STOCHASTIC MEDICAL REASONING AND ENVIRONMENTAL HEALTH EXPOSURE

STOCHASTIC MEDICAL REASONING AND ENVIRONMENTAL HEALTH EXPOSURE

George Christakos
San Diego State University, USA

Jin-Feng Wang
Chinese Academy of Science, China

Jiaping Wu
Zhejiang University, China

Imperial College Press

Published by

Imperial College Press
57 Shelton Street
Covent Garden
London WC2H 9HE

Distributed by

World Scientific Publishing Co. Pte. Ltd.
5 Toh Tuck Link, Singapore 596224
USA office: 27 Warren Street, Suite 401-402, Hackensack, NJ 07601
UK office: 57 Shelton Street, Covent Garden, London WC2H 9HE

British Library Cataloguing-in-Publication Data
A catalogue record for this book is available from the British Library.

STOCHASTIC MEDICAL REASONING AND ENVIRONMENTAL HEALTH EXPOSURE

Copyright © 2014 by Imperial College Press

All rights reserved. This book, or parts thereof, may not be reproduced in any form or by any means, electronic or mechanical, including photocopying, recording or any information storage and retrieval system now known or to be invented, without written permission from the Publisher.

For photocopying of material in this volume, please pay a copying fee through the Copyright Clearance Center, Inc., 222 Rosewood Drive, Danvers, MA 01923, USA. In this case permission to photocopy is not required from the publisher.

ISBN 978-1-908977-49-6

Typeset by Stallion Press
Email: enquiries@stallionpress.com

Printed in Singapore

To friendship

Contents

Preface	xi

Chapter 1. Medical Sciences in the Age of Synthesis — 1

1.1 Professional Practice and Stochastic Medical Reasoning 1
    1.1.1 Synthesis in medical sciences 1
    1.1.2 Environmental health and geomedicine 5
    1.1.3 On ancient Greek and Chinese medicine 9
    1.1.4 Decision-making in conditions of *in situ* uncertainty 12
    1.1.5 A brief note on logical thinking in ancient Greece and China 14
    1.1.6 Enter stochastic medical reasoning 16
1.2 Health: The Fundamental Roles of Space–Time and Uncertainty 19
1.3 Abstract and Concrete Modes of Thinking 23
1.4 Issues of Sound Medical Decision-Making 25
    1.4.1 Key elements of a medical investigation 26
    1.4.2 Reflection, recognition primed decision and robust decision methods 30
    1.4.3 Algorithmic medical decision-making methods 33
    1.4.4 Does expert knowledge translate into expert judgment? 34
1.5 Medical Dialectics and Knowledge Synthesis: An Outline . . . 36

Chapter 2. Reasoning Amidst Uncertainty — 39

2.1 When "To Know" Means "To Be Uncertain Of" 39
    2.1.1 Common medical reasoning errors 39
    2.1.2 Physician's language and metalanguage 42
    2.1.3 The notion of knowledge base 44

2.2	\multicolumn{2}{l}{The Space–Time Domain of Stochastic Medical Reasoning}	47
	2.2.1	Location–time coordinates and metric	47
	2.2.2	The spatiotemporal random field model	51
2.3	\multicolumn{2}{l}{*In Situ* Logic and Uncertain Mind States}	53
	2.3.1	How much a health care provider does not know: Ontic and epistemic uncertainty	54
	2.3.2	Appreciating case individuality and legal disputes	56
	2.3.3	Case communication uncertainty: Entitled to their own opinions but not to their own facts	58
	2.3.4	Formal vs. *in situ* logic	60
	2.3.5	From state of nature to state of mind (assertion)	63
	2.3.6	Ranking of assertion forms	71
	2.3.7	The Three Qs of the triadic case formula	74
	2.3.8	Uncertainty factors: A review	76
2.4	\multicolumn{2}{l}{SMR's View of Medical Connectives: Beyond Drug Digestion}	79
	2.4.1	Conversational (dialogical) connective interpretation	80
	2.4.2	Content-dependent vs. content-independent connectives	83
2.5	\multicolumn{2}{l}{Natural Laws and Scientific Models}	85
	2.5.1	Infectious disease and human exposure modeling	86
	2.5.2	Medical syllogism and the justification of professional assertions	89
	2.5.3	Reconstructing Chinese arguments in terms of Greek syllogisms	94
	2.5.4	Revisiting content-dependent and content-independent assertions	96
2.6	\multicolumn{2}{l}{Substantive Conditionals in Medical Thinking}	99
	2.6.1	The notion of content-dependent conditional	100
	2.6.2	Paradoxes of mainstream logic	105
	2.6.3	Conditionals and metalanguage	114
	2.6.4	Conditionals and natural laws	117
	2.6.5	Over-extending and extrapolating	119
2.7	\multicolumn{2}{l}{The Object Language–Metalanguage Connection}	121
	2.7.1	Relations between states	122
	2.7.2	Combinations of medical inferences and derivative assertions	131
	2.7.3	Levels of justification and uncertainty	133
	2.7.4	Does postmodern decision analysis make sense?	138

Chapter 3. The Role of Probability — 141

3.1 How Much Understanding is Sufficient in Medical Investigations? 141
    3.1.1 Medical assertions and partial understanding 141
    3.1.2 On rationality and belief 144
    3.1.3 Knowledge theory revisited: Platonism, context and continuity in medical thinking 145
    3.1.4 Concerning medical expertise 147
3.2 Space–Time Probabilities of Medical Cases 148
    3.2.1 Common probability interpretations in health care practice 148
    3.2.2 Probability of a case assertion (mind state) 152
    3.2.3 Basic probability rules 155
    3.2.4 Probability interpretations in object language and metalanguage 157
    3.2.5 Body of evidence and medical interventions 161
3.3 Probabilities of Medical Conditionals 164
    3.3.1 Standard logical relations and inference rules 164
    3.3.2 Choosing a conditional probability form 166
    3.3.3 Stochastic truth tables: A second look 176
    3.3.4 More on probability calculation: Is there a probameter? 181
3.4 Stochastic Medical Inferences 186
    3.4.1 From standard to stochastic syllogisms 187
    3.4.2 Premise strengthening, internally consistent and uninformative inferences 199
3.5 Probability, Uncertainty and Information of Diagnoses or Prognoses Sets 205
3.6 Diagnosis Ranking and Symptom Confirmation Strength ... 211
    3.6.1 Quantitative case parameters 212
    3.6.2 Principles of medical practice and their quantitative expressions 215
3.7 The Trouble with Medical Probability 217
3.8 Translating Medical Assertions into Probabilistic Terms 222
3.9 Space–Time Reasoning Dynamics 229
    3.9.1 Changes in assertions and substantive conditionals .. 229
    3.9.2 Probability dynamics and hypothesis confirmation ... 235
    3.9.3 The case of non-monotonic medical reasoning 238
3.10 Medical Syllogisms Involving Likelihood Ratios 240

3.11	Summing Up: Checking the Validity of Medical Arguments	242
3.12	Self-Referential Medical Assertions and Cognitive Favorability	244
3.13	Not Just a Set of Guidelines	247

Chapter 4. Space–Time Medical Mapping and Causation Modeling   249

4.1	Techniques With a "Health Warning"	249
4.2	Space–Time Disease Mapping	250
	4.2.1 Objectives of medical mapping	251
	4.2.2 The fundamental mapping equations	253
	4.2.3 The insight behind the BME–SIR equations	256
	4.2.4 A study of French flu	258
4.3	Modeling Space–Time Infectious Disease Spread	264
4.4	Space–Time Causation Revisited	268
4.5	Medical Causation in the *SMR* Inference Setting	271
	4.5.1 Defining the problem	272
	4.5.2 The role of KB and the interpretation of probabilistic causation	277
	4.5.3 Causation: Epistemic vs. non-epistemic	280
	4.5.4 Some remarks regarding the form of the causation conditional	283
	4.5.5 Stochastic causal inferences	285
	4.5.6 The role of secondary case attributes	287
4.6	Causation in Terms of Integrative Space–Time Prediction	289
4.7	Causation Justification and the Dualistic Opposition	291

Chapter 5. Looking Ahead   293

5.1	An Ibsenian Transformation	293
5.2	*SMR* and Divergence of Rationality in Medical Thinking	296
5.3	Challenges Emerging from the Incompleteness Principle and Unanticipated Knowledge	298
5.4	Information Technology-Based Medical Reasoning	302
5.5	Social and Cultural Dimensions of Medical Thinking	304
5.6	Quod Iacet Ante?	308

Bibliography   315

Index   331

Preface

The validity of critical reasoning steps carried out during or on the sidelines of a clinical case, public health survey, medical study, human exposure, population risk assessment or disease mapping often does not receive sufficient attention or even feature on the investigator's radar. For example, the technical complexity of a human exposure experiment may overshadow the logical assumptions made when moving from one phase of the experiment to the next; or the study of population risk assessment may focus on analytical and computational matters, whereas methodological and cultural factors are neglected. This book hopes to help health investigators structure their thinking so that they avoid logical mistakes and argumentation pitfalls and, at the same time, gain new insights about clinical reality, and improve awareness of the environment and context within which one's thinking takes place.

A central thesis put forward in the book is that *in situ* medical reasoning extends beyond clinical procedures and technical guidelines (checklists), and is viewed as a synthesis of theoretical and empirical considerations in science, epistemology, sociology and stochastics.[1] Stochastics has given medical reasoning a vocabulary on which to base its own methodology. Remarkably, while mainstream formal (or discursive) logic relies on a set of rules (norms) by means of which a form of orderly closed-system reasoning can be applied, cognitive limitations, partial understanding, and environmental and cultural factors affect *in situ* reasoning and the inferences that people draw in real world situations. Thus, the book is about stochastic medical reasoning (SMR) in a space–time synthesis context, which is reasoning based on the integration of different knowledge sources under conditions of *in situ* uncertainty and health status heterogeneity. This is medical reasoning that overlaps with

---
[1] The term "stochastics" refers to the mathematical study of phenomena that vary in space–time under conditions of uncertainty.

logic and rationality, but is considerably more than that. To some extent, the methodological underpinnings of this kind of reasoning are closer to a neo-Kantian "substantive unity" of reason (its original form was considered by Immanuel Kant; Kant, 1922) rather than to the Hobbesian calculative reason (i.e., reasoning seen as a broader version of generalized computation; Malmesbury, 1839) or to the Habermassian conception of reasoning as strictly formal or procedural (Habermas, 1995). Accordingly, since *SMR* is substantive rather than normal, it encompasses the observation of medical cases, the prediction and control of future disease-related events and attributes, and the testing of health hypotheses, as well as novel ways of seeing real medicine and interpreting things (test results, diagnoses, prognoses, treatments). Otherwise said, *SMR* views medical thinking as a non-egocentric process that does not merely incorporate logical or rational thinking modes, but goes considerably beyond that in the domains of insight, creativity, imagination and the unconventional. Over time, this methodology and the related technology have gradually proceeded from a research and standardization context to a concrete and productive *in situ* setting.

A brief survey of the book follows: Chapter 1 looks at medical reasoning from a synthetic perspective, the term "medical" being viewed in the broad sense that includes clinical medicine, environmental health and geomedicine. Chapter 2 introduces the basics of medical syllogism, and is concerned with the fundamental distinction between medical language and metalanguage in interviewing, decision-making, and communication. Chapter 3 focuses on the introduction of probability in medical thinking, and on techniques testing the validity of diagnoses in realistic conditions of time pressure. The vital role of uncertainty in medical decision support is constantly emphasized throughout the book. The value of a sound decision analysis lies not merely in its ability to generate disease diagnoses but also in the responsible examination and rigorous assessment of the uncertainty surrounding diagnostic tests and hypotheses. The criteria of disease etiology (causation) and the methods of space–time disease evolution and spread discussed in Chapter 4 apply in a wide range of problems in medicine, environmental health, and geomedicine. A central thesis of this book is that a rigorous and efficient way for health care professionals to understand the underlying logical structure of their everyday language is to learn how to put it into the *SMR* terms and methods introduced in Chapters 2 to 4. Concluding, Chapter 5 presents a few thoughts about what potentially lies ahead in medical thinking and decision-making.

All chapters include a considerable number of examples and real case studies to help the readers gain adequate insight about the theory and methods of medical reasoning in conditions of uncertainty. It has been said that people who do not know history do not have future, and people who do not know philosophy do not have vision. Accordingly, most chapters of the book discuss examples from medical practice where philosophical logic could play a key role. One of these examples is the crucial distinction between "knowledge of the real medical case" and "knowledge of a physician's mental model of the case", which is both a crucial clinical diagnosis issue and a deep philosophical problem (specifically, belonging to the philosophical field of epistemology). That is, an investigator should distinguish between "science's limits" and "scientist's limits", which can have a profound effect in setting priorities in clinical practice and medical research. When discussing a view, notion or method in the book, our aim is not necessarily to defend that view, notion or method, but to state that the view exists, as shown in various medical sources. Therefore, the view, notion or method should be accounted for rather than simply dismissed.

As noted above the concern of real medical reasoning and clinical judgement is not limited to the domain of deterministic logic, in the sense that it can deal with questions that are unanswerable within the deterministic domain. This is made possible because *SMR* avoids the Gödelian issues of unprovability (undecidability) that characterize deterministic logic. Indeed, the latter employs strictly deductive reasoning, whereas *SMR* uses a mode of thought that works under the uncertainty conditions characterizing public and private medical practice. In the same milieu, a situation encountered by medical decision-makers is that the notion of content-independent formal reasoning captures the essence of many health statistics techniques. This happens despite the fact that a health care provider's powers of noticing, perceptual grasp and understanding depend upon recognizing what is salient and the capacity to respond to the substantiveness of the particular medical case. The above considerations show that the need to reinterpret mainstream logic in a realistic medical setting is a result of the desire to retain, as far as possible, substantive interpretations of medical decision-making and exposure assessment. Ideally, a physician is trying to develop reasoning in such a way that the justification of a medical assertion embodies content-dependent arguments (scientific principles, empirical evidence etc.) rather than content-independent ones.

As intellectual provocation constitutes an invaluable component of creative thinking and decision-making, it underlies many developments and discussions throughout the book. A steady stream of evaluations, considerations and challenges to ideas, mainstream methods and "common sense" notions is a necessary component for improvement in medical reasoning and clinical judgment.

In a similar setting, otherwise intelligent health care providers who lack a special mathematical aptitude may argue that certain parts of medical reasoning (e.g., those dealing with the study of space–time disease distributions and epidemic matters) require some level of quantitative sophistication. Also, it has been argued that quantitative studies purporting to deal with health and medicine are essentially dealing with mathematical analysis, formal logic, probability theory, statistics, and so on. Even if this is valid to some degree, it would surely be a big mistake to overlook the great transforming effect the introduction of stochastic mathematics has upon medical decision-making. In reality, most of *SMR* inference can be mastered with reasonable effort, and it is a versatile tool that produces a plethora of useful results in medical decision support. *Inter alia*, *SMR* can radically improve clarity of clinical argumentation, rigor of diagnostic expression, and power of prognostic information.

Lastly, the design of the book cover is a visualization of the book's perspective on the integration of medicine with logic, and its historical roots. Remarkably, although the evolution of Greek medicine and Chinese medicine followed rather remote paths (geographically and conceptually), still there were significant similarities. Many experts cite the view that the medical theories and clinical practices of ancient Greek and Chinese medicine had interesting similarities to each other, and they separately produced particular merits of themselves. The idea of synthesizing medicinal observation with logical rigor was deeply rooted in Aristotle's thinking, who grew up under the influence of the Hippocratic tradition of medicine, whereas the semantics of valid inference were studied in Mohist *Canons*. The book cover's design emphasizes the above facts by presenting a synthesis of images relevant to ancient Greek and Chinese medical scholars and pioneers of logical thinking.

The authors would like to express their gratitude to Ms Tasha D'Cruz for the dedication with which she managed the book's production.

GC, JFW, JPW
San Diego, Beijing, Hangzhou

Chapter 1

Medical Sciences in the Age of Synthesis

1.1. Professional Practice and Stochastic Medical Reasoning

If the 15th century (medieval times) was the age of *Belief*, the 16th century (Renaissance) the age of *Adventure*, the 17th century the age of *Reason*, the 18th century the age of *Enlightenment*, the 19th century the age of *Ideology* and the 20th century the age of *Analysis*, then, naturally, the 21st century should be the age of *Synthesis*. Most of today's real world problems cannot be solved within the boundaries of a single scientific discipline. Instead, these problems have an essential multidisciplinary structure, which means that their successful study transcends disciplines and requires a goal-directed synthesis of concepts, data, techniques and thinking modes from all these disciplines.

1.1.1. *Synthesis in medical sciences*

In a developing environment of synthesis, medical sciences[1] cannot be an exception. In fact, the idea of synthesizing medicine with logical rigor was deeply rooted in *Aristotle*'s thinking, who grew up under the influence of the *Hippocratic* tradition of medicine. Aristotle valued both medical observations and techniques (he documented many of Hippocrates' medical achievements), but he argued that logic should be carefully used

---

[1]The term "medical sciences" is considered in the broad sense that includes *clinical medicine* (science of diagnosing, treating or preventing disease and other damage to the human body or mind), *environmental health* (assessment and control of those environmental factors that can potentially affect health) and *geomedicine* (effects of the environment on the spatiotemporal distribution of population health and disease spread).

to confirm hypotheses. This synthesis was promoted by other medical thinkers, including *Avicenna*, who incorporated Aristotelian logic and *Galen*'s teachings into medical diagnosis and treatment.

Synthesis is a *process* whereby logic, critical reflection, medical knowledge, clinical experience and environmental awareness are combined in evaluating multiple objectives (diagnoses, prognoses, treatments), while accounting for the patient's situation (signs, symptoms, pre-existing conditions). From their routine practice, physicians know that a patient's symptoms, signs and diagnostic test results may be associated to more than one disease, and that often it is not possible to distinguish between them with certainty. Clinicians need to synthesize the medical information they obtain during a patient's interview and physical examination into an opinion about the patient's state (a clinician can then conclude that a patient is not diseased, or that more information is needed by means of additional diagnostic tests, or that enough is known to suggest a specific treatment). Remarkably, the diagnosis of a disease may require the synthesis of information obtained with the help of theories and tools having their origins in non-medical scientific fields. Also, the interest of large-scale health studies, like epidemic spread or population exposure in a region of the world, clearly exceeds the domain of the respective disciplines. Decision support tools that advance the quality of medical practice and patient care[2] are often the direct or indirect product of developments in different sciences and technologies, such as modern physics, molecular and cellular biology, nanotechnology, engineering (mechanical, nuclear, electrical and electronic) and biomaterials or tissue engineering. Integrating knowledge from different disciplines is needed in order to provide medical researchers and clinicians with the most reliable tools to make the necessary *measurements* and improve their *methodological* quality (the degree to which the measurement process matches the case objectives, the likelihood that a measurement instrument will generate unbiased results) (Arrivé et al., 2000; de Vet et al., 2011).[3] In view of the above and similar considerations, researchers and clinicians routinely form teams that develop *styles* of professional practice with the goal of providing high-quality care, including communication mechanisms, protocols for transforming knowledge and experience, ways of

---

[2] For example, by improving clinicians' decisions involving tests and treatments that are also more personalized, or constructing more accurate and more rapid diagnostic techniques.

[3] A review of medical decision-making methods is presented in a later section of this chapter.

doing things (including guidelines assessing the expertise of team members) and shared evaluation of their performance.

As most *health care providers*[4] would admit, the state of affairs in public health is quite heterogeneous. Human *genetics* is, of course, a crucial factor of quality health care, however, it has become increasingly clear that people's *environment* can also provide important contextual information, medically speaking. An increasing number of studies have clearly shown that where people live and work are intrinsically intertwined with their health (Wright et al., 1982; Burke et al., 2003; Yoon et al., 2003; Eggleston, 2009; Hovell et al., 2002; Davenhall, 2012). There is plenty of evidence that certain of the pollutants people are exposed to can serve as crucial precursors to respiratory and circulatory illnesses, some kinds of cancer, and, in some cases, heart diseases. The impact of breathing bad air in many of the places people have lived will follow them wherever they go during their entire lives. Also, *lifestyle* factors, including diet, smoking, working and exercise habits are key determinants of human disease — accounting for perhaps 75% of most cancers (Sharpe and Irvine, 2004). *Diet*, in particular, can play a major role in people's health. The readers may be surprised to hear that there exist populations around the world that surpass highly developed Western countries in health and longevity without having much of a medical system in place. This includes populations in which a healthy diet has a greater impact on children's health than the availability of a medical insurance plan.

Example 1.1: Public health investigators discovered that during the 1950s only the children of wealthy families in the Philippines developed *childhood liver cancer*. This was because only these families could afford peanut butter and, during this time period, much of the peanut butter in the Philippines was contaminated with a mold called aflatoxin that is a highly carcinogenic substance; the aflatoxin started the cancer tumors and dairy consumption contributed to their rapid growth (Campbell, 1967).

Another example of the lifestyle's impact on modern populations is the increasing numbers of thyroid cancer cases in certain parts of the world.

Example 1.2: In the plots of Fig. 1.1 one can see the dramatic increase of *thyroid cancer* in Hangzhou city (Southeast China) as a result of lifestyle change. Over the years, the number of male patients with thyroid cancer

---

[4]The term includes physicians, physical practitioners, optometrists, chiropractors, dentists, nurse practitioners and physician assistants.

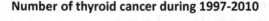

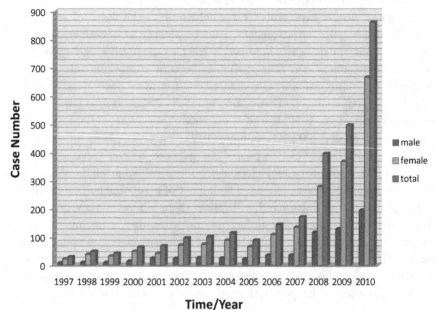

Figure 1.1: Thyroid cases during the years 1997–2009 in Hangzhou city (HZ), Southeast China.

presented a slow growth, whereas for the women patients the growth rate was much faster (the total number of female cases is more than three times larger than that of males). The thyroid growth rate fluctuation is very severe. It exhibited negative growth in 1998 and 2005, however, in 2007 the growth rate initiated a sharp increase, 131.6%. Furthermore, the number of disease cases changes with age. High incidence populations were those of 31–40 years old (24.2%), 41–50 years old (28.1%) and 51–60 years old (20.9%). Clearly, the population of 31–60 years old had the highest percentage of thyroid cases (about 73.2% of the total number of cases). In addition, the improving economic conditions in the region contributed to the increased number of reported thyroid cases (larger number of people have access to better health services). It is of utmost significance that the medical records of the various hospitals and health centers be automatically updated as new findings like these ones become available concerning health risks due to different factors (diet, sex, income).

In view of the above considerations, it is unfortunate that most health care professionals rely solely on the data collected from standard clinical procedures and medical guidelines (physical examination, lab tests, patient interviews etc.), neglecting the fact that during their lives people accumulate undetected environmental exposures, experience *socioeconomic* adversities and changing lifestyles and face unseen risks that can play a key role in making them sick. There is abundant evidence that lower socioeconomic status often impedes the management of chronic illness due to barriers to health care (transportation problems, limited health literacy, language barriers, inadequate number of local care providers, absent health insurance; e.g., Alexander et al., 2005). Currently, physicians do not possess systematic procedures to translate the rich bases of environmental health and geomedical data into information that could benefit their patients (Riley et al., 1978; Al-Jahdali et al., 2003; Walker et al., 2003; Furuva, 2007). This book then adds its voice to the call to appreciate the value in quality care and patient safety of the synthesis of *internally* generated health care records and the increasing amount of *externally* available health data. The convergence of two powerful elements — environmental conditions and human health factors — should increasingly drive medical judgment and decision-making.

1.1.2. *Environmental health and geomedicine*

As already noted, medical science, as viewed in this book, does not include solely clinical medicine, but also extends into the fields of "environmental health" and "geomedicine". *Environmental Health*, in particular, is the branch of medical science that is concerned with all aspects of the natural and built environment that may affect human health (Tiefelsdorf, 2007; Dhondt et al., 2011; Moeller, 2011). This includes physical, chemical and biological factors external to a human being, and all the related factors impacting behaviors (usually excluding behavior not related to environment, as well as behavior related to the social and cultural environment and genetics). *Geomedicine*, on the other hand, is a term initially used in the early 1930s as synonymous to geographical medicine (Zeiss, 1931). A more comprehensive definition was suggested in the late 1970s as "the science dealing with the influence of the ordinary environmental factors on the geographical distribution of health problems in man and animals" (Läg, 1978, 1990). A similar definition views geomedicine as "the branch of medicine dealing with the influence of climatic and environmental conditions on health" (Dorland, 2007).[5] One readily observes that the two disciplines, environmental health and geomedicine, have certain things in common.

The daily routine of environmental health and geomedical scientists is often to synthesize, for instance, what they have learned in the morning meetings from their study of a region's chemical contamination or radiation hazard with what they have learned in the afternoon panels about human exposure, in order to reach some conclusions concerning population health effects. Generally, this decision-making process seeks to untangle the available literature on environmental chemistry, biophysics, toxicokinetics, physiology, disease spread, population dynamics, biostatistics and other relevant fields in an effort to derive optimal predictions of population risks in the actual world.

Example 1.3: Figure 1.2 is a summary of an environmental health study of the impact of spatiotemporal *ozone* (O_3) exposure distribution on population health of human in eastern US (including New York City and Philadelphia) during the year 1995 (Christakos and Kolovos, 1999). The analysis started with O_3 exposure distributions ($E(s,t)$,[6] Fig. 1.2) producing the input to toxicokinetic laws of burden on target organs and tissues ($B(s,t)$) that were linked to human effect ($H(s,t)$) via the burden-response curve which, in turn, were integrated with relationships describing how health damage was distributed across population ($\Psi(s,t)$). Different Ψ-maps were obtained for different population cohorts (groups of people according to age, sex, physiological characteristics, pre-existing medical condition) and during different days. These $E-B-H-\Psi$ relationships offer a biophysical blending of the environmental exposure and biological processes that affect public health.

Natural uncertainties and human variations were taken into consideration in terms of random field models. The stochastic perspective offered a deeper epistemic understanding and the development of improved models of spatiotemporal human exposure analysis. Also, it explicitly determined the knowledge sources available and developed logically plausible rules

---

[5] A more recent definition of geomedicine is as the field that "uses modern information technology to deliver information on a patient's potential environmental exposures into the hands of the clinician while they are in the examination room" (Davenhall, 2012). This definition, which is limited to one's exposure while in the examination room, is rather unsatisfactory for the purposes of this book.

[6] Here s denotes geographical coordinates and t is time (details in Section 2.2 and references herein).

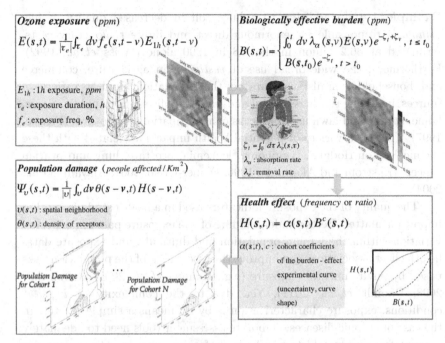

Figure 1.2: Outline of an environmental health study in an eastern US region during the year 1995.

and standards for data processing and exposure map construction. The approach allowed the horizontal blending of scientific findings about O_3 exposure reality in the New York City–Philadelphia region, which led to informative spatiotemporal maps of O_3 exposure and effect distributions, and an integrative analysis of the whole risk case.

Possible exposures of interest to environmental health include indoor and outdoor air pollutants, dumping of chemicals on the ground or into stormwater drain, soil fertilizers, bisphenol A, radiofrequency fields, animal waste products and mosquito breeding grounds. Exposures have been linked to a variety of health problems like respiratory diseases (asthma, emphysema, bronchitis, pulmonary fibrosis, pneumonia and tuberculosis), different types of cancer, neurotoxicity, neural tube defects, malaria, West Nile virus and dengue fever (Boezen et al., 1998; McConnell et al., 1999; Ahlbom et al., 2004; Hunter et al., 2004; Reynolds et al., 2005; Gauderman et al., 2007). The above environmental health effects are frequently linked to considerable *monetary costs*.

Example 1.4: In addition to a huge toll in decreased quality of life, *asthma* is responsible for enormous direct and indirect monetary costs, estimated at $6.2 billion in the US in 1990 alone (Weiss et al., 1992). Furthermore, the widespread use of *pesticides* in agriculture, commerce and households entails exposure to the population from a variety of sources, including residues in food and water, applications to public spaces, home, garden and lawn use and occasional occupational exposures (Fenske, 1997). There is a variety of potential health impacts associated with these chemicals (non-Hodgkin lymphoma, leukemia, prostate, lung and ovarian cancer) (Ekstrom and Akerblom, 1990; Fenner-Crisp, 2001; Alavanja et al., 2004).

The quality of the exposure indicators used in a health study is another important matter. Better understanding of the exposure pattern, exposure variation within the human population and diurnally, and exposure data-health effect associations can improve the assessment of the population risks posed by environmental exposure (e.g., Haining et al., 2007; Wiwanitkit, 2008; Dhondt et al., 2011). Yet, due to the complexity of *in situ*[7] conditions, exposure characterization is by no means a simple matter. In the case of chronic diseases, elaborate assessment tools need to adequately account for exposure history. The tools for assessing long-term exposure include several types of scientific biomarkers, statistics techniques and suitable questionnaires. In many cases (e.g., pesticides; see Alavanja et al. (2004)) the quality of exposure indicators are considerably better when collected prospectively, that is, prior to disease onset. This is because most environmental exposures do not leave "foot prints" indicating exposure duration and intensity, and because time and disease can influence recall of exposures as well as interpretation of biological markers.

Many geomedical studies deal with the influence of natural factors on the geographical distribution of problems in human and veterinary medicine, including the potentially harmful effects of soil pollution on human and animal health (Oliver, 1997, 2004; Deckers and Steinnes, 2004; Selinus et al., 2005). Other studies have focused on potentially harmful impacts on human and animal health related to trace element deficiencies in soils that are frequently reported in agricultural crops and livestock (these impacts are important in a global perspective) (Alloway, 2005; Gupta and Gupta, 2005; Andersen, 2007; Steinnes, 2009). This is the case of the following example.

---

[7]The term "*in situ*" means "in a real world environment" or "in its natural place".

Example 1.5: Geomedical studies have shown that the Keshan disease (*KSD*) and the Kashin–Beck disease (*KBD*) are closely related to *selenium* (*Se*) deficiency in geoecosystems (soils, food grains, animal and human hair, human blood etc.). In particular, Tan *et al.* (2002) illustrated the relations in terms of summary statistics indicators between (*a*) total *Se* in Chinese soil and (*b*) *KSD* and *KBD* in disease-affected regions vs. disease-unaffected regions in China. As it turned out, in the *KSD* and *KBS* affected regions the *Se* concentrations both in cultivated and natural soils are lower than in disease unaffected regions.

In sum, nowadays a patient's medical records include a vast collection of clinical data, physical exams, observations, lab results and disease diagnoses, but in most cases they lack any information about the accumulated exposure risks, environmental health impacts, geomedical factors and socioeconomic factors. Nevertheless, patients deserve to be better served by the large amount of environmental and geomedical data that is becoming available at an increasing rate, and physicians need to improve their ability to utilize this information in efficient ways in their clinical practice. It is anticipated that health care professionals, in addition to their clinical data and health care standards, will increasingly use environmental health and geomedical records to aid diagnosing, treating and preventing certain diseases and, on occasion, making recommendations to patients on places to live and work. This is why the reasoning concepts and methods discussed in this book apply in all the above fields of medical sciences and are not limited to clinical studies.

1.1.3. *On ancient Greek and Chinese medicine*

A historical note may be appropriate at this point.[8] Undoubtedly, Western medicine and its Greek roots have truly historic origins. *Hippocrates*, father of medicine, was the first to introduce the notion of "physics" and to transform theocratic medicine into rational medicine. He used a *holistic* approach for diagnosing and treating human disease that combined scientific knowledge with drug therapy, diet programs, physical/mental exercises and emotional compassion and prayer (Porter, 1997; Orfanos, 2007). Hippocrates' detailed records $I\pi\pi o\kappa\rho\alpha\tau\iota\kappa\acute{\eta}$ $\Sigma\upsilon\lambda\lambda o\gamma\acute{\eta}$ (*Corpus Hippocraticum*), consisting of several volumes enriched and completed by his students and scholars, became the standard reference for the entire

---

[8]This is an extremely brief note on ancient Greek and Chinese medicine. The interested readers should consult the voluminous literature on the subject.

antique world for medicine (Adams, 1939). Hippocratic medicine was practiced for several centuries around the Mediterranean during the Roman Empire, until the Middle Ages and even later (Jones, 1953; Weisser, 1989). *Dioscurides*, a physician and botanologist, wrote $Περί\, 'Yλης\, Ιατρικής$ (*De Materia Medica*), a five-volume encyclopedia about herbal medicine and related medicinal substances. This fundamental work, which focused on the preparation, properties and testing of drugs, became the most central pharmacological work in Europe and the Middle East for more than 16 centuries.

Zhang Zhongjing, one of the greatest Chinese physicians, is perhaps most famous as the author of *Shanghan Za Bing Lun* (*Treatise on Cold Pathogenic and Miscellaneous Diseases*). This work contains many formulas, including gynecological remedies for infertility problems, disorders during pregnancy, prevention of miscarriage and post-partum weakness. Contemporary to Zhang Zhongjing was *Hua Tuo*, another renowned physician. His medical techniques and therapies are considered milestones in Chinese medical history. They include techniques of externally extruding heart, mouth-to-mouth artificial respiration and anesthesia. Moreover, Hua Tuo is considered the originator of surgery in China. The readers may be also aware of *Huangdi Neijing* (*The Inner Canon of Huangdi* or *Yellow Emperor's Inner Canon*), which was perhaps the most important medical text in ancient China (it has been treated as the fundamental doctrinal source for Chinese medicine for more than two millennia). Some experts compare it in importance to the *Corpus Hippocraticum*.

In general, when comparing medical developments in Greece and China one should keep in mind two essential observations. One should not compare things out of context and in separation, whether they are concepts, values, techniques, individuals, or groups of individuals. And, because their settings in some cases may be different, they can also have different implications. One looks for similarities with different significances, or different means that lead to analogous ends, as well as contrasts, trying to learn how different cultural circumstances push medical institutions and ideas in different directions.

Example 1.6: Table 1.1 lists a few remarkable associations and similarities between the Greek and Chinese perceptions of the functions of certain internal organs. As pointed out earlier, these associations and similarities have deeper intellectual and cultural roots and should not be considered out of the relevant contexts. This approach is followed throughout the book when different aspects of the two traditions are compared.

Table 1.1: Similarities between the Greek and Chinese perceptions of organ functions.

Organ	Function — Ancient Greek Medicine	Function — Ancient Chinese Medicine
Kidney	Νεφρός, from abbreviation of words νερό (water) and φως (light, fire).	Stores primordial *yin* (water) and *yang* (fire), energy source of entire body.
Heart	Most prominent part of ψυχή (soul).	Emperor, houses the *shen* (spirit).[1]
Liver	Ἧπαρ, from the word επαίρομαι (boast) expressesing an urge towards growth and expansion.	Expresses the primeval power *Yuanshen* associated with the urge towards expansion and growth.
Spleen	Σπλήνα, because is dark and wet like a σπήλαιον (cave).	Related with dampness and its main function is to transform/transport it.
Lung	Πνεύμων, from the word πνέω (blow) meaning wind movement.[2]	Governs breathing and *qi*.

[1] Heart is associated with the primeval power *Hun* (fire).
[2] From the word πνέω also come the words πνοή (breath) and πνεύμα (spirit).

Another domain in which the two cultures followed parallel paths was *herbal medicine*. Two comparable works of herbal medicine cited by the experts are Dioscurides' *Περί την Ιατρικήν* and *Shennong Chang Bai Cao* (*Divine Farmer's Materia Medica*) in ancient China.

Example 1.7: In these works every herb is analyzed at two levels (Tilikidis, 1999). One should notice that, although these levels of analysis are found in both works, they do not agree about the healing properties of all herbs. The first level analyzes the herb's character based on its features (hot, cold, damp, dry, astringent, acrid, taste etc.). The second level investigates the healing properties of the herbs, based on their effect on various parts and organs of the human body (e.g., cathartic, diuretic, stimulating, demulcent, aphrodisiac, cicatrizing, emetic, cholagogue, expectorant, haematogogue). Dioscurides cites about 950 herbs (from the plant or mineral kingdoms), whereas the *Shennong Chang Bai Cao* cites about 350. Several researchers have found that many of them are essentially identical.

In recent years, certain scholars have become increasingly critical about science-based differences between the two medical paradigms. For example, Wang (2012) maintains that although in ancient times the medical theories and clinical practices of both Greek medicine and Chinese medicine were quite similar to each other, and they separately produced particular merits of themselves, due to the lack of support to natural philosophy in ancient China, the progress of medicine, with its original native qualities for thousands of years, only showed an increase of clinical experiences, rather than scientific reformation of its essences. Therefore, these scholars argue, Chinese medicine should also receive scientific "baptism" as Greek medicine. In this way, the valuable medical experiences of Chinese medicine can be picked up for wide application, and its great historical achievements can be revealed for later pondering. We will revisit the subject in later parts of the book.

1.1.4. *Decision-making in conditions of* in situ *uncertainty*

It is widely recognized that decision-making plays a central role in every human exposure field and clinical domain where medicine is practiced (from the anatomic pathology laboratory to the intensive care unit). When practiced adequately, medical decision analysis can help both physicians and patients to cope with disease or injury uncertainties.

Undoubtedly, *health judgment* and *medical decision-making* have turned into interdisciplinary affairs that have attracted attention from many different research disciplines outside medicine, including psychology, sociology, philosophy, computer science, artificial intelligence, mathematics, logic and statistics. These disciplines are distinguished by their theoretical stance and methodology, which varies from the experimental to the axiomatic. Each of them yields important insights into the way health care professionals think, how they behave under pressure and what is the best response in these cases. The synthesis of these insights, which are both factually- and belief-based, underlies interdependent decision-making in clinical practice and medical research. For instance, the synthesis of clinical medicine, geomedicine and environmental health (related to molecular, biological, genetic, psychosocial, physical, chemical and other environmental attributes) can bring a comprehensive approach to health care and patient safety. Synthesis is often necessary even within the same domain of expertise, since a care provider who is fluent in one sub-field may be less fluent in another sub-field of the same domain. Also, physicians must have a solid grasp of the decision analysis tools in order to implement them adequately in active clinical environments, and they must possess considerable synthesis skills to integrate these tools into medical student and residency curricula.

Decision support includes the synthesis of decisions regarding the need for further diagnostic testing (that may be, e.g., burdensome or costly) with therapy decisions (e.g., surgery with morbidity and mortality risks). Typically, such decisions are binary (e.g., 0 and 1), quantitatively expressed in terms of probabilities, and requiring clinically relevant decision thresholds.

Yet synthesis in a decision-making setting is by no means a trivial matter. Generally, committees produce medical *guidelines* (reference points) and *checklists* that are assumed to come out of consensus among experts. These guidelines and checklists include algorithms, decision trees and graphs, which have been put together by the committees on the basis of "best" evidence or "optimal" practices. Nevertheless, many experienced care professionals have noticed that the guidelines frequently are not the result of a careful and balanced synthesis of viewpoints of experts who do not necessarily agree with each other, but rather the opinion of a forceful personality dominating the committee's decision-making and the way the guidelines are finally written. And, as the distinguished physician Jerome E. Groopman (2010) remarked, "if the physician is under financial pressure to follow the guidelines, his or her interests and those of the patient might no longer be in alignment".

Truly, under certain circumstances medical guidelines and clinical checklists may provide useful information in a structured way; nevertheless, the way they are frequently implemented is restricting rather than expanding diagnostic accuracy and disease understanding. Which is why physicians and other health care providers need to consciously and methodically apply logical thinking in conditions of uncertainty and time pressure, and to engage their patients in a narrative (listen to their concerns, prompt the patients in an open-ended way, collect first-hand information, and show empathy).

As a matter of fact, the interest in the above and similar matters arises from a wider point at issue: space–time synthesis in conditions of *in situ* uncertainty. Indeed, the challenge to the health care provider is multi-thematic and multi-objective. More specifically, some important questions emerge:

(1) What is the medical content that links space and time to real world conditions?
(2) What features and values should a care professional attribute to this point of view in the context of real medicine?
(3) How can intelligences develop within a professional culture, and how can they be mobilized in various *in situ* settings?
(4) What should be the role of guidelines in the dynamic conditions of clinical practice?

(5) When care providers encounter different thinking styles (clinical judgment, diagnostic reasoning, critical reflection, rational argumentation, dialogue), how can they avoid misconstruing the logic and objectives of the different styles?
(6) How much of present day medicine is a social service rendered by professionals in personal relationships with their patients, and how much is pure business?

One should ponder carefully what is at stake in this challenge to the realistic study of medical cases and health systems, in general.

1.1.5. *A brief note on logical thinking in ancient Greece and China*

Undoubtedly, this section does not provide any detailed study of the ancient Greek and Chinese logic and language. Yet, it aims at pointing out that some obvious comparisons are possible when notions, assumptions, thinking modes and argumentation techniques turn up in both cultures, as well as showing that intriguing conclusions can be drawn when their settings are different, thus having different implications in medical reasoning and inference.

In the West, logic essentially started with *Aristotelian logic*, which has had an unparalleled influence on the history of Western thought (Smith, 2012).[9] Aristotle's logical works contain the earliest formal study of logic that we have. It is therefore all the more remarkable that together they comprise a highly developed logical theory, one that was able to command immense respect for many centuries. Even nowadays several studies on the application of the very techniques of mathematical logic to Aristotle's theories have revealed a number of similarities of approach and interest between Aristotle and modern logicians (Lear, 1980). According to many experts (Graham, 1967; Hansen, 1998), technically, classical China had not worked out any systematic logical system. It used, however, some important logical laws and rules, known in Greek logic (e.g., non-contradiction, *modus ponens*). In fact, Wang (2012) argues that the medical theories and clinical practice of ancient Greek and Chinese medicine had significant similarities and that they also produced meritorious results working in separation from

---

[9] Readers may appreciate that in the Hellenistic period, *Stoic logic*, in particular the teachings of *Chrysippus*, was also very influential (unfortunately, Chrysippus' works have not survived).

each other. Other authors (Sivin, 1995) have cited the view that there was a profound complementarity of Greek logic and Chinese *semantics*.[10]

Mozi (Master Mo, a contemporary of Confucius) is credited with founding the *Mohist* school, whose remarkable *Canons* dealt with issues of valid inference and conditions of correct conclusions (*inter alia*, they contained an approach to logic and argumentation that stressed rhetorical analogies over mathematical reasoning, and is based on the *three fa,* or methods of drawing distinctions between kinds of things). In 4th century BC, the best known among the Mohists were *Hui Shi* and *Gongsun Long.* The Canons began with a long list of definitions considered as standards of moral, psychological, geometrical and logical terms. Examples of definitions of logical terms include the term "All" meaning "none not so"; and the term "Some" meaning "not all". Another example may illustrate some similarities between the Aristotelian and the Mohist schools.

Example 1.8: The Mohist Canons argued that to each name we associate an "is this" and an "is not" (e.g., "is it or is it not the case that....?") The "is not" generates an opposite for each name and marks the point of distinction. Thus, in a distinction/dispute (*bian*), one party will always be right (Chinese doctrine portrays disagreements as arising from different ways of making the distinctions that give rise to opposites). This argumentation can be seen in the context of the Aristotelian *principles* of logic. Given a statement and its negation, Aristotle's *non-contradiction* principle asserts that at most one is true (not both), whereas his *excluded middle* principle asserts that at least one is true (not neither); taken together they assert that exactly one statement is true. Mohists also analyzed *antinomy*, known in Greek logic as $\alpha\nu\tau\iota\nu o\mu\iota\alpha$. It is worth noticing that the logical inference underlying *Lao-Tsi's* ethical arguments is an illustration of the *chain rule* of formal logic.

In certain parts of the Canons there was a shift of interest from names to sentences and to the deduction of one sentence from another. Although the Chinese did not analyze deductive forms, the Mohists noticed that the formal parallelism of sentences does not necessarily entitle us to infer from

---

[10] Briefly, logic is concerned with the forms of thought and its expression, whereas semantics deals with the signification and meaning of words. Semantics is what was discussed, e.g., by Chinese people known as *Ming-Chia. Syntactics*, on the other hand, deals with the rules governing the arrangement (syntax) of words and symbols in an assertion (the rules are distinct from the meanings which words convey).

one in the same way as from another, and they developed a procedure for testing parallelism by the addition or substitution of words.

Example 1.9: According to the Canons, while "asking about Mr. Shi's Hepatitis E disease implies asking about Mr. Shi", on the other hand, "disliking Hepatitis E disease does not imply disliking Mr. Shi".

When studying ancient Chinese arguments many investigators find it appropriate to focus on the admissibility of the premises and the conclusion rather than on whether they are true or false (as is the case in ancient Greek logic). In general, ancient Chinese thinkers do not seem to make what from a Greek logic viewpoint may be considered logical mistakes, although, on occasion, they can commit epistemological fallacies (Chmielewski, 1962, 1963a, 1963b, 1965a, 1965b, 1966, 1968, 1969). This is an observation that one might find appropriate to relate to the notion that truth is more epistemological than logical, as many scholars maintain. In the pragmatic spirit of ancient Chinese philosophy, one finds more often in medical texts the kind of reasoning modes that are persuasion arguments rather than logical proofs. Before leaving the topic, it should be noticed that, no doubt, the logical heritage of the later Mohists is the most developed piece of ancient Chinese logic. Unfortunately, for several centuries the rule of *Legalism* repressed the Mohist line of investigation. The study of logic in China was revived following the transmission of Buddhism in China, which introduced the Buddhist logical tradition. The importance of the more detailed Mohist work came to light in modern times.

1.1.6. *Enter stochastic medical reasoning*

The readers may appreciate the less known fact that Aristotle's father was a prominent physician. This fact may be, at least in part, responsible for Aristotle's inculcation of the thought of ancient Greek scholars concerned with the study of health and the natural elements (Modell, 2010). Remarkably, Aristotle valued medical observation, but he also argued that logic should be carefully integrated with observation to confirm medical hypotheses. He applied his approach in the study of animals and the human body. This led to a broad range of medically- and biologically-related works, most of which, unfortunately, did not survive. In works such as $Περί\ Ζώων$ (*On Animals*) and the lost $Ανατομών$ (*Dissections*),[11] he actively dissected

---

[11] It is estimated that three quarters of Aristotle's works have been lost.

organisms and recorded his findings (Lennox, 2011). Aristotle is properly recognized as the originator of the scientific study of life.[12] Before Aristotle, very few of the Hippocratic works are both systematic and empirical, and their focus is exclusively on human health and disease. By contrast, Aristotle placed the *empirico-logical* investigation of living things at the center of the theoretical study of nature. Aristotle deepened the fourfold theory of the elements with anatomic and physiologic observations, he laid the foundation of comparative anatomy, and he established embryology on a scientific foundation by his direct studies of the chick embryo.[13] Aristotle was the first to correct the erroneous views of the time regarding blood vessels, which were thought to arise from the head and brain. His description of the hectocotyli arm was about 2000 years ahead of its time, and widely disbelieved until its rediscovery in the 19th century. The corpus of medically-related works by Aristotle had a direct influence on subsequent Greek thinkers (like Galen) and on medieval Islamic and Western scholars (including William Harvey and Charles Darwin). Aristotle's ideas continue to influence modern medical thinkers in the West through his writings' enduring impact on medical researchers balancing scientific and more personalistic approaches to medicine, and also gaining the attention of medical ethicists (Modell, 2010).

The Aristotelian influence is also present in the line of thought leading to the threefold thesis of this book:

(a) the space–time study of medical cases is not merely a technical subject (experimental, analytical, computational), but one that transcends different *theories* of knowledge (epistemologies) and *domains* of knowledge (physical, biological, cultural);
(b) medical reasoning should be based on a well-grounded *fusion* of clinical observation and logical rigor (which may include identification of different systems of logic contained in Greek and Chinese traditions); and
(c) a medical study should appreciate the *process* of problem solving (allowing health care professionals to think about the disease as well as about the patients in front of them), not just its outcomes and quick results.[14]

---

[12] Charles Darwin considered him as the world's greatest natural scientist (Malomo et al., 2006).
[13] His preformation theory of embryonic development survived until the 17th century.
[14] A significant obstacle to a meaningful synthesis is the shortsighted perspective that opposes every inquiry that is not translated to instant monetary profit. Among the vital domains of inquiry that have suffered from this worldview is basic research. It is

Underlying this threefold thesis, *inter alia*, is the strong belief that guidelines alone are not the solution to medical decision-making (as is the situation with the guidelines for operating an industrial apparatus or a mechanical equipment).

The views outlined above fit well with recent trends towards a broad knowledge synthesis that is a dynamic process fusing medical sciences and spatiotemporal stochastics[15] with content- and context-dependent logics. The need for this synthesis has led to the development of *stochastic medical reasoning* (*SMR*), in which decision-making is an undertaking illuminated from different angles (scientific, logical, philosophical, empirical) and uncertainty management seeks a balance between the substantive and the technical. In the synthetic milieu, *SMR* emphasizes theory-based[16] and evidence-supported thinking, as opposed to mainly rhetorical, descriptive and *ad hoc*. This includes the *quality assessment* of medical assertions and health care decisions on the basis of the process by which they were generated and by their ability to explain new clinical data and ill-understood symptoms. *SMR* properly emphasizes the role of *quantification* in medical investigations. Rigorous quantification can keep investigators from going too far astray. If the description of a diagnosis or treatment is purely qualitative and verbal, experience shows that it is easy to bend the meaning of words and to modify conceptual relationships to make a medical case appear to be fully understood when, in fact, it is not.

At the heart of *SMR* inference lies the notion of *health status*, and the possible ways it could be measured. Generally, one distinguishes between the health status of an individual and that of a population. Individual health status is measured by a care professional who examines the patient and decides about the presence or absence of illness, its severity (e.g., life-threatening or of mild severity), disease risk factors and overall health. In addition to the objective health measures above, the adequate assessment of an individual's health may benefit from the individual's subjective perceptions (e.g., biophysical functioning, pain or discomfort, emotional well-being and overall impression of health). Population health, on the other

---

infinitely saddening that the greed generation has no interest for long-term investment on knowledge that will actually give fruits for the future generations rather than "here and now".

[15] The readers are reminded that the term "stochastics" refers to the mathematical study of phenomena that vary in space–time under conditions of uncertainty (e.g., Christakos and Hristopulos, 1998).

[16] Theory should not be confused with mere opinion.

hand, is determined by aggregating data collected on individuals. However, while the individual status can be placed along a continuum from perfect health to death (once the definition of optimum health for the individual is agreed upon), this is not possible in the case of entire populations. Since it is rather unlikely for an entire population to die, one cannot define the population-level equivalent of death. Similarly, since it is highly unusual for an entire population to share the same health standards, one cannot define the population-level equivalent of optimum health status. In the absence of comprehensive or absolute measures of the actual health status of a population, the average life span, the prevalence of preventable diseases or deaths and the availability of health services currently serve as measures of indicators of health status. In such a framework, judgments regarding the level of health of a particular population are usually made by comparing one population to another, or by studying the trends in a health indicator within a population over time. Some commonly used measures of population health status are: morbidity measures (incidence rate, prevalence) and mortality measures (death rate).

SMR inference assumes a rational individual with adequate cognitive skills and professional background (say, a physician, an epidemiologist, an environmental scientist or a quality care professional). This includes individuals capable of following the consequences of what they know as far as they logically and cognitively lead them. The term "assertion" refers to beliefs, statements and claims about disease diagnoses, prognoses and treatments in the course of clinical and medical practice. The term "case (or health) attribute" may refer to any quantitative characteristic of the health status of an individual case or of a population group. Among others, the term includes body temperature, heart rate, blood pressure and glucose levels, degree of cellular dysplasia/anaplasia, circadian rhythmicity, melatonin, tumor malignancy, cholesterol and hemoglobin levels, and disease incidence, prevalence and mortality.

1.2. Health: The Fundamental Roles of Space–Time and Uncertainty

As the field of medical decision-making matures, one hallmark of its progress is that it eventually begins to appreciate its own boundaries. This includes a certain number of problems that have not been overcome as far as rigorous decision-making is concerned. *Inter alia, space–time dependence* and multi-sourced *uncertainty* are two key notions that seem to have been neglected or insufficiently studied in the medical and public health literature.

Real world health phenomena hold space–time dependencies and relationships among each other in a way that reflects the underlying mechanisms (biophysical, environmental, social). The fundamental relationship between space–time and the laws governing whatever exists within it can be developed in a general *SMR* setting: space–time, which is the container of every natural system, is also the arbiter of whatever this system may be. An initial realization of the fundamentally spatiotemporal structure of the notion of health status was that it may differ considerably from a person in one place and time to a person in another place and time and also in the same person from time to time. Analogous is the case with entire populations. Subsequently, these variations have been associated with differences in physiologic functions and changing environments. People's health clearly depends on the environment in which they live (the air they breathe, the water they drink, where and how the food they eat is grown). Accordingly, where and when a person has lived must be considered as part of the context in which clinical decision-making occurs. In symbolic terms, at any specified location–time coordinates $p = (s, t)$, where s denotes location and t is time, a *case attribute* will be generally denoted as CA_p. In light of the discussion above, CA_p may denote disease symptom DS_p, disease incidence DI_p, human exposure HE_p etc.

Furthermore, in medical practice disease diagnosis is an uncertain affair, for a variety of reasons, including the fact that many times a symptom array rarely constitutes a definite proof of the presence of a disease, physicians frequently interpret symptoms in different ways, and they are not fully confident of the statements they make.

Example 1.10: At the regional population health level, a typical reasoning situation is the study of the distribution of an infectious disease.[17] This is basically a twofold process (Christakos and Hristopulos, 1998):

(1) a structural space–time aspect that includes an adequate appreciation of the disease's dependence structure (biophysical laws, epidemic models, population dynamics and cultural features); and

---

[17]Such as diseases that are caused by microbes (bacteria, viruses, fungi and protozoa) and that spread. They vary from the common cold, ear infections, tonsillitis and the flu (influenza) to pneumonia and mononucleosis.

(2) an uncertainty aspect in the physician's thinking that relates to incomplete knowledge about certain disease features (symptoms, diagnosis, causes, trends, spread, frequency and duration).

These two aspects are the principal drivers for the determination of space–time change in the medical decision and health care setting, acting in synergy, interacting and exchanging information, when appropriate.

Technological advances in $TGIS^{18}$ have brought together health care providers with investigators from a wide range of disciplines (environmental science, physics, ecology, sociology, economics, statistics), which has increased the need to adopt a rigorous spatiotemporal perspective to disease and health. This being the case, it is rather inexplicable that certain health studies neglect fundamental aspects of the space–time domain and the associated uncertainty of disease spread, with no regard to sound science, and the production of rather unrealistic results.

Example 1.11: Disease transmission between susceptible individuals in a geographical region has been studied before without the rigorous consideration of space and related interactions in the proposed disease model. *Exotic infection* modeling (Roberts, 2004) involves questionable simplifications such as temporally unchanging the number of susceptibles, neglecting key parameters of disease variation (without any real justification) and turning a blind eye that ignores the evidence before it, whereas the consideration of vital space–time dependencies and interactions is notably absent. What is left is a simplistic temporal analysis (of what is actually a fundamentally spatiotemporal phenomenon), which assumes a linear model approximation limited to a few isolated micro-locations.

In some other cases, the use of medical decision support tools in conditions of uncertainty has been empirically supported, yet reasoning mechanisms underlying the results are poorly understood and, as a consequence, the tools are not easily generalized to a range of health care decisions (Reyna, 2008). For reasons that are not clear, results that were obtained in one disease setting have not been possible to replicate for other settings. Since spatiotemporal structure and multi-sourced uncertainty are two basic aspects of everyday practice, an appropriate model should account for these aspects as well as for their relationships and interactions.

---

[18] Temporal Geographical Information Systems.

This is of particular importance, e.g., in the resolution of documented medical disagreement that may involve a certain degree of departure from rigorous scientific thought. The adequacy of this resolution often makes the difference between mere treatment and definite cure.

Example 1.12: Characterization of space–time *disease spread* due to natural causes is an essential matter, for it provides the necessary background for the study of two crucial components of scientific inquiry: disease causation (etiology) and prognosis. Different methods of knowledge acquisition are characterized by different sources and degrees of uncertainty. When the study investigators themselves search for possible answers to a series of symptoms, the associated uncertainty is of another kind and degree than when these investigators rather rely on experts (specializing in certain aspects of a disease). In turn, the above types of uncertainty may differ from that of studies where many independent investigators search for the truth in a specified domain. Some study participants may propose concepts of "evidence" that are based on assumptions incompatible with those scientists make when they speak of and offer evidence for hypotheses (hypothesis is a term that here refers to any kind of inference made by a health care provider concerning symptoms, diagnoses, treatments, causes, traits, categories, explanations, stereotypes). And all the above uncertainties are carefully distinguished from uncertainties (and even obscurities) characterizing studies in which a privileged group of investigators is given the exclusive right to search for the truth in the given health domain.

As we will see in more detail in the next chapter, uncertainty frequently emerges from different sources, such as incomplete medical knowledge, conceptual errors, lack of adequate tools and techniques, and biases of various kinds (experimental procedures, subjective assessments, professional prejudices).

Example 1.13: A case of interest is the study of the potential health effects of *radioactive repositories*. Such a study must combine information of various types and levels of scientific and logical justification: natural analogues with long time-scales but ill-defined boundary conditions; better constrained but with medium time-scale field experiments; and less realistic with short time-scales but fully controlled laboratory tests (Toth, 2011). In all the above cases, the relevant methodology must possess capacities in multi-uncertainty source identification and modeling.

In the end the most important medical decisions and clinical actions are not merely procedures or prescriptions but the judgments from which all other aspects of clinical medicine flow. Yet, since the study of health care is a multi-thematic process, naturally a combination of closely linked abstract (conceptual) and concrete (case-specific) issues emerge in the medical decision-making setting. This is the topic of the following sections.

1.3. Abstract and Concrete Modes of Thinking

Do medical practitioners duly appreciate deep and broad thinking in conditions of uncertainty? Undoubtedly, the type of thinking that directs decisions is related to the phase of one's professional development. In a presentation given at the 1943rd Stated Meeting of the American Academy, Jerome E. Groopman (2010) remarked that medical students and Harvard house staff "were not thinking deeply or broadly about the patients under our care" and, he continued,

> "I stopped myself because I realized that to teach these young doctors to think better I had to know how I thought as a physician. And I realized that despite all my training at prestigious institutions no one had ever really taught me to think; and at times (many times) I did not think deeply or broadly".

Many readers would agree with Groopman that physicians ought to know *why* they get it right and *what* accounts for the times *when* they get it wrong (e.g., when they misdiagnose). Accordingly, it is imperative that care providers are well aware of their *mode of thinking* when making decisions about diagnoses, prognoses and treatments.

We continue with some observations. Sooner or later, most health care professionals are confronted with a familiar kind of disciplinary tension that can take at least three formulations:

Experimentalists (who criticize modelers for their complicated theoretical constructs that are, supposedly, of no practical use) vs. *modelers* (who complain that the medical experiments are inadequate or even false representations of scientific theory);

Particularizers (who argue that theorists neglect critical complexities in their eagerness to assimilate a medical case of interest into their favorite theoretical setting) vs. *generalizers* (who complain that particularizers are so immersed in idiosyncratic detail that they miss the big theoretical picture); and

Formalists (who accuse substansivists for developing conceptions of the *in situ* medical case that lack mathematical rigor and consistency) vs. *substansivists* (who complain that formalists exhibit excessive adherence to recognized forms and symbolisms rather than substance and medical content).

This situation is probably one of the first domains of medical thought to which the logic of synthesis could be fruitfully applied. A noteworthy feature of the synthesis embodied in health studies is that it proposes the fusion of two thinking modes:

(1) *abstract* mode (reflecting on concepts, attributes and relationships separate from the specific medical case characterized by these attributes and relationships); and
(2) *concrete* mode (focusing on specific details of the case under consideration, on empirical issues and on technical data analysis).

Discussing similarities and differences between that which is unfamiliar and distant (i.e., abstract) and that which is familiar and close to home (i.e., concrete) can help an investigator grasp the deeper meaning of the actual case. The careful blending of abstract and concrete thought plays a key role in a professional's reasoning. It helps formulate the appropriate framework of a health problem within which a solution will be sought. Becoming an expert in a medical field requires that the physician's knowledge is twofold: from abstract to concrete, and from explicit to implicit. The two modes of thinking and their interconnections are dynamic, and the degree of the connections is predicated on the space–time domain of the study but also on the health care provider (e.g., different doctors are likely to experience varying connections or degrees of connections between the thinking modes; Groopman, 2007). Accordingly, mode blending offers an explicit way of structuring the physician's reasoning and problem solving.

In view of these and similar considerations, medical data processing needs to be internally *consistent* (free from logical contradictions), *concrete* (generate accurate diagnoses and results that agree with medical reality) and *abstract* (account for an ever larger variety of cases). A sophisticated formulation of a health problem and its solution is useful when it helps the professional gain insight concerning vital facets of the actual phenomenon and is applicable in practice with considerable efficiency. When medical decision models satisfy neither of these two conditions, the professional may pose the question: what is the point of developing complex technical formulations that are neither sharp and theoretically insightful nor flexible in handling realistic medical cases and health systems?

1.4. Issues of Sound Medical Decision-Making

At this point, we briefly describe the lay of the land in medical decision-making. Several studies relating general decision theory to medical issues have appeared in the literature during the last three decades (Dawson and Cebul, 1992; Peng and Hall, 1996; Vickers and Elkin, 2006; Karnon et al., 2010). One notices an increased interest in quantitative methods, analytical techniques and an improved understanding of the theory–practice interface (e.g., methods that prescribe how decisions should be made and those that describe how decisions are actually made). In a few words, medical decision-making and its related technology gradually proceeds from a research and standardization context to a concrete and productive *in situ* setting. Yet, implementing decision-making in medicine proved to be considerably more difficult than developing it. Formulating answers for individual patients' dilemmas using decision analysis techniques is often costly, time consuming, and requires special expertise. Also, many experts argue that a consensus approach toward diagnostic decision-making has not yet emerged, with the overall diagnostic error rate remaining unacceptably high (Croskerry, 2009).

It is true that clear-cut statements and decisions are frequently desirable in medical practice and public health planning. It will be helpful to have a few examples to refer to as the list below:

(1) Does an individual suffer from a specific disease or not?
(2) Does the environment affect the presence of a symptom or not?
(3) Has an individual been exposed to a hazardous environment?
(4) Should a physician administer a drug or not?
(5) How can one avoid recommending diagnostic or therapeutic interventions that are no longer considered essential?
(6) What is the speed of disease spread and what factors can affect it?

Yet, in many cases of clinical judgment and medical decision-making, definite answers to the above questions are not a pragmatic possibility. As noted earlier, physicians are rarely completely sure of the assertions they make, they frequently disagree with each other, and the symptoms do not always provide definite proof of the presence of a disease, whereas diagnosis and prognosis are inherently uncertain. Diagnostic errors due to poor clinical judgment and decision-making are recurrent and usually under-appreciated (Croskerry and Nimmo, 2011). In fact, although the true overall prevalence is unknown, it is estimated to be in the order of 10–15% (Schiff et al., 2009). Accordingly, the ways in which health care providers think is an extremely important part of providing safe health care. Let us continue with a review of the key elements of an investigation.

1.4.1. Key elements of a medical investigation

In the 5th century BC, by examining the condition of his patients, Hippocrates carefully documented all physical demonstrations of their diseases and their individual complaints. His approach today is called the *clinical symptomatology* of a patient's condition. Subsequently, Hippocrates used the well-known formula for defining disease by description, and a series of common clinical terms are routinely used nowadays (e.g., alopecia, apoplexia, erythema, exanthema, lepra, leuke/leukoderma, lichen, melancholia, oedema, pachyderma, poliosis and psora/psoriasis). In modern medicine, among the key elements of a typical medical investigation are disease symptoms, disease diagnoses, diagnostic procedures and standards and case prognoses.

Disease symptoms (DS) constitute the actual facts of the medical case (nausea, pain in the heart, bodily fever, a dry mouth that has no thirst, hands covered in warts, cold hands alternating with cold feet, burning sensation, memory weakness). The factual feature of DS, once more, calls the physician's attention to the substantial viewpoint that care providers are entitled to their own opinions but not to their own facts. Very few DS can become too exclusive and many DS can become confusing. Hence, in practice the aim is to extract a reasonable number of DS with the maximum significance and relevance to the case.

Disease diagnosis (DD), often simply termed diagnosis, is concerned with the process of determining or identifying a possible disease from its symptoms. A DD can take many forms, which are generally uncertain and provisional. It may be concerned with naming a disease, dysfunction or disability. It can be a brief summation or an extensive formulation, occasionally taking the form of a story or a metaphor. In fact, professionals are familiar with many different types of diagnoses, including those listed in Table 1.2. DDs used for prognostic or therapeutic recommendations are evoked after they are assessed for their adequacy in explaining existing

Table 1.2: A list of *DD* types with their descriptions.

DD type	Description
Clinical	Made on the basis of medical signs and patient-reported symptoms, rather than diagnostic tests
Laboratory	Based on laboratory reports or test results, rather than the patient's physical examination.[1]
Radiology	Based primarily on medical imaging results.[2]
Prenatal	Specific diagnosis work done before birth.
Self	Identification of medical conditions by the patient oneself.[3]
Principal	Most relevant to the patient's chief complaint or need for treatment.
Admitting	The reason why a patient was admitted to the hospital.[4]
Discharge	Recorded when the patient is discharged from the hospital.
Differential	Identifies all possibilities connected to signs, symptoms and lab findings, and then rules out them until a final determination is made.
Exclusion	A medical condition whose presence is not established with complete confidence from either examination or testing, and is made by elimination of all other reasonable possibilities.
Dual	Diagnosing two related but separate medical conditions.[5]
Remote	Diagnosing a patient without being physically in the same place as the patient.
Computer-aided	Allows the computer to diagnose the user to the best of its ability.[6]
Nursing	Rather than focusing on biological processes, it identifies people's responses to situations in their lives.[7]

[1] For example, a proper diagnosis of infectious diseases usually requires both an examination of signs and symptoms, as well as laboratory characteristics of the pathogen involved.
[2] For example, greenstick fractures are common radiological diagnoses.
[3] Very common and typically accurate for everyday conditions, such as headaches, menstrual cramps and headlice.
[4] Which may differ from the discharge diagnoses.
[5] The term almost always refers to a diagnosis of serious mental illness and substance addiction.
[6] Health screening begins by identifying the part of the body where the symptoms are located. For example, the computer cross-references a database for the corresponding disease and presents a diagnosis.
[7] For example, a readiness to change or a willingness to accept assistance.

clinical findings (positive, negative, normal) and for their pathophysiologic reliability (checking the reasonableness of causal linkages between clinical events; Kassirer et al., 2009). Interestingly, *DD* may be influenced by non-medical factors such as power, ethics and financial incentives for the health care providers or the patients.

Diagnostic procedure (*DP*) is a method used to obtain a diagnosis for a particular disease (physical examination, lab test, biopsy, autopsy, surgery etc.). The actual *DP* is a cognitive process during which the physician assimilates various data sources to reach a conclusion about the case. The initial diagnostic impression may be broad involving a category of diseases rather than a specific disease or condition. Subsequently, the physician uses follow-up tests and procedures to get more data to support or reject the initial diagnostic impression and narrow the diagnostic possibilities down to a more specific level. Often in real practice, diagnostic hypotheses are quickly generated with only minimal clinical data and then used as a framework for further focused information gathering. Subsequent assessment and possible modification of these hypotheses involves probabilistic and causal reasoning modes (see also Kassirer (2010)).

Diagnosis procedure *standard* (*DP std*) is a term used to define the actual disease status against which the results of a new diagnostic test are currently compared. In other words, a *DP std* is a sort of a "gold standard" that is currently considered to generate a definitive diagnosis of a disease. For illustrative purposes, Table 1.3 displays a list of *DP std* that are widely considered as definite diagnostic procedures that will confirm whether or not an individual has the disease. Some of these procedures are quite invasive and this is a major reason why new diagnostic procedures are being developed.

Table 1.3: A list of common diseases and the corresponding *DP std*.

DD	DP Std
Coronary stenosis	Coronary angiography
Breast cancer	Excisional biopsy
Prostate cancer	Transrectal biopsy
Myocardial infarction	Catheterization
Coronary stenosis	Coronary angiography
Lynch syndrome	Amsterdam criteria
Multiple sclerosis	McDonald criteria
Systemic lupus erythematosus	American College of Rheumatology criteria
Strep throat	Throat culture

Part of *DP* is the *a priori* identification of the case's key biomedical notions. Different physicians may chose different sets of key notions — a choice which is critical in correct case diagnosis.

Example 1.14: In the case of *hyperkalemia*, the key biomedical notions are glomerular filtration rate, transtubular potassium gradient (*TTKG*) and anion gap. The correct implementation of these notions by a nephrologist is cited in the following (McLaughlin *et al.*, 2010):

> "I want to look at the urine handling of potassium. TTKG is 10.1 — so the principle cell is doing what it should be doing in the face of hyperkalemia, although I note that the urine chloride is low — so there may be some reduction in the distal delivery of sodium to the cortical collecting duct. Creatinine is 112 — so there is enough glomerular filtration for it to be able to filter potassium and lead to its excretion. The likely explanation for hyperkalemia in this setting is due to a shift. So what might cause the shift? Bicarb is 12; she has a normal anion gap acidosis. This would lead to the buffering of hydrogen ions and displacement of the potassium from the ICF to the ECF causing hyperkalemia which, given the lack of evidence to support cell lysis or any other cause, would suggest that this is the diagnosis here."

McLaughlin *et al.* (2010) found that diagnostic success increases with the number of key biomedical notions involved. The improvement was more distinct in novice physicians than in experienced ones (biomedical knowledge augments performance in a novice, but it may be redundant in experienced physicians). Novices' diagnostic success equaled that of experienced physicians when both used the complete set of key biomedical notions of a case. The crucial role of biomedical knowledge in the diagnostic performance of novices presumably reflects the dearth of their clinical knowledge. Experienced physicians were able to supplement their biomedical knowledge with clinical knowledge, which made them less dependent upon biomedical knowledge, thus performing better than novices despite applying the same biomedical knowledge. Another explanation is that experienced physicians actually applied more biomedical knowledge than novices, but did so indirectly, through its encapsulation within clinical knowledge.

Case prognosis (*CP*) is concerned with the prediction of the course of a disease based on the preceding medical investigation elements (*DS*, *DD*, *DP*). More specifically, the complete *CP* includes the expected duration of the disease, its function and a description of the disease course (e.g., progressive decline, intermittent crisis, or sudden, unpredictable crisis). For certain diseases prognostic scoring is used to predict disease outcome. A *Manchester score*, e.g., is an indicator of prognosis in small-cell lung cancer. For *non-Hodgkin lymphoma*, the international prognostic index has been developed to predict patient outcome. Also, prognostic indicators are used in drug-induced liver injury and an exercise stress test is used as a prognostic indicator after myocardial infraction.

From a research perspective, *DD* and *CP* constitute a similar challenge: the clinician has some information and wants to know how this relates to the true patient state, whether this can be known currently (*DD*) or only at some point in the future (*CP*). Interestingly, when applied to large human populations, *CP* estimates are frequently quite accurate: that "a medication has an 85% probability for curing a disease" means that in a group of 100 people with the disease being given the same medication (with all other conditions being the same) approximately 85 people would be cured and 15 not cured. However, one cannot say beforehand that, say, Mr. Volonte who is picked out from a group of people who take the medication will turn out to be one of the 85 people who would be cured.

1.4.2. *Reflection, recognition primed decision and robust decision methods*

In principle, medical decision-making is an *empirico-logical* process[19] in which care professionals use their knowledge and experience to make rational and informed decisions (evaluating a symptom, making a diagnosis, choosing a treatment, generating a prognosis). Just as in other scientific fields, medical decision-making has advanced from the narrow limits of mere empiricism to the broader realm of rationalism. Naturally, the quality of decision-making schemes varies widely. Some schemes ignore contextual information essential to conceptual retrieval, some give no consideration to semantic interoperability and even render little inferencing capabilities. Basically, one could distinguish between two major classes of medical decision-making methods:

(1) purely *empirical* or *intuitive* methods, and
(2) *analytical* methods based on a *rationally structured* methodology.

In clinical, geomedical and environmental health cases where the stakes are high, the time pressure is considerable and the *in situ* conditions

---

[19]The readers may recall that Aristotle was the one who initiated the empirico-logical investigation of living things (Section 1.1.6).

are uncertain, many practitioners seem to prefer the methods of *class* 1 rather than those of *class* 2, not always without a cost in quality care. In a professional's daily routine the "quick and dirty" solution is sometimes to resort to so-called *common sense* judgments and *intuitive* inferences (usually based on simple perceptions of the health situation). Real experience, however, provides convincing evidence that this kind of approach is often unreliable and leads to false conclusions with far reaching consequences (poor health risk assessment and emergency management, waste of precious resources; Ioannidis, 2005). Yet the same commonsensical mistakes are often made in the handling of medical cases, which calls to mind the ancient story of the *Danaides*.[20]

Taking into consideration such concerns, the *dual-process* approach of decision-making (Croskerry, 2009) suggests a combination of the intuitive (heuristic) mode (1) and the analytic (systematic) mode (2) described above. In this setting, well-calibrated reasoning requires the physician to be in the right mode at the right time. The intuitive mode (1) is instinctual, reflexive and rather effortless, characterized by first impression, rapid response to information, and reflexive rules of thumb. Diagnosis aspects belonging to mode (1) include symptom detection and hypothesis generation. Although intuitive clinical reasoning in many cases produces valuable and accurate results, due to its inherent features (quickness, lack of computation, little or no awareness or active thought), it can be influenced by the context of the moment (time pressure, uncertainty conditions, emotions), and is frequently prone to cognitive errors. The analytic mode (2) is a deliberate and mindful process that consciously considers case alternatives. It is slower than mode (1) since it requires considerable cognitive work and is solidly based on science, logic, causal links and probability calculus. Mode (2) is usually activated when the case pattern (patient's symptoms, clinical and laboratory results) is not clear and does not fit a recognizable clinical picture. Diagnosis aspects belonging to mode (2) include hypothesis testing, differential diagnosis and diagnosis confirmation. The analytic mode is less likely than the intuitive mode to be error prone. Based on a rigorous rational it is capable of checking and overriding the first impressions of rapid case recognition, if necessary.

Another important dimension of medical decision is related to public health *finances*. In a recent paper in the *Journal of the Americal Medical Association (JAMA)*, Berwick and Hackbarth (2012) estimated that as

---

[20] Creatures condemned by the gods to pour water endlessly into a leaking jar.

much as 30% of all health care spending is totally wasted. In a related note, Crosson (2012) observed that doctors' decisions account for about 80% of health care expenditures. In such a crucial decision-making setting, the rigorous incorporation of transparent information and credible data regarding the relative value and risk of the available medical interventions (diagnostic and therapeutic) can help reduce unnecessary tests, scrupulously avoid superfluous procedures, optimize the appropriate allocation of resources and improve quality care. Gradually removing barriers to decision-making will advance the knowledge status and thinking style of health care professionals in the best interest of the patient (achieving and maintaining a high-quality life for the population).

A health care provider needs to make some distinctions. There are cases in which *critical reflective* thinking is an essential component of good clinical reasoning and judgment. The daily routine of care providers requires that they examine thoroughly the assumptions underlying medical decisions and, if necessary, be prepared to question the validity of assertions and even facts about the disease at hand. In such cases, critical reflection skills enable one to gain a sense of salience (what to pay attention to) that informs one's powers of perceptual grasp. Reflection skills, however, may not be particularly useful in cases in which clinicians must decide and act quickly, while at the same time avoid patient injury. This example is from a study by Benner *et al.* (2008).

Example 1.15: Working under pressure to figure out a patient's alterations from the well-established tenets of typical *human circulatory systems*, a clinician may not afford the time to critically reflect on that well-grounded understanding. Another situation in which critical reflection may not provide what is needed for a clinician to act is the lack of adequate understanding of the differences between female and male circulatory systems and the typical pathophysiology linked to heart attacks. Current understanding is based upon multiple, taken-for-granted starting points about the general nature of the circulatory system, in which case reflection is of limited use and one rather needs to resort to science-based medical reasoning and judgment.

Some practitioners favor the so-called *recognition primed decision (RPD)* method (Klein, 1998): fitting a set of indicators into the practitioner's experience and quickly arriving at a satisfactory course of action, frequently without weighing alternatives. The *RPD* is another model of how practitioners would make quick, effective decisions when faced with complex medical situations. The physician should be able to generate a possible course of action, compare it to the case constraints and select the first course of action that is not rejected. RPD has been found to function relatively well in conditions of time pressure and incomplete information. Among the *RPD* limitations is that it requires practitioners with extensive experience in correctly and quickly recognizing the salient case features, especially in unusual or misdiagnosed circumstances (e.g., when enumerating symptoms, decision-making should make due reference to those symptoms that are indicative of danger, or even of fatal issue). On the other hand, proponents of *robust decision* methods (*RD*) claim that by formally integrating uncertainty into medical argumentation in a way that eliminates as much uncertainty as possible given the available resources, *RD* leads to the best possible decision with an acceptable risk level (Ullman, 2006).

1.4.3. *Algorithmic medical decision-making methods*

Most *algorithmic decision-making* (*ADM*) methods of medical diagnosis are the result of intensive collaboration between physicians and mathematicians (Miller, 1994; Niknam and Niknam, 2008). One should keep in mind that most of these methods rather provide explanations of how health care providers should make decisions than of how they actually make decisions. Also, the methods often demand a large number of restrictive and even unrealistic assumptions and a large amount of complex calculations.

There are diagnosis cases in which the implementation of the *Bayesian networks* theory has been efficient in medical practice (Kahn *et al.*, 1995; Nikovski, 2000; Lucas, 2001). Basically, a Bayesian network is a graphical model that encodes the joint probability distribution for a large set of variables. Among the limitations of the approach is that the computation of the large Bayesian networks is hard to handle requiring either large amounts of valid data or numerous approximations and assumptions, they are not expressive enough for many real world applications assuming a simple attribute-value representation (i.e., each problem instance involves reasoning about the same fixed number of attributes, with only the evidence values changing from problem instance to problem instance) and that the numbers, types and relationships among entities usually cannot be specified in advance and may have uncertainty in their own definitions. These limitations may impede the application of Bayesian networks to complex medical cases.

Among the disadvantages of *fuzzy reasoning* techniques are: the problem of adopting an appropriate fuzzy compositional rule of inference for making diagnostic decisions; the selection of possibility distributions parameters is often highly subjective; it is not easy to express the fuzzy inference model to a closed form; it is limited to applications that are based on crisp rules; and finally it may depend to a considerable degree on the selection and the quality of the initial set of rules (Vasilescu et al., 1997; Yao and Yao, 2001; Innocent and John, 2004; John and Innocent, 2005; Tsipouras et al., 2007). The *case-based reasoning* approach of medical decision-making (Yearwood and Pham, 2000; Hsu and Ho, 2004 and Holt et al., 2005) relies on the assumption that health care providers frequently relate the current case to similar ones they encountered in the past. Among the weakness of this approach to medical diagnosis is that the assumption is not always valid in professional practice, and when it holds it is difficult to produce an appropriate metric of case similarity (under what conditions two cases are sufficiently similar, what is the quantitative sense of "sufficient" etc.).

1.4.4. *Does expert knowledge translate into expert judgment?*

Methods producing intuitive or commonsensical assessments are usually organized around the potentially misguided assumption that "expert knowledge translates into expert judgment". The ostensible reason to hope that a physician's judgment is better than someone else's lies in the degree and substance of the physician's expertise. Experts know, recall and perceive more than novices do, and to become an expert requires dedication, focus, effort and meticulous self-assessment. Moreover, the potentially serious limitations of commonsensical, "quick and dirty" solutions is that decision-makers are confronted with a diverse body of uncertain information, in which case their reasoning is often logically inconsistent and contradictory, and they are frequently unaware of the hidden variables of the situation (including the incentives at work). Another essential reason contributing to the unreliability of many medical decision methods is the manifold nature of large-scale phenomena, such as epidemics and population dynamics. In these cases, it is much more difficult to derive "quick and dirty" solutions since a deeper study of the phenomenon involves many different knowledge sources, the analysis of which often requires investigators from different disciplines.

Example 1.16: The space–time maps of the spread of the *Black Death* in 14th century Europe (Christakos *et al.*, 2005) were the outcome of a process that involved a manifold of databases, including hospital data, ecclesiastical documents, court rolls, chronicles, guild records, testaments, church donations, letters, edicts, tax records, financial transactions, land desertion patterns, tombstone engravings, historical documents and artistic creations (paintings, poems, etc.). At the same time, the contributing study investigators maintained their different perspectives: while a historian sought to describe, explain and interpret what happened during the 14th century Black Death, the medical investigators aimed to understand how and why the Black Death epidemic happened. Gummer (2009) studied the devastating effects of the plague in the British Isles based on an insightful synthesis of historical data, sociological perspectives and scientific modeling.

Increasingly sophisticated mathematical and statistical methods are developed to handle quantitative data from varying sources, whereas language logics are used in the case of qualitative evidence. In professional practice, the *SMR* methods balance theory-based and evidence-supported decision-making that enables quality assessment of medical assertions in terms of the process by which they are generated and the ability to explain new data and previously ill-understood symptoms. These rationally structured methods suggest a context- and content-oriented decision process that is capable of embedding natural languages and non-formalized knowledge for feature extracting, reasoning and causation inference. Health situations such as the Black Death epidemic of Example 1.16 are characterized by the need that the contributing investigators make informed assertions to which they appropriate a level of justification and a degree of uncertainty (e.g., maintaining a belief that the plague was a viral disease is something the investigator must justify). These assertions take into account the fact that the available information sources transcend several disciplines and consist of both quantitative and qualitative data of varying credibility.

SMR notions and methods are particularly useful when they enable health care providers to test rigorously what are previously accepted as intuitively valid results and demonstrate the occasional unreliability of common sense in medical practice. As we will discuss in the following chapters, in many cases this crucial task is accomplished in practice by translating an intuitive medical argument in terms of the corresponding probability (or uncertainty) functions, and seeing whether or not the resulting formulation is logically and scientifically sound. The approach includes quantitative tools to aid medical decision-makers structure the clinical, exposure, epidemiological and economic evidence bases, and gain valuable insight to assist them in making better decisions. Basically, *SMR* is trying to develop reasoning in a way that retains, as far as possible, substantive interpretations of case assertions and health assessments, whereas medical decision-making embodies content-dependent arguments (scientific principles, empirical evidence) rather than content-independent ones. Lastly, *SMR* inference methods take into account the fact that understanding the real phenomenon is often a multi-thematic and multi-disciplinary affair, which is based on interpenetration, discovering connections and drawing analogies with other phenomena and even between seemingly remote topics. In a metaphorical setting, an investigator may study medical reality as an art critic comments about a painting by constantly referring to other paintings associated with different space–time contexts (e.g., noticing that a tree or a shade in an 18th century Dutch painting has certain similarities with those of 17th century paintings of the Italian school). The readers may appreciate the fact that in the Black Death epidemic (Example 1.15) the investigators gained valuable information from paintings of the time,[21] a fact that supports the knowledge-theoretic view that art is really tacit experiential knowledge.

1.5. Medical Dialectics and Knowledge Synthesis: An Outline

We started this chapter with the thesis that the 21st century will be the Age of Synthesis and, in light of the discussion of the previous sections, will close the chapter with a further reinforcement of this. As many argue, there is a "cultural turn" from thinking in disciplines towards the promotion of intellectually rigorous interdisciplinary practice. As regards the objectives of this book, *medical dialectics* is seen as a framework of logical debate in which differing views concerning disease diagnosis are debated, not with the idea that some of them should be necessarily declared to be true, but with the aim of resolving the differences between them, and so coming closer to the true diagnosis. Clinical arguments and human exposure notions often unfold via a succession of contradictions and resolutions; a thesis (concerning a

---

[21] See, e.g., Pieter Bruegel's (the Elder) 16th century paintings.

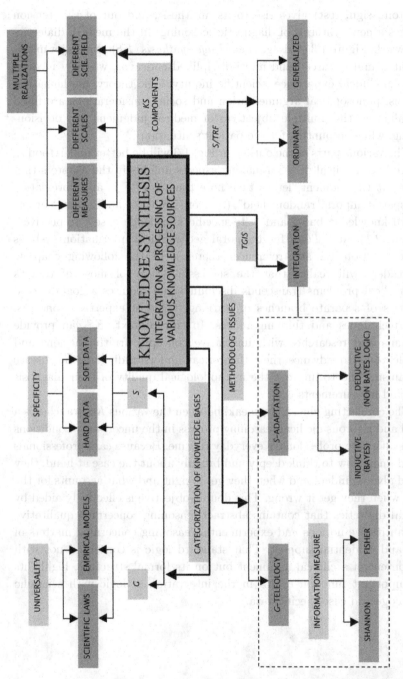

Figure 1.3: An outline of knowledge synthesis in medical sciences and its various components.

symptom, sign, test) gives rise to its antithesis, and out of this tension emerges a new synthesis of diagnostic reasoning. In the medical dialectics framework, Figure 1.3 envisages *knowledge synthesis* (*KS*) as a major multi-thematic, multi-sourced and multi-disciplined endeavor, whereby medical research, clinical experience, scientific inquiry, logic theory, mathematical analysis, philosophical argumentation and social considerations are fused in evaluating the multiple objectives of medical judgment and decision-making, while accounting for the patient's situation.

The various parts of the outline in Fig. 1.3 will be better understood as the study of medical decision-making aspects unfolds in the chapters that follow. At the moment, let us take note that *S/TRF* is an abbreviation for "spatiotemporal random field" (Section 1.8.5), *G* and *S* for "core or general knowledge base" and "site-specific knowledge base", respectively (Section 2.1), and *TGIS* for temporal geographical information systems (earlier Section 1.2 and references therein). In the following chapters the readers will realize that the successful study of most of today's real medical problems transcends disciplines and requires a goal-directed synthesis of separate branches of learning, fields of expertise, concepts, data, techniques and thinking modes. In this context, *KS* can provide clinicians and researches with innovative concepts, fruitful notions and reliable tools to advance their theoretical understanding of the disease mechanisms and to improve the methodological quality of their diagnosis tests and measurements.

The connecting glue, so to speak, between the various *KS* branches is a sound and rigorous medical reasoning process in the uncertainty conditions of a health care professional's everyday routine. Because care professionals should know how to think deeply and broadly about the case at hand, they should always understand when they get it right and what accounts for the times when they get it wrong. This double-objective is effectively aided by medical dialectics that contains abstract reasoning concerning qualitative or quantitative notions and experimental reasoning concerning matters of facts and evidential support. That standard logic is concerned not with the content of a clinical argument but on its formal structure, highlights the important function of *KS* in the integration of logic with scientific knowledge and case-specific data.

Chapter 2

Reasoning Amidst Uncertainty

2.1. When "To Know" Means "To Be Uncertain Of"

This chapter focuses on four major components of medical reasoning, namely, language, space–time, logic and uncertainty. But before proceeding any further, it may be appropriate to review some of the most common medical errors that are related to these components, to a larger or smaller degree.

2.1.1. Common medical reasoning errors

As noted in the previous chapter, the state of affairs in public health is quite heterogeneous and uncertain. In many organized societies the existence of a medical plan is the key to high-quality care. In other populations a healthy diet has an equal or even larger impact on people's health than the availability of a medical insurance plan. And there are cases in which a clean environment determines the population's quality of life. This being the situation in public health affairs, among the key objectives of medical reasoning is the extraction of high-quality disease symptoms (DS), on the basis of which the disease diagnosis (DD) determines or identifies a possible health problem, and allows an effective case prognosis (CP), i.e., the prediction of the disease course. Complex trade-offs exist between potential benefits and risks of different diagnostic tests and case treatments, and cognitive errors in clinical reasoning. Physicians typically work under tremendous time pressure, in conditions of uncertainty and with limited data available to them (Montgomery, 2006). Yet, important medical documents include almost nothing about *thinking* or *reasoning* (Kohn et al., 1999). Instead, they view the hospital primarily as a large factory, focusing on systems, techniques and procedures, and trying to pinpoint where on the "assembly line" the physicians usually fail (the latter includes misinterpreted guidelines, spread of infections due to poor hand washing, neurosurgeons operating on the wrong side of the brain or orthopedists amputating the wrong limb). Unfortunately, such studies do not explain why a considerable percentage of patients are misdiagnosed. This happens because the errors causing misdiagnoses, delayed or never-made diagnoses, mistreatments etc. are all *medical reasoning errors*. Errors may be due to disease manifestations that are not sufficiently noticeable, omitting possible diagnoses from consideration, giving too much significance to certain diagnosis aspects, dealing with a rare disease with symptoms suggestive of many other conditions or the condition having a rare presentation. Case misjudgments may be due to physicians' starting points (miscalculated *a priori* disease likelihood, overemphasized significance of positive tests) or their recollection of cases with similar characteristics. Physicians may jump to erroneous conclusions with limited information at their disposal, or they may judge prematurely that they possess a working diagnosis. As they themselves will admit, clinicians often start treatment while they are uncertain about the actual state of the patient (Sox *et al.*, 2013). Among the cognitive biases that are most prevalent and especially relevant to clinical reasoning are cognitive disposition to respond (CDR) and affective dispositions to respond (ADR) biases. As discussed in Croskerry *et al.* (2003, 2010) over 50 CDRs and 12 ADRs have been identified in the relevant literature. Jenicek (2011) has expanded the list of cognitive biases to over 100. These errors and biases are amplified when the physicians continuously concentrate on checking off boxes of close-ended questions and, at the same time, fragmenting the patient to fit the structure of the electronic record shown in their computer screens (Hartzband and Groopman, 2008). It then comes as no surprise that medical error and unaccounted uncertainty are increasingly recognized as constituting a major cause of mortality and morbidity. A 2004 investigation by the Hearst media corporation reported that preventable medical mistakes and infections were responsible for about 200,000 deaths in the US each year, a quite remarkable figure. Hence, as was emphatically noted in the previous chapter, it is very important that clinical practitioners are fully aware of their reasoning mode when making decisions that involve diagnoses, prognoses and treatments.

Interestingly, it was the laboratory confirmation of cognitive errors in psychology that made it possible for medical researchers to identify and classify cognitive errors in every stage of the diagnostic process (Kassirer and Kopelman, 1989; Redelmeier, 2005). Yet, despite the early recognition of cognitive errors and the fact that many of these errors can lead to life-threatening situations, due attention to them has been only a recent endeavor (Kassirer, 2010).

Human reasoning, in general, is the ability to abstract the essential elements of a situation and to understand the concepts and methodology underlying one's thinking process:

(1) Why one reasons in a particular manner (consistency of argumentation).
(2) What is the reliability of the available information sources.
(3) Why a physician has made the right decision.
(4) What accounts for when a physician made the wrong diagnosis.

Yet, if "to know" means "to be certain of", the term is of little use to health care professionals who wish to be undogmatic. *SMR* (Section 1.1.6) is rigorous and insightful thinking in the uncertainty conditions of clinical practice and *in situ* human exposure (associated with the investigator's epistemic or cognitive condition) and their space–time variability (tied to ontic aspects of a disease). In this setting, uncertainty analysis can serve various purposes, such as ascertaining the value of collecting additional information and ordering diagnostic tests, and assessing confidence in a chosen course of medical action to support the physician's decision better. It may be worth noticing that *SMR* differs from the systems-based thinking process (used, e.g., in business administration; Dettmer, 2007). *Inter alia*, the emphasis is on substantive reasoning based on rich semantics (scientific meaning, *in situ* case variation, conceptual uncertainty) rather than merely on formal systems theory and techniques. Moreover, *SMR* inference takes into account the fact that health care providers are thinking while they are acting, whereas systems theory assumes that the decision-maker has the time availability to study the data thoroughly and analyze all the decision steps in a systematic manner.

Another kind of error is sometimes made when quantitative analysis is part of medical thinking. The example below shows that careful consideration of the assumptions underlying a quantitative formulation is needed when one attempts to move from an empirical case-specific *rectal temperature–heart rate* relationship to a more general mathematical formulation.

Example 2.1: A group-specific study has shown that heart rate (HR, beats/min) and post-work rectal temperature (RT, in °C) satisfy the regression relationship (Edholm et al., 1962 and Tanaka et al., 1979):

$$RT - 0.0167HR = 35.97.$$

It is a common approach in scientific modeling that an investigator seeks to go beyond group-specific evidence and obtain a more general expression that applies in other cases as well. For mathematical convenience, let $X = RT - 0.0167HR$. Further formal manipulations of the regression relationship involve the steps:

[i] $35.97X = 35.97^2$.
[ii] $X^2 - 35.97^2 = X^2 - 35.97X$.
[iii] $(X + 35.97)(X - 35.97) = X(X - 35.97)$.
[iv] $X + 35.97 = X$.
[v] Last equation holds for $X = 35.97$, which gives $35.97 + 35.97 = 35.97$.

This is obviously false.

The rather straightforward procedure yielding the equation in *step* [v] overshadows an error made when moving from one *step* to the next. In particular, the term $(X - 35.97)$ in both sides of *step* [iii] cannot be canceled because the term is equal to zero.

The didagma of Example 2.1 is that a careful consideration of the validity of the underlying reasoning and mathematical calculations is needed when one attempts to move from a specific quantitative formulation of a medical case to a general one.

2.1.2. Physician's language and metalanguage

Much of what was different between Greek and Chinese medicine began with the medium of expression in each culture: *language*. Among other authors, Matalene (1985) has identified distinct systems of logic in Chinese and Greek language. Because of conventions established in ancient Greece, Western languages are governed by specific rules and conventions that determine correct expression or writing. In the West, physicians subscribe to Aristotle's (logical) dictum,

"state your case and prove it",

and they search for premises and conclusions connected by inductive or deductive reasoning. Chinese language is logically constructed and expressed, but the "hierarchy of culture, language and rhetoric" is logically different than in the West (Hansen, 1983).

A clinical practitioner recognizes that in its everyday practice, an essential component of medical inference is the language it employs. The tangles that physicians involve themselves in often arise out of misconceptions about the way that language works. Broadly speaking, the practical need for language consists of both communicating the generated medical statements and helping oneself in remembering and reconstructing previous clinical test results. Given these two practical functions, the more precise the language and syntax of clinical assessment and medical decision-making are, the higher the quality of the provided care. The significance of language and syntax in critical reasoning and medical judgment is illustrated in the following case study (Redelmeier et al., 1995).

Example 2.2: An emergency admission was described in detail to a group of 148 medical students. Half of the students were presented with the following statement:

> "obviously, many diagnoses are possible given this limited information, including CNS^1 vasculitis, lupus cerebritis, intracranial opportunistic infection, sinusitis and a subdural hematoma";

whereas the other half of the students were presented with the shorter statement:

> "obviously, many diagnoses are possible given this limited information, including sinusitis".

Subsequently, the two groups were asked to decide whether they would recommend ordering a CAT^2 scan of the head. Logically, since the two statements describe the same situation, there should be no difference between the students' responses. However, it turned out that 20% of the medical students recommended a CAT scan in response to the short statement, whereas 32% did the same in response to the long statement.[3]

While the notion of language is understood in the broad sense that includes statements and theories in linguistics, mathematics, physics and biology, the *metalanguage* (after-language) is generally a form of language

---

[1] Central nervous system.
[2] Computed axial tomography.
[3] This seems to contradict *support theory* (Tversky and Koehler, 1994), according to which the possibility of sinusitis should be larger when it is the only specified diagnosis than when it is accompanied by other specified diagnoses.

used to describe another language, the *object language* (original language). While object language is a language that speaks about nature and refers to a system of events and case attributes (body temperature, heart rate, blood pressure, cholesterol levels, cancer incidence, plague mortality), metalanguage is a system of assertions (statements, beliefs, propositions, claims) about these events and case attributes. As such, metalanguage is used when one speaks about symbols linked to events and attributes. "Validity," "truth" and "belief" are properties of assertions about health attributes, not of attributes themselves. The support theory, e.g., assigns probabilities not to events (object language) but rather to descriptions of events (metalanguage).

Example 2.3: The statement

"the patient's tumor is malignant"

belongs to the object language, whereas Dr. Wu's assertion

"I believe that the above statement is valid"

belongs to the metalanguage. Also, the statement

"the patient's body temperature (BT) rhythms will be much higher tomorrow"

belongs to object language, whereas Dr. Wu's assertion

"I expect that 'patient's BT rhythms will be much higher tomorrow'"

is part of the metalanguage.

As most professionals would readily admit, the language–metalanguage association is implemented in various contexts of their everyday routine, in some of which without being explicitly spelled out.[4]

2.1.3. *The notion of knowledge base*

At this point, it is appropriate to bring to the reader's attention some important convention and terminology. In medical sciences, what is called a *knowledge base* (KB) may include data, information and knowledge.

---

[4] As we shall see in a following section, one of these contexts is *probability theory*.

These three constituents may be considered in relation to each other in the hierarchy:

$$data \rightarrow information \rightarrow knowledge \rightarrow understanding.$$

In this hierarchy, *data* is the raw material of information without context, that is, a piece of data has no meaning unless the context is understood; data needs to be transformed to information. Accordingly, *information* is the result of processing and manipulating data in ways that add to the knowledge of the person receiving it. In other words, information adds sense to data; it is descriptive and can be presented in a wide variety of forms including text, images and sounds; and it provides answers to the types of questions that begin with words such as, *who, what, when, where, how many*. Information is that commodity capable of yielding *knowledge*, which is instructive rather than descriptive, and can provide answers to questions that begin with words such as, *how, how to*. Then, knowledge is a multifaceted concept with multilayered meaning (the history of philosophy since the classical Greek period is regarded as a neverending search for the meaning of knowledge). The readers may find it interesting that in certain cases, the distinction between data, information and knowledge reflects only relative judgments, that is, on occasion what is one person's knowledge may be another's raw data. For instance, what scientists think they know about the merits of flu vaccination is merely data to health administrators and policy makers. The highest level of the hierarchy is *understanding*, which is generative, i.e., it gives rise to creative insights and requires intuition.

A medical study relies on the construction of a KB with semantically rich information about the case and further contextual information. For future use, let S denote the *case-specific* or *specificatory KB*, which includes new data (hard or exact; soft or uncertain), case-specific information (unique elements that are particular and peculiar to the specific case) and the deliverance of clinical or field experience. This sort of KB expresses the polymorphisms of actual clinical material, and is disclosed to health care providers in a certain context, through their practical encounters with experimental results and observational data, and by communication with patients (listening to a patient's story, a physical examination). On the other hand, G denotes the established and accepted *core* or *general KB* (biophysical laws, medical training, scientific theories, phenomenological models, well-grounded expertise), which plays a vital role in critical reflection, reasoning and judgment. For example, the clinicians' skill in providing high-quality care depends upon their ability to think, reason and judge, which can be limited by the lack of adequate *G-KB*. Physicians rely heavily on core knowledge to enable them, before seeing the patient, to identify the most likely diagnoses and assign a chance of occurrence to each one of them. As such, core knowledge is a fundamental prerequisite of successful clinical argumentation, inference and decision-making. *G-KB* obtained outside physician *A*'s direct experience (e.g., book knowledge, published results) is linked to *A*'s semantic memory, whereas *S-KB* is linked to *A*'s episodic memory involving traces that represent databases of past experience. For example, an *S-KB* may be associated with 100 traces of malaria and 10 traces of cholera in *A*'s episodic memory, whereas the corresponding *G-KB* would consider unique representation of malaria and cholera in *A*'s semantic memory. Moreover, it is possible that although *A* may have never seen a patient with influenza (i.e., no record of malaria exists in *A*'s episodic memory), the physician would be able to diagnose a patient with malaria because knowledge of malaria and its associated symptoms were learned in medical school and are represented in *A*'s semantic memory.

In daily practice, the incorporation of core *KB* in disease diagnosis may be intentional or incidental. Several studies have shown that the use of core biomedical theory augments disease diagnostic performance (Lesgold et al., 1988; Gilhooly et al., 1997). Other studies maintained that expert performance in clinical medicine is not explained by experts simply processing more data (Schmidt and Boshuizen, 1993). McLaughlin et al. (2010) found a trend towards improved diagnostic performance when experienced physicians used core knowledge, but they also noticed that this may reflect the fact that the cases considered had helpful clinical information. Biomedical knowledge is sometimes redundant, but in cases where clinical information is limited, unhelpful or on occasion misleading, even experienced physicians rely upon biomedical knowledge to yield a correct diagnosis.

While mainstream (formal or discursive) logic relies on a set of rules (norms) by means of which a form of orderly closed-system thinking can be used, environmental and cultural factors affect *in situ* reasoning and the inferences drawn in the public and private medical practice. This leads to the study of reasoning as a non-egocentric individualism process (Christakos, 2010), which does not merely incorporate logical or rational thinking modes, but often goes considerably beyond that in the domains of insight, creativity, imagination and the unconventional. In concise terms, health care practitioners rely on the interplay between object language and metalanguage, critical and creative thinking, and a consciousness of the spatiotemporal case domain.

2.2. The Space–Time Domain of Stochastic Medical Reasoning

As was pointed out in Chapter 1, people's health may vary from place to place and from time to time due to prominent physiological and environmental variations. Medical data — such as cholesterol and hemoglobin levels, body temperature and heart rate, and blood pressure and glucose levels — differ from an individual to another and diurnally. Furthermore, environmental factors — such as weather, climate, hazardous events (accidents, natural disasters), adverse psychological status (mental illness, shock, anger) and socioeconomic conditions (living standards, income, inequality, lifestyle) — also affect population health from place to place and in a dynamic manner (as a function of time).

2.2.1. *Location–time coordinates and metric*

We already touched upon the spatiotemporal features of health in the previous chapter. Here we look at the subject in a more technical manner. Since space–time synthesis is a fundamental methodological component of real world problem-solving, it is implied that the mathematics used by physicians practicing *SMR* inference to represent a medical case and its associated attributes cannot be considered aspatial and atemporal.[5] There are two aspects of medical reasoning that are worth noticing: one intuitive and one technical. Intuitively, it is argued that nature should be describable in a space–time milieu, and does not really care which space–time concept people use. In this conceptual setting, the main features of disease evolution (speed of spread, geographical variation) must be properly assessed in order to take effective measures to confine the disease (regional distribution of medical supplies, prevention and treatment centers).

Example 2.4: Epidemiologists know that geography and time are key factors in any effort to prevent disease spread. For many known epidemics, the first few weeks are critical if the disease is to be adequately treated. In the case of *poliomyelitis*, for example, there is little hope of marked good

---

[5] As is the structure of formal mathematics in the sense of Hilbert.

being done by any treatment that started after the nerve cells have been damaged.

In view of the realization that space and time are conceptually interrelated (e.g. the conception of time by the metaphor of a geometrical line), it is plausible to search for some sort of *technical unification* of space and time. Technically, the scientific analysis of health systems that vary across large space and time scales (e.g., epidemics, infections) requires the introduction of the notion of a domain equipped with a

$$location-time\ coordinate\ set\ \boldsymbol{p} = (\boldsymbol{s}, t),$$

where $\boldsymbol{s}$ denotes the spatial location vector and t denotes time; and the notion of

$$location-time\ metric\ \Delta\boldsymbol{p} = \boldsymbol{p}_i - \boldsymbol{p}_j,$$

(*space–time distance*) between any pair of points in the domain. Assessing the composite space–time *dependence* structure of disease spread relies both on $\boldsymbol{p}$ and the relevant $\Delta\boldsymbol{p}$. The meaning of the term "relevant" here depends on the *in situ* conditions (e.g., an epidemic law is intimately connected to a specific metric, and the observation scale determines population exposure to a noticeable extent). In symbolic form,

$$\Delta\boldsymbol{p} = \varepsilon(\boldsymbol{h}, \tau),$$

where $\boldsymbol{h} = \boldsymbol{s}_i - \boldsymbol{s}_j$, $\tau = t_i - t_j$, and ε is a function determined by the available KB, which may include laws of nature, scientific models and empirical evidence. Different biomedical laws can lead to the determination of different metrics associated with distinct location–time variations of the disease.

Consider a case attribute, say population disease incidence $DI_{\boldsymbol{p}}$, which spreads geographically and diurnally. An adequate space–time distance for many disease distributions is the *Riemann–Minkowski* metric

$$\Delta\boldsymbol{p}^2 = \sum_{i,j=1}^{n} \varepsilon_{ij} h_i h_j + 2\tau \sum_{i=1}^{n} \varepsilon_{0i} h_i + \varepsilon_{00}\tau^2, \qquad (2.1)$$

where the coefficients $\varepsilon_{00}, \varepsilon_{0i}$, and $\varepsilon_{ij}(i,j = 1,\ldots,n)$ depend on the available KB, as before. Technically, Eq. (2.1) is a non-separable location–time metric of disease distribution. Consider now the space–time covariance of disease spread, $c_{DI}(\Delta\boldsymbol{p})$, where $\Delta\boldsymbol{p}$ is of the form (2.1). The covariance describes the distribution of disease attributes (incidence, prevalence, mortality, transmissivity). Elementary mathematical manipulations yield the general set of equations obeyed by the disease covariance and the associated space–time metric

$$\frac{\frac{\partial}{\partial h_i} c_{DI}(\Delta p)}{\frac{\partial}{\partial h_j} c_{DI}(\Delta p)} = \frac{\sum_{j=1}^n \varepsilon_{ij} h_j + \varepsilon_{0i}\tau}{\sum_{i=1}^n \varepsilon_{ij} h_i + \varepsilon_{0j}\tau}$$

$$\frac{\frac{\partial}{\partial h_i} c_{DI}(\Delta p)}{\frac{\partial}{\partial \tau} c_{DI}(\Delta p)} = \frac{\sum_{j=1}^n \varepsilon_{ij} h_j + \varepsilon_{0i}\tau}{\sum_{i=1}^n \varepsilon_{0i} h_i + \varepsilon_{00}\tau} \qquad (2.2\text{a–b})$$

($i, j = 1, 2, \ldots, n$). In other words, Eqs. (2.2a–b) establish a noticeable connection between knowledge of the disease's space–time dependence structure (as expressed by its covariance function) and the geometry of the disease domain (as expressed by its metric). Using the metric of Eq. (2.1) in conjunction with a known epidemiological law or empirical health model that describes the evolution of a disease can lead to the derivation of a space–time metric that is biophysically appropriate. This is illustrated next.

Example 2.5: Suppose that the following law describes the space–time change of the covariance function c_{DI},

$$\left(\frac{\partial}{\partial h} - \frac{h}{v^2\tau}\frac{\partial}{\partial \tau}\right) c_{DI}(h, r) = 0, \qquad (2.3)$$

where v is an empirical parameter of the speed of disease spread. A metric that is mathematically consistent with the disease law (2.3) is $\Delta p^2 = h^2 + v^2\tau^2$, which, of course, is a special case of the metric (2.1) for $n = 1$, $\varepsilon_{00} = v^2$, $\varepsilon_{01} = 0$ and $\varepsilon_{11} = 1$. This location–time metric is not only a function of space and time, but also of the speed with which disease spreads along a direction.

There is a strong connection between the natural laws that govern the evolution of an infectious disease (biomedical laws, population dynamics) and the corresponding space–time geometry. And like all the important concepts, that of "space–time" is studied in a more scientific manner within the *in situ* realm than as a reflection about its data-based usage,[6] since the former allows a better interpretation of the concept and a deeper understanding of its meaning. These space–time features are highly

---

[6] For instance, arbitrarily assuming a convenient separable space–time metric, and computing the covariance based on the available, yet uncertain dataset.

consequential in the solution of large-scale health problems (see Chapter 4). This numerical example looks into some of these features.

Example 2.6: Using a regional population mortality (PM_p) dataset and the covariance model $c_{PM}(h) = e^{0.3|h|}$, $|h| = (h_1^2 + h_2^2)^{1/2}$, routine implementation of data-driven spatial statistics or geostatistics techniques (kriging estimation; Lin et al., 2011) generated the map of Fig. 2.1a that represents the regional PM_p distribution. Now assume that the scientific study of the situation, which goes beyond the available dataset, showed that the true metric of the phenomenon is $|h| = |h_1| + |h_2|$, in which case kriging the same dataset and the same covariance form leads to the new map of Fig. 2.1b. Clearly, the PM_p distribution of Fig. 2.1a is very different from that of Fig. 2.1b, which means that the naïve usage of quantitative techniques can lead to a false PM_p representation.

The above is a simple yet sufficiently illustrative example of false results obtained when, on the altar of ill-conceived practicality, a health study systematically ignores the essential interplay between location–time and the *in situ* phenomenon of interest.

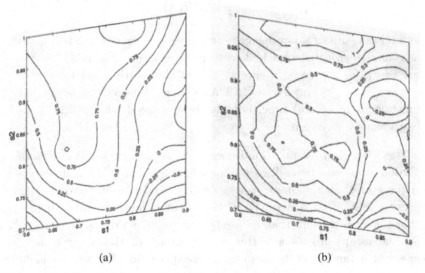

Figure 2.1: Population mortality (PM, %) distributions obtained using the same dataset and the same covariance model with metrics: (a) $|h| = (h_1^2 + h_2^2)^{1/2}$, (b) $|h| = |h_1| + |h_2|$.

2.2.2. The spatiotemporal random field model

To account for its threefold nature (space–time structure, logical pattern, and uncertainty), *SMR* inference favors the mathematical representation of medical case attributes in terms of the

spatiotemporal random field (S/TRF) $CA_{\boldsymbol{p}}$ at location−time points $\boldsymbol{p}$

(2.4)

(Christakos and Hristopulos, 1998).[7] Our readers may recall that the crucial role of the symbol $\boldsymbol{p}$ in a health study is the ability to build and organize relevant medical content that links place and time to real world clinical conditions and population exposure risks. Specific realizations (possible worlds) of $CA_{\boldsymbol{p}}$ across space–time are denoted by $CA^{(i)}(i = 1, 2, \ldots, n)$ and, in general, different probabilities are assigned to these possible worlds (e.g., different probabilities are assigned to influenza incidence rates at different locations of a region and time periods) (Choi *et al.*, 2008). By definition, each $CA^{(i)}$ must be compatible with the biomedical and relevant *KB*s and logically consistent. The uncertainty of a care professional's subjective belief about a case is higher (which is reflected in an increased number of possible worlds) than the uncertainty of a rational assertion (which permits fewer possibilities).

Example 2.7: In principle, disease spread is a fundamentally spatiotemporal phenomenon, the commonly used purely temporal models being merely convenient simplifications of the actual situation. Case attributes that can be mathematically represented as *S/TRFs* are the geographical and temporal distributions of *HIV* susceptibles ($S_{\boldsymbol{p}}$), infecteds ($I_{\boldsymbol{p}}$) and clinical cases (R_p). Similarly, the $S_{\boldsymbol{p}}$, $I_{\boldsymbol{p}}$ and $R_{\boldsymbol{p}}$ may represent local infectious disease rates of susceptible, infected and recovered population fractions within a spatial area around location $\boldsymbol{s}$ and at time t. Generally, these *S/TRFs* are functions of three factors: the population fraction that migrates during the specified time period, a kernel that controls population movement and the probabilities of recovery and transmission (more technical details are discussed in Section 2.5.1). For visualization purposes, Figs. 2.2 and 2.3 present two real case studies. Specifically, Fig. 2.2 displays maps of the *hand–foot–mouth disease* (*HFMD*) spread in

---

[7]Readers may recall (Section 1.2) that $CA_{\boldsymbol{p}}$ may represent different *case attributes*, like disease symptom $DS_{\boldsymbol{p}}$, disease diagnosis $DD_{\boldsymbol{p}}$, population incidence $PI_{\boldsymbol{p}}$ and population exposure $PE_{\boldsymbol{p}}$.

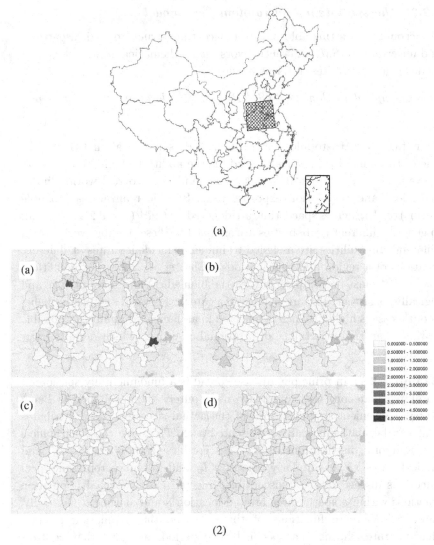

Figure 2.2: (1) Study region and its location in China. (2) HFMD rates (cases per 10,000 people) in study counties across four selected weeks: (a) $t = 5$, (b) $t = 10$, (c) $t = 15$ and (d) $t = 20$.

145 Chinese counties with relatively higher disease incidences over a period of 20 weeks during the years 2008–2009 (Angulo *et al.*, 2013). Lastly, the maps of Fig. 2.3 display the geographical distributions of plague mortality in India at different time periods (Yu and Christakos, 2006).

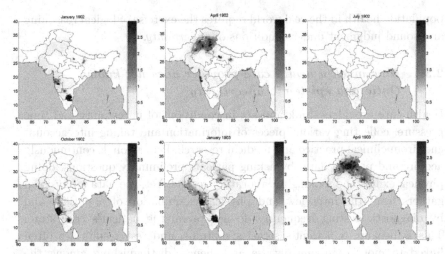

Figure 2.3: Space–time maps of bubonic plague mortality (%) in India during the years 1902–1903.

Some readers may appreciate that the set of possible worlds of the *S/TRF* model reveals a certain *link* between medical reasoning and standard epistemic logic (Fagin *et al.*, 1995), a logic that explicitly deals with knowledge. Specifically, epistemic logic deals with the possession of, or access to information, and also focuses on aspects of object language, whereas *SMR* is concerned with the context- and content-dependent mathematical representation of the medical decision process. In this sense, when physicians apply *SMR* to clinical and medical practice they assume an initial epistemic state characterized by the possible worlds of the relevant case *S/TRF*. As more data is collected during the reasoning process and an improved understanding of the case is established, the reliability of the physician's cognitive state increases.

2.3. *In Situ* Logic and Uncertain Mind States

Core activities of real-time medical decision-making (patient-centered communication and resolution, utility assessment, cost effectiveness) depend on assumptions regarding human judgment and cognitive condition. An important lesson of experience that a health care professional's thinking takes to heart is that no matter how well founded a claim may be, its truth cannot be established with absolute certainty. Hence, in addition to an adequate understanding of the relevant space–time domain, studying population health in the real world requires the exercise of logical reasoning and sound judgment under conditions of *uncertainty*.

2.3.1. How much a health care provider does not know: Ontic and epistemic uncertainty

Health care providers think and act in conditions of uncertainty and time pressure, collecting various pieces of information and taking into account cues from clinical practice and medical research. Professionals continuously acquire and process data, by asking patients preliminary questions, doing physical exams and looking at laboratory results. Yet, the data a physician gathers from the interview and physical examination of a patient and by diagnostic testing frequently do not reveal the patient's true state. This is because a patient's signs, symptoms, and test results are often linked to more than one disease and, hence, distinguishing among the possibilities with absolute certainty is not feasible. For illustration, consider a patient's complaint about "nocturnal substernal chest pain for one month". As Sox *et al.* (2013: p. 13) notice this complaint may initially elicit a considerable number of possible diagnoses, including lung diagnoses (pneumonia, pulmonary embolus, tuberculosis, tumor, pneumothorax), bony thorax (rib fracture, muscle strain, costochondritis), esophageal diagnoses (reflux esophagitis, esophageal spasm, esophageal rupture, esophageal cancer), cardiac diagnoses (coronary artery disease, aortic stenosis), referred pain diagnoses (cholecystitis, peptic ulcer, pancreatitis, cervical arthritis), pericardial or pleural diagnoses (pericarditis, pleuritis), mediastinal (mediastinitis, tumor), or great vessels disease (dissecting, aneurysm). Spatial and temporal variations in population health generate uncertainties that have an essential effect on medical studies. Medical uncertainty is linked to the polymorphism of most diseases, the complex manner diseases evolve in different patients over time and the polyglot way in which individual physicians retain biomedical knowledge and clinical information.

These and similar concerns emphasize the need for rigorous quantification of *in situ* uncertainties such as:

(1) *Ontic* uncertainty attributed to natural heterogeneities and intrinsic variations (genetic, biological and environmental factors, cultural and psychological deviations, technical fluctuations of medical instruments, differences among laboratories). Many of these variations are beyond complete control, yet their impact may be manageable to a certain degree.

(2) *Epistemic* uncertainty associated with a health care provider's cognitive state, incomplete knowledge, skills and experience, as well as on documented medical disagreement and available resources.

Physicians and epidemiologists alike are aware of the fact that in many cases one cannot make decisions with certainty regarding diagnosis, treatment and prognosis, in which case the notion of uncertainty could provide a reasonable quantitative measure of how much a professional does not know about the actual case conditions. Uncertainty and incomplete knowledge can cause health care professionals to generate *absurd* statements (see Section 3.2.2). Furthermore, the need for an improved understanding of the theoretical, experimental, scientific and practical aspects of medical thinking in conditions of real world uncertainty, frequently draws on insights from the various disciplines where it is currently studied (logic, psychology, cognitive science). As a matter of fact, many health care providers argue that the current lack of a coherent scientific theory of medical reasoning and behavior explanation, as opposed to behavior and judgment idealizations, imply that clinical approaches and decision support tools are based on mere speculations (Reyna, 2008). This lack of theory often hampers the soundness and generalizability of the decision tools. The above is well-known material, and the goal here is to set up an appropriate framework in which to place the medical reasoning process.

Example 2.8: Relevant to the view of many experts concerning the lack of a coherent reasoning theory in conditions of uncertainty is the inappropriate interpretation of medical guidelines. One problem is that while guidelines are mere recommendations, they are frequently used as strict rules by the medical profession. The conversion of recommendations into rules can have serious drawbacks, both scientific and cultural. The readers may recall that guidelines are based on imperfect evidence and frequently they are not developed by a rigorous and objective process (see Section 1.1.4).

Decisions concerning disease treatment or cure typically involve a number of steps, during which a patient may encounter several observers and instruments: considering the patient's history, recording essential data (weight, blood pressure, heart rate), physical examination and further investigations (x-rays, electrocardiogram, blood glucose measurements). Variations in these steps contribute significantly to clinical practice uncertainty, which, when a sound theory is absent, can lead to false diagnosis, treatment and prognosis.

Example 2.9: As they furnish their minds with the requisite data for recognizing the nature and possible cause of the diseases that physicians are called upon to treat or remedy, awareness and due assessment of the relevant disease symptoms, their groupings and connections are of vital importance to them. Accordingly, when women in their 30s are complaining of headaches, an expert neurologist would initially consider three possible disease diagnoses (Cordingley, 2005):

$DD^{(1)}$: "migraine",
$DD^{(2)}$: "medication-overuse",
$DD^{(3)}$: "none of the above".

The last option includes any other possibility like sinus disease, tension-type headaches, arthritis of the neck or jaw-joints, or tumors. Moreover, even before seeing the patient — say, Ms Nicolis — and based on previous experience, a physician assumes that for each one of the three possible diagnoses the odds that it will turn out to be correct is about 33%. This number offers a measure of a physician's *initial* uncertainty about the true diagnosis. During subsequent examination and supplemental tests, the initial uncertainty may undergo a series of upward and downward adjustments according to what Ms Nicolis has to say and what does or does not turn up on her physical examination and tests. For example, the original set of possibilities may be replaced with a new set,

$DD^{(1)}$: "migraine",
$DD^{(2)}$: "medication-overuse",
$DD^{(3)}$: "sinus disease",
$DD^{(4)}$: "tension-type headache",
$DD^{(5)}$: "none of the above".

Hence, diagnosis is a dynamic and sequential process characterized by different levels of uncertainty. Eventually, the physician's goal is to focus on one diagnosis by eliminating as much as possible of the associated uncertainty.

2.3.2. *Appreciating case individuality and legal disputes*

Continuing on the subject, a physician often seeks to evaluate and synthesize different and sometimes conflicting core knowledge and case-specific evidence sources. Most clinicians would admit that change confronts them every day with new situations, some of which may be particular to the specific case. Such unique circumstances deserve individual consideration, which means that adequate decision-making is a matter of "situational appreciation", i.e., reading and appreciating the *individuality* of the case and properly integrating it into the physician's decision process. Moreover, synthesis is made necessary given that, as every physician knows, the findings of one medical study frequently do not match the conclusions of another study. Here are some rather typical examples.

Example 2.10: Swedish findings emphatically opposed the amyloid hypothesis that dominated *Alzheimer's* disease research for decades, suggesting that it is, in fact, neurons' inability to secrete beta-amyloid that is at the heart of pathogenesis in Alzheimer's disease (Tampellini et al., 2011). In India different studies suggested levels of *hypertension prevalence* that varied by orders of magnitude. In Italy studies contradicted each other regarding the effect of diabetes, smoking and blood pressure in *population mortality*.

Remarkably, in some cases questioning the established view can lead to major findings that may influence the status of a medical field.

Example 2.11: For many years, patients with *peptic ulcer* (stomach or duodenal ulcer) could spend 20 years or more on antacids that counter stomach acidity. Their life was difficult with certain restrictions (avoiding alcohol, spicy foods and the like). However, in the early 1980s, the physicians Barry J. Marshall and Robin Warren suggested that peptic ulcer was not a permanent condition but rather an infection that could be treated accordingly. Most experts immediately dismissed the idea as ludicrous. Nevertheless, it was soon proved that Marshall and Warren were right.[8] Nowadays, a patient can be cured with the help of antibiotics within a week.

Medical uncertainty due to incomplete understanding plays a major role in *legal disputes* and *lawmaking*, as well. It is common knowledge that the courts have been repeatedly criticized for their inadequate consideration of medical uncertainty.

Example 2.12: In a widely-publicized case (Grawert, 2009: pp. 390 and 406),

> "[Justice] Kennedy relied upon Congress's apparent ability to regulate despite medical uncertainty... in stating that congressional findings

---

[8]In fact, they were awarded the 2005 Nobel Prize in Medicine.

should be upheld when the regulation is rational and in pursuit of legitimate ends... this position erroneously lets Congress manufacture a scientific controversy and then take its own side in the ensuing farcical debate... The precise point of this paper is that the 'uncertainty' was improperly evaluated, amounting to a constitutional loophole".

In sum, a rigorous assessment of medical uncertainty allows both physicians to produce their diagnoses, and policy makers to propose health legislation based on critical thinking and sound science. At the community level, medical uncertainty characterizes population health risks and the significant effects of various preventive and moderating interventions.

2.3.3. *Case communication uncertainty: Entitled to their own opinions but not to their own facts*

Undoubtedly, everyone is entitled to their own opinions but not to their own facts. This viewpoint may be the key to an adequate *communication* between interested parties (health care professionals, patients, policy makers). Adequate medical decision-making, in turn, relies on the communication between health care professionals, as well as between professionals and patients. For one thing, it is widely recognized that *medical journaling* is an important method of recording and tracking the disease condition. For instance, diabetics who journal their day-to-day health-related information (blood glucose, diet, exercise and medications) to self-manage their condition can prolong their lives, gain a better understanding of their disease and communicate more effectively with health professionals.

The domain of communication includes time, location and the relationships between the parties involved. Communication is dynamic, and can involve various kinds of data, information and knowledge organized in different ways, ranging from simple text and static images to complicated diagrams, mathematical equations and animated graphics. Epistemically one may distinguish between three possibilities in medical communications: case items that are known and agreed by all, those that are known but not agreed and those that are known to none. Different physicians can have different perceptions about which communication function[9] should

---

[9] Generating impartial case information, expressing expert views or establishing guidelines, providing instructions about health care, influencing patients' behaviors, referring to the nature of the interaction etc.

be dominant in a particular case and, hence, should most influence the character of the communication; or they may have radically different opinions about what information should or should not be included in the communication concerning the same case. Part of the data exchanged via communication may be qualitative (oral history including information from those medical practitioners who were part of important cases). This kind of data may be valuable in practice, assuming that the data can be put in a sound theoretical context. As a matter of fact, it is this context that distinguishes qualitative clinical research from journalism (Hannay, 2002).

Communication uncertainty is argued to be due to *vagueness* and *ambiguity*,[10] which can arise from conceptual, presentational and linguistic factors. If vagueness and ambiguity characterize the communication between health care professionals or between professionals and patients, similarities and differences can be uncertain, e.g., vagueness and ambiguity can mask ontological and semantic uncertainties. When vagueness or ambiguity exist in communication, different health care professionals can use the same terms and phrases but mean different things, without those differences being apparent to others. In this setting, syntactic[11] ambiguity arises from case assertions that may be parsed in more than one way, whereas semantic ambiguity arises if the same words in the same communications elicit either different cognitive states or different emotional states.

One distinguishes between three principal sources of vagueness and ambiguity: first, the conceptual source encompasses distinct frames of reference used by health care providers, patients etc. (note that concepts exist in the mind as abstract entities independent of the terms used to express them). Second, the presentational source encompasses different contexts and styles that can affect communications between physicians, as well as between physicians and their patients. Third, the linguistic source refers to different characters, symbols and phrase structures, which are used by clinicians when speaking and communicating with others. No doubt the reliability of communication will be compromised if the same communication is understood differently by different care professionals.

---

[10] Generally, vagueness characterizes cases of which the meaning cannot be determined in any context, whereas ambiguity characterizes the existence of at least two specific meanings that make sense in a particular context.

[11] As noted earlier, syntactics deals with the rules governing the arrangement (*syntax*) of words, symbols, equations etc. in an assertion (these rules are distinct from the meanings words convey).

Experience shows that interacting medical professionals usually distinguish between two types of communication uncertainty. The usual ontic uncertainty (Section 2.3.1) involves different professionals who interact and have different conceptualizations about what kinds of entities are relevant to the medical case of interest, what kinds of interactions these entities have and how the entities change as a result of these interactions (ontic uncertainty exists when similarities and differences between different professionals' ontologies are not certain). This uncertainty may be more closely related to conceptual vagueness and ambiguity than to linguistic vagueness and ambiguity, and it is likely to exist when phenomena are inherently unpredictable. Semantic uncertainty, on the other hand, is of an epistemic nature, involving different professionals in the same interactions giving different meanings to the same term, concept, method, assertion and action. Similarly, semantic uncertainty exists when similarities and differences between the meanings given by different professionals to the same terms, concepts and assertions are not certain. There are several conceptual and linguistic factors related to this kind of uncertainty. Remarkably, conceptual factors can be more important in semantic uncertainty than presentational and linguistic factors. In sum, ontic and semantic uncertainties can lead to intractable misunderstandings between health care professionals, as well as between professionals and patients.

2.3.4. *Formal vs.* in situ *logic*

At this point, our readers may appreciate a very brief regression concerning the development of *formal* or *classical logic*.[12] Aristotle's definition of logic was as new and necessary reasoning, Boole made logic as clear as algebra and Frege developed a system that is considered the best starting point for the investigation of issues of truth, meaning, thought and language. It was Russell who tried to apply higher logic to human affairs, and without him doing the "dirty work", Gödel may not had been able to ask certain crucial questions that led to important developments.

For formal logic,[13] the axioms of a theory are the starting points for the reasoning process — all that is asked of them is that they are logically compatible. Formal logic is used with great success in mathematics. Yet,

---

[12] For a review of elementary logic notions and techniques, the reader is referred to the many volumes available in the relevant literature (Carnap, 1962; Bennett, 1987; Mendelsohn, 1987).

[13] Also called classical, standard, deterministic or mainstream logic.

underlying this kind of logic is the rather unrealistic assumption of a domain free of the error and confusion that actually plague the world of material reality (Mikler et al., 2006). As noted earlier, in medical practice, different kinds of uncertainty characterize actual cases in various space–time scales, including epistemic, ontic, conceptual and technical uncertainty. Furthermore, research in artificial intelligence (involving cognition, psychology and computer science) has shown that the assumption that all knowledge can be represented in terms of symbols and processed by formal logic is profoundly wrong.

In light of the above and similar considerations, when logic is used in public and private medical practice, rather than formal logic, a more adequate kind of *in situ* logic is sought that emphasizes the central role of both uncertainty and pattern. We will revisit the subject in Section 2.4. Among other things, *SMR* is distinguished from formal logic by the interpretation of the term

it is valid

of the classical setting (closed-system formal logic) as,

it is substantiated[14] *in uncertain real world conditions*

(open-system *in situ* logic). Recall that assertions that state the truth or validity of object language statements (e.g., description of symptoms) belong to the metalanguage, which, in a sense, is one level higher than object language. In the following, the symbols

"$\neg$", "$\wedge$", "$\vee$", "$\therefore$"

will denote, respectively, the *logical connectives* of *negation, conjunction, disjunction* and *entailment* (in the broad sense). As we shall see in Section 2.4, *SMR* encompasses the re-interpretation of logical connectives in a metalanguage setting that facilitates a substantive justification of assertions involving these connectives. To make serious progress in medical decision-making and health planning, one must distinguish and capture the fundamentally different consequences that arise when the *content-dependency* (*CD*) of *in situ* SMR replaces the *content-independency* (*CI*) of standard logic.

Example 2.13: In studies of medical causation (etiology), the investigator may choose to replace the concept of "proof" with that of "causality",

---

[14] Or, *established by substantive means*.

meaning that one can arrive from the available data to disease prognosis via a series of causal arguments that are content-dependent, internally consistent and scientifically sound. This distinction and its effects on the investigation are discussed throughout the book.

Any health care professional knows that applying formulas may not be good enough when one is faced with really hard problems under conditions of *in situ* heterogeneity, real-life uncertainty and time pressure. On the other hand, one cannot ignore fundamental knowledge sources expressed in quantitative terms (biomedical laws, epidemic models, environmental determinants, population dynamics) (Daley and Gani, 1999; Kolovos et al., 2011). Although mainstream *purely data-driven* analysis (*PDD*) is a useful tool in empirical studies, it often overlooks the significance of *knowledge synthesis* in real conditions of uncertainty. SMR appreciates the value of synthesis driven analysis, two key features of which are content-dependence and semantics. Both features are ignored by *PDD* analysis (Lee and Rogerson, 2007; Chien and Bangdiwala, 2012).

Example 2.14: A physician's cognitive burden is to figure out if and how the set of medical guidelines concerning a disease correspond to the particular patient, Ms Nicolis. An important issue is that the *PDD* analysis that generated the guidelines for a disease involves rather homogeneous populations of individuals that may or may not have enough in common with Ms Nicolis (who may also have other medical problems). Accordingly, while thinking about the disease, the physician should also be thinking about the individual. These are issues of studied and sound medical reasoning rather than straightforward implementation of guidelines and recommendations. In the preceding Example 2.9 it is a matter of rigorous *SMR* inference to understand if and how Ms Nicolis fits the published data, and then to find out from her what her goals, values, priorities and preferences are.

All the above facts are often overlooked by health statistics that focus on *PDD* techniques at the expense of important semantics (retrieving and processing irrelevant data, masking substantive features of data and failing to identify all relevant documents in the case *KB*s). In our view, the "*SMR* vs. *PDD* analysis" debate is passé in favor of the former. *SMR* renders rich inferencing and reasoning capabilities, and a framework to integrate heterogeneous *KB*s. It deals with contextual aspects of meaning, involves a science-based and semantically rich logical process that synthesizes, and on occasion reconciles, varied and sometimes conflicting medical information and clinical evidence. When viewing logic as a medical decision support tool, one must carefully choose what use to put it to. If a care provider starts from the wrong premises and relies on poor *KB*s, logic can be the executioner's handmaiden or a fool's ideal accomplice.

Concluding this section we notice that, although different cultures may have varying epistemologies, there is no such diversity in logic. When it comes to an evaluation of arguments, one always relies on logic defined as a systematized set of valid inference schemes. Depending on the circumstances, one can decide whether a given inference in a specified language is conducted according to valid logical rules. For example, one may find that a large number of inferences in medical texts originated in ancient Greek medicine, compared with a relatively smaller number in the ancient Chinese medical tradition. But this does not necessarily constitute proof that one culture uses a different logic than the other.

2.3.5. *From state of nature to state of mind (assertion)*

As noted earlier, it is important to distinguish between different kinds of core knowledge and not mix them indiscriminately in medical decision-making. The distinction can be made, e.g., on the basis of the authority of their results and conclusions, and the reliability of the methods they employ. These are two essential indicators of the *epistemic strength* of a knowledge source. A possible ranking of *G-KB* on the basis of their epistemic strength is as follows (in order of decreasing strength):

natural laws, scientific theory, ad hoc *empirical models.*

Similarly, for the epistemic strength of *S-KB*:

hard data, soft or uncertain information, secondary evidence;

see also Eq. (2.9) below.

For a health care professional practicing *SMR*, chance and necessity constitute an integrated whole; *G-* and *S-KB* are blended in terms of a decision process that emphasizes meaning and has well-defined goals; the cognitive situation usually includes *KB*s that are incorporated into the professional's thinking mode in a coherent and consistent manner; the structure from which a medical decision formalism is abstracted is often richer than the formalism itself and the *S/TRF* modeling of a case requires the development of a suitable metalanguage that describes the object language.

Before processing a particular health dataset using quantitative methods, the health care provider encounters a number of crucial questions in need of answers. Such preliminary questions include:

(1) What kinds of medical data are studied and how were they obtained?
(2) Are the available medical records semantically and pragmatically rich?
(3) Is the data acquisition process (via experimentation, observation, surveillance) consistent with the underlying scientific theory?
(4) What is the population from which health data is drawn?
(5) What is the context relative to which the data is studied?
(6) What is the target of the clinical inferences based on this data?
(7) What are the boundaries of the study?

Naturally, data need to be converted into useful information before they can be used. This is achieved by relating data to medical knowledge, scientific theory, sound experience and other evidence. In this way, data can aid physicians to make accurate case diagnoses, judgments and conclusions (which could be used in other contexts or for different purposes as well). A simple example: the fact that little Maria's body temperature is 39.8°C remains merely a piece of data, but when it is related to the medical knowledge that a temperature above 39°C is considered high fever, then one understands that the child needs medical care.

The study of a medical case is not merely technical (computerized guidelines, statistical data analysis, laboratory instrumentation) but substantive too, including better theory and sound science enabling the construction of improved instruments and medical tests. In fact, one should always keep in mind that *there is nothing so practical as a good theory*.[15] Yet, health care professionals sometimes take for granted the validity of the datasets presented to them and the adequacy of the representation of the actual case these sets offer, without examining the theoretical consistency of the measurement process or questioning the relevant experimental conditions. As a result, assertions (beliefs, estimates, predictions) concerning medical diagnosis and therapy can be incomplete

---

[15]There is some connection between features of practical rationality and epistemic features and relations, which entails that the obtaining of the latter has a practical dimension.

and uncertain, even if highly sophisticated technology has been used in the process.

With regards to scientific argumentation, an obvious implication is that it is appropriate that the health care professional uses statements that allow a range of possibilities reflecting one's cognitive state such as,

CA_p is *believed to be* valid at location–time coordinates $p = (s, t)$,
CA_p *ought to be* valid at p,
CA_p is *eventually* valid at p,
CA_p is *likely to be* valid at p. (2.5)

The above real *in situ* possibilities make sense in the "multiple possible worlds" setting of the clinical case introduced by the *S/TRF* theory (Section 2.2.2) which, and this is important in medical decision-making and exposure assessment, also offers a rigorous method to quantify these case possibilities. Accordingly, *SMR* involves a logic of knowledge and belief (this is a property that it shares with epistemic logic) in which the health care providers are inquiring and active: they possess core and specificatory *KB*s, can expand or revise some of them, can test the internal consistency of the *KB*s, follow a sound methodology, decide and act according to certain rational rules, and interact and share *KB*s.

Example 2.15: When studying health phenomena that vary in space–time, an expert capitalizes on the premise that statements of the form

"expert *knows* that the Reed–Frost model[16] applies in the Black Death epidemic",

or

"expert *believes* that the daily population exposure to O_3 in Anqing city during 2012 was over 0.11 *ppm*"

have features that admit a systematic study (use an axiomatic–deductive scheme to view knowledge claims in a manifold of statements concerning clinical inferences and medical decisions). In the above sense, medical inference may be seen as a modal approach based on "possible world" semantics and thus grounded in a realistic ontology. This is due to the

---

[16] Abbey (1952).

fact that the "actual world" plays a central role in modal logic (as it does in S/TRF theory): the actual health state is the point of evaluation of a formula (in a similar way that the actual space–time disease distribution is the evaluation point of the S/TRF realizations).

The view cited by many scholars is that we are currently experiencing the fourth revolution as regards the role of human agency in the world. The *materialistic* perspective that focuses on biophysical objects and processes has been replaced by the *informational* perspective, in which the key role is played by a physician's cognitive condition about biophysical objects and events. Assume that physician A studies a case attribute CA_p distributed in space–time (disease event, diagnostic proposition, environmental exposure, epidemic speed). *Mutatis mutandis*, the perspective change above is reflected in SMR's switching focus from the deceptively certain statement of object language,

$$CA_p \text{ occurs} \tag{2.6}$$

(associated with the often unknown or unknowable features of the health system[17]), to *states of mind* of the uncertain form of metalanguage,

$$\text{expert } A \text{ asserts that "}CA_p \text{ occurs", denoted as } a_A(CA_p) \text{ or } CA_p\dot{:}_{KB}. \tag{2.7}$$

Let us look more carefully. While (2.6) expresses uncertainty about the occurrence of CA_p, expression (2.7) is of the cognitive state type (2.5) considered earlier. In particular, in (2.7), the $a_A(CA_p)$ represents a *general assertion* made by physician A about CA_p, which is associated with different levels of comprehension but not necessarily certainty. In other words, the statement (2.6) refers to the *in situ* CA_p itself and belongs to language, whereas the assertion (2.7) refers to A's cognitive situation concerning the space–time occurrence of CA_p that belongs to metalanguage. Metalanguage entails systematic, technical knowledge of the ways in which the resources of language and semiotic systems (signs, symbols, plots) are deployed in meaning-making.

Example 2.16: If the case attribute is *tumor malignity*, TM_p, statement (2.6) expresses a certainty about the tumor, independent of the physician's

---

[17]CA_p, e.g., refers directly to a biomedical entity.

cognitive condition; whereas (2.7) describes the physician's own view about the tumor's condition. That is to say, expressions (2.6) and (2.7) refer to different things. Unlike (2.7), statement (2.6) says nothing particular about states of mind concerning the case (beliefs, normative views, insights). As such, (2.6) can be valid without (2.7) necessarily being so in the real world.

When convenient, the symbolism "$a_A(\cdot)$" may be used interchangeably with the symbolism "$(\cdot)\vdots_{KB}$", where, as before, KB denotes the knowledge base that provides the epistemic support for A's assertion. In relation to expression (2.7), for many care professionals it is intuitively valid that $a_A(\neg CA_p)$ amounts to $\neg a_A(CA_p)$, which can also be shown to be formally valid in terms of probability (Example 3.47 later). Lastly, the crucial role of the space–time symbol p used in the CA_p is the ability to build and organize relevant medical content that links place and time to real world health conditions and exposure risks. Nevertheless, herein for simplicity the subscript p will be occasionally omitted (yet implicitly assumed) when this omission does not affect the clarity of the presentation.

For future reference, readers should remember that in many cases there is a correspondence between the rules researchers and clinicians employ in object language and those used in metalanguage expressions. This is a consequence of the relation between "assertion" and "mind state".[18] To make an assertion a_A about CA_p is to say that the health care provider A is able to construct a mental state that refers to *in situ* characteristics of CA_p (depending on the construction, of course, the assertion may possess various degrees of validity). In some other cases, though, this correspondence is not valid; e.g., an inference that is valid in object language may be not so in metalanguage.

A famous line in the *Talmud* begins, "Ayn kemach, Ayn Torah", i.e., without bread there is no learning. Health care providers need *resources* to do science and to educate patients. But the present day focus on money in medicine is so intense that an entire dimension of medicine is being negated as a result. Resources characterize mind states, which, in turn, exert a considerable influence on the *epistemic strength* of a physician's assertions concerning a case. Because of issues having to do with limited resources, incomplete evidence, education level, cognitive abilities, attention

---

[18]Mind states can process complex information, generate sophisticated inferences and solve complicated real world problems. Hence, when developing a theory of reasoning (*SMR*) one should have a basic understanding of these states.

spans, memories and time, health care providers may possess a limited understanding of a medical case. The inability to form a semantically rich mind state can make care providers resort to "quick and dirty" solutions, commonsensical tricks, rules of thumb and heuristics, some of which are non-epistemic. To avoid these complications, general assertions may assume different case-specific forms, depending on the physician's cognitive situation and acumen about the case of interest and the available resources; symbolically,

$$a_A = k_A, \quad \text{or} \quad r_A, \quad \text{or} \quad s_A, \quad \text{or} \quad b_A, \quad \text{or} \quad g_A, \qquad (2.8)$$

where the term "asserts" (a_A) may denote a specific mind state in which a health care provider A:

(1) *knows* (k_A), understood in a veracity sense (adequate resources assure a cognitive status in which if A knows that a case occurs, then the case must occur);
(2) *rationalizes* (r_A), considered in a rational thinking sense (supportable objectively on the basis of the existing evidence);
(3) *sustainably believes* (s_A),[19] viewed in a sustainable belief sense (although not necessarily fully objective, not completely guided by A's state of mind);
(4) *believes* (b_A), assumed in a personal belief sense (subjectively formed given the resources available, depending on A's acumen and state of mind) or
(5) *guesses* (g_A), considered in the sense of mere supposition (with a lower degree of justification assigned to it, as facilitated by limited resources).

Several possibilities can emerge in medical practice. It is possible that the same KB is available to different physicians A, B or C, but that they interpret it in different ways; e.g.,

$$\vdots_{KB} \equiv b_A, \quad \vdots_{KB} \equiv r_B, \quad \vdots_{KB} \equiv k_C,$$

i.e., given the same KB different physicians may form different mind states, depending on their acumen, biomedical knowledge, experience with similar situations and stage of professional development. On the other

---

[19]In other words, an agent A expresses the concept of justifiable belief by saying A *has reason to believe that* p instead of A *believes that* p.

hand, the KBs potentially available to A may be associated with different a_A-forms; e.g.,

$$\vdots_{KB_1} \equiv g_A, \quad \vdots_{KB_2} \equiv b_A, \quad \vdots_{KB_3} \equiv s_A, \quad \vdots_{KB_4} \equiv r_A, \quad \text{and} \quad \vdots_{KB_5} \equiv k_A.$$

Which one of the above assertion forms most adequately describes the particular cognitive state of a physician concerning the medical case is a function of a number of factors. So, developing an adequate a_A-form depends on one's ability to recognize what is salient and the capacity to respond to the particular case (the physician's powers of perceptual grasp of the real issues).

Example 2.17: A health care professional A has the sustained belief that the daily $PM_{2.5}$ mass concentration in Beijing during the period May 10, 2010 to December 6, 2011 was in the range 5–450 μg/m^3 (Wang *et al.*, 2012), i.e.,

$$s_A(PM_{2.5} \in [5, 450 \,\mu\text{g/m}^3]),$$

if this result is determined on the basis of (a) the KB available to A and (b) the substantive links (logical, physical) between the relevant KB and the result, and, also, (c) A is aware of the way this is done.

The various a_A-forms above are associated with different uncertainty levels. We will have to say more about this later; at the moment, let us illustrate the concept with one example.

Example 2.18: In behavioral analysis terms, the uncertainty of a care professional's subjective belief is higher, allowing more possible worlds, than the uncertainty of a rational assertion that permits fewer possibilities. Physician A's subjective belief regarding the expected patient's body temperature (BT),

$$b_A(BT \in [37, 41°\text{C}]),$$

allows a larger set of possibilities (all numbers between 37 and 41) than does physician B's rational assessment of the patient's temperature obtained on the basis of the existing KB,

$$r_A(BT \in [38.3, 39.1°\text{C}]),$$

which only includes temperatures between 38.3 and 39.1°C. Interestingly, the reverse uncertainty ranking is obtained when the above assertions are considered in certain technical terms (Section 2.3.8).

After we introduce the notion of probability, in Section 3.3.2 we will describe some quantitative methods for choosing an appropriate a_A form. At this point we should bring to the readers' attention two of the real world studies that have attempted to assess case uncertainty based on some cognitive processes.

Example 2.19: The oral cancer study by Yu *et al.* (2010) analyzes the uncertainties of standardized oral cancer mortality from low population areas. The ten-year data was used to assess the degree of uncertainty of yearly-recorded mortality at the sustained belief (s_A) level. The results showed that the systematic consideration of data uncertainty can improve the rigor and estimation accuracy of the spatiotemporal distribution of oral cancer in Changhua (Taiwan). Furthermore, Yu and Wang (2010) estimated the $PM_{2.5}$ retrospectively during the period that no $PM_{2.5}$ observations were available. The $PM_{2.5}/PM_{10}$ ratio was used as an indicator of human activity. The investigators assumed that human activity was relatively stationary during the period in which they performed retrospective $PM_{2.5}$ estimation. This was also an assumption associated with the s_A cognition state of the investigators supported by local knowledge that was gained from living in Taipei during several decades (space–time maps of $PM_{2.5}$ values in the past were derived based on data and inferences from local experts).

When required in medical practice, useful relationships can be established between different assertion forms, e.g., for a disease diagnosis, DD,

$$g_A(DD) \equiv \neg r_A(\neg DD)$$

(see also Eq. 2.29). More precisely, "physician's guess is esophageal disease" is equivalent to "it is not true that the physician rationalizes that it is not esophageal disease". Lastly, one should keep in mind that a physician's epistemic uncertainty about a disease may also be a function of practical ontic factors: *resources*, *costs* and *benefits* of disease treatments. If the assertions of physicians A and B about a diagnosis have the same high epistemic strength, then that diagnosis is warranted enough to be a reason why each of the two physicians has to choose a treatment (or, if the strength is low, the assertion is not warranted enough to be a reason why either A or B has to make this choice). Yet, it is possible that, although the physicians' assertions about a diagnosis may have the same epistemic strength, only A can use that diagnosis as a reason for choosing a specific treatment due to practical reasons (e.g., A has the necessary resources, whereas B does not).

2.3.6. Ranking of assertion forms

As noted earlier, the epistemic strength of the assertion k_A is greater than that of r_A, which in turn is greater than that of s_A etc. This kind of ranking of the various assertion forms can be expressed as

$$k_A \therefore (r_A \therefore (s_A \therefore (b_A \therefore g_A))).^{20} \tag{2.9}$$

In S/TRF terms, the smaller the number of possible worlds (realizations) in which the CA is considered valid, the higher the epistemic strength of the assertion $a_A CA$. Interestingly, if k_A is based on unassailable theory (not to be confused with mere opinion) or on extensive empirical evidence concerning the same circumstances, $k_A CA$ may be considered equivalent to CA (in which case, the symbols $k_A CA$ and CA could be used interchangeably). No doubt, the appropriate characterization of an assertion in terms of k_A, r_A, s_A, b_A, g_A is a crucial component of medical investigations, which could help a health care provider to avoid viewing as certain knowledge what is, in fact, an assumption or a belief.

Example 2.20: That *gastric ulcer* is caused by acidity was a perspective that was considered for many years as possessing knowledge (k_A) strength, whereas the appropriate characterization should have been b_A, which has a lesser epistemic strength. The latter characterization was fully justified when it was later established that in many cases ulcer is caused by a bug (*Helicobacter pylori*) upsetting the stomach and upper intestinal tract (see also Example 2.11).

As was noticed on various occasions, distinct assertion forms, *KB*s and decision-making skills are associated with different developmental stages of a health care professional. At the novice's stage, professionals rely primarily on *G-KB*; their clinical behavior is rather limited and inflexible, following "context-free" instructions; and their assertions are usually of a lower

---

[20]That is, that A knows that CA occurs entails that A rationalizes that CA occurs, that A sustainably believes that CA occurs, that A believes that CA occurs and that A guesses that CA occurs.

justification level (g_A, b_A), hesitating to attempt higher-level ones. With time and experience, the novice could start formulating action guidelines, but still does not have in-depth experience with similar health cases. At the competency stage, health care professionals have developed significant S-KB, in addition to the existing G-KB. They can form s_A assertions with greater confidence, making conscious decisions and viewing their actions in terms of goals and plans, and they begin to recognize patterns and the essence of clinical cases faster and more accurately. At the proficiency stage, health care professionals can use their considerable S-KB (based on their encounters with similar patient populations) together with higher-level assertions (r_A and k_A) more frequently to organize and understand cases spontaneously and as wholes rather than as unconnected aspects; yet they still rely on analytical thinking to choose an action. At the expertise stage, professionals take advantage of their superior skills (including assertions that are usually of the highest level) and their rich G- and S-KBs, acting naturally and making fast, intuitive and fluent decisions. In addition to his/her developmental skills, the physician's assertion state itself (a_A-form) may also change during a medical case (g_A may be upgraded to b_A, or r_A may be reduced to s_A etc.).

Example 2.21: The comprehension of a disease by physician A may initially be expressed as b_A, then turn into an r_A and finally into a k_A state. Matters may be different when A's clinical belief is demonstrated only in the current situation and does not rely on well-grounded tenets. This happens when A is faced, say, with the issue of how to suction a very fragile patient whose oxygen saturations are observed to sink too low.

If the medical assertion a_A includes instructions on how to get to disease event or case attribute CA, then a_A may be seen as a proof of CA's validity, i.e. $a_A \equiv k_A$ may be appropriate. This condition is related, to some extent, to the notion of *intuitionistic* logic that is also based on a notion of proof (van Dalen, 2001). Formally, intuitionistic logic can be succinctly described as a part of classical logic, in the sense that all formulas provable in the former are also provable in the latter. Interpretationally, intuitionistic logic is characterized by the strict rendition of the basic phrase "there exists" of its traditional counterpart, classical logic, as "there can be constructed". As a consequence, the logical connectives are seen as instructions on how to construct a proof of a medical statement that includes these logical operators. Noticeably, certain authors find it instructive to distinguish between the phrases:

- *it is valid* (exists or is true) of classical logic vs.
- *it is constructed* of intuitionistic logic vs.
- *it is substantiated in an uncertain environment* of stochastic logic.

It is worth revisiting the topic in subsequent parts of the book.

In a relevant note, although the mind state in the earlier expression (2.7) is epistemic, there is no sense in talking about epistemic states in the absence of any ontic (actual or real) states of health. *SMR* inference incorporates the distinction between ontic and epistemic states in an essential way in its semantics. Note that "epistemic" is not necessarily a purely subjective notion, in the sense of happening only in the mind of a physician without any connection to an actual health situation (in the same way that there is no knowledge without things to be known, and there is no software without hardware). As a result, the intended interpretation of a mind state in (2.7) is to offer a representation of a physician's overall cognitive or information condition concerning an actual disease, which is why in all of the above cases the a_A is *relativized* with respect to the KB available. If a test result suggests that a disease occurs, the physician A may assume to know, rationalize or simply believe it to be the case, depending on the epistemic situation (logic mode, education, intelligence, experience).

In addition to physician A's cognitive abilities, how much is *at stake* in a medical case could also have a direct impact on which one of the assertion forms in Eq. (2.8) should be chosen. Due to the possibility of very serious surgery, A's assertions regarding the clinical case may need to be at the level k_A or r_A. What is at stake can have a direct impact on whether something counts as hard evidence to A in the particular case. People, on the other hand, will probably be more reluctant to grant that physician A knows the specified case (i.e., A's assertion about the case is of k_A form) in high-stakes situations than in low-stakes situations. Another reason for the difficulties of implementing sound and rigorous reasoning is due to the increasing *financial* stakes associated with present-day medicine. In particular, many scholars have cited the view that the health care system has been turned into a profitable industry (McGlynn *et al.*, 2003). In the past, health care providers were required to spend time with patients and to think in an open, expansive way. For economic reasons, such an open-ended practice is no longer available. Instead, today's physicians are increasingly conditioned to think of their professional time as money, i.e., they are subject to the so-called "relative value units" (doctors must strictly account in dollars for their time).

Lastly, the credibility of a health study fundamentally depends on the physician's context awareness being aware not only of one's own, often uncertain, state of mind, but also of one's mind state being properly communicated and adequately assessed by other participants in the case (investigators, experts, juries).

Example 2.22: Consider a court case as follows:

(a) a forensic expert searches a fingerprint database with two different options ("match" and "no match"), and then asserts that "match" is the appropriate characterization of the situation;
(b) the expert transfers this result by testifying in court;
(c) after receiving the expert's assertion, the jury members produce a judgment (e.g., decide whether the fingerprint left at the crime scene belongs to the suspect or not) and
(d) this judgment is one of the inputs to be taken into consideration during the jury's larger deliberations.

The process a–d produces a number of assertions made by the different participants of the court case: assertions in *steps a–b* are based on scientific findings and their interpretation by a forensic expert; subsequent assertions by the jury members (*step c*) may vary depending on a number of factors (e.g., a scientist among the jury members may assign more weight to the expert's assertion than a non-scientist); and the court's final judgment in *step* d is a certain synthesis of all previous assertions.[21] It is rather obvious that all the above assertions and their synthesis are not characterized by the same degrees of justification and reliability: some assertions may be justified as knowledge, some others as rational claims and yet some others as merely personal beliefs.

2.3.7. *The Three Qs of the triadic case formula*

In light of the analysis so far, the readers agree with the view that medical reasoning and clinical inferencing, essentially, is a response to the *Three Qs* proposed by *Hermagoras*:[22]

$$Q^3: Quid \text{ (What)} - Quis \text{ (Who)} - Quomodo \text{ (How)}.$$

---

[21] Naturally, different kinds of synthesis may be possible, depending on the jury.
[22] Hermagoras (Ερμαγόρας), of Temnos, was an ancient Greek rhetor (1st century BC).

The above lead to the fundamental *triadic case formula* of medical reasoning, which may be visually represented as follows:

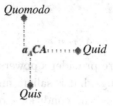

In more detail, the triadic consists of:

What is the object of study: One or more case attributes CA.
Who studies CA: Health care professional A.
How A expresses him/herself: Assertion form $a_A(g_A, b_A, s_A, r_A, k_A)$.

The triadic case formula is the basis of much of stochastic medical reasoning and inferencing in conditions of uncertainty.

Medical thinking involves logical functions of triadics that are consistently developed during the medical decision-making process, and have the general form

$$\Phi[a_A CA],$$

where the case vector $\boldsymbol{CA}$ may involve a set of individual attributes $\{CA_i\}, i = 1, 2, \ldots, n$. For illustration, some examples of Φ-functions are,

$$a_A((CA_1 \vee CA_2) \vee (\neg CA_1 \vee CA_3) \to (CA_2 \vee CA_3)),$$
$$\boldsymbol{CA} = \{CA_1, CA_2, CA_3\};$$
$$a_A(CA \in \Theta) \wedge a_A(M_{CA}),$$

where Θ is an interval of values, e.g., $\Theta = [38.3, 39.1°C]$, and M_{CA} is a biomedical law or a scientific model.

Interestingly, Φ-functions are also the uncertainty, $U[a_A CA]$ (Section 2.3.8), and probability, $P[a_A CA]$, functions (Chapter 3).

The triadic case formula is essential for training the mind to think in the substantive logic setting. To truly and thoroughly describe a sign or a symptom in terms of the triadic can strengthen the care providers' thinking style, empower their medical choices and lead to more awareness of their clinical decisions. It also trains the mind to see the relationships between decisions and consequences, a type of thinking that is particularly valuable in a professional's everyday routine. By subsequently comparing what has been described in terms of multiple diagnoses, stating the value of whatever has been described and compared and evaluating one's prognosis, the care provider has a scope and view that will lead to advances in medicine and quality care.

2.3.8. Uncertainty factors: A review

As noted earlier, a health care provider's powers of noticing or perceptual grasp depend upon recognizing what is salient and the capacity to respond to the particular medical case in conditions of real world uncertainty. Distinct levels of uncertainty could characterize a physician's *in situ* assertion of a medical case. Reviewing the discussion in previous sections, we could sum up the uncertainty notion, $U[a_A CA]$, as a function of three main factors:

(1) health care provider A,
(2) case attribute CA and
(3) assertion form $a_A(g_A, b_A, s_A, r_A, k_A)$ based on the relevant KB.

A care provider A creates $a_A CA$ and assigns to it an uncertainty measure $U[a_A CA]$, which then A may interpret in two possible ways: as a more or less subjective notion reflecting care providers' views and feelings, or as a technical notion at the syntactic level (see, also, technical interpretation of the relevant notion of information, Section 3.5). To get a better grasp of these notions and their close interrelations, let us examine an example.

Example 2.23: Physicians A and B consider the same medical case (say, they examine a patient suffering from a certain disease). Suppose that following the medical interview A's diagnoses set is

$$\{DD^{(1)}, DD^{(2)}, DD^{(3)}\},$$

whereas B's is

$$\{DD^{(1)}, DD^{(2)}\}.$$

The goal of both physicians has been to narrow the list of possible diseases that could be the cause of the patient's situation and focus physical examination and diagnostic testing on a small number of possible diagnoses. Justifying their own diagnoses,

> "physician A feels more uncertain (less confident) than physician B about the disease situation, which is why A's diagnoses set includes more possibilities than that of B's".

For example, rigorous medical testing can reduce the uncertainty of diagnostic reasoning, allowing the physician to select the kind of antibiotics that are most likely to treat an infection. This view of uncertainty is consistent with many informal epistemologies in which it is typically stipulated that the smaller the set of scenarios (diagnoses) a physician considers possible, the smaller the physician's uncertainty. On the other hand, technically

"it is less uncertain that A will turn out to be correct than B, since A's diagnoses set allows more possibilities about the true diagnosis than B's".

Yet, A's diagnoses set has a lower value (it is less informative) than B's set, for the same reason. In other words, physician A may develop an uncertainty interpretation about $a_A C A$ in a subjective or a technical manner. From this two-sided perspective, the more uncertain A is about CA the larger the set of possibilities allowed by $a_A C A$, whereas the higher the number of the possibilities allowed by $a_A C A$, the lower the technical uncertainty one assigns to $a_A C A$.

The two (*a priori*) interpretations of uncertainty lead to two opposing rankings in health care practice: in the case of the technical uncertainty interpretation of A's assertion, the following ranking seems appropriate[23]:

$$U[g_A CA] < U[b_A CA] < U[s_A CA] < U[r_A CA] < U[k_A CA]. \qquad (2.10)$$

On the other hand, in the case of the expert's subjective uncertainty interpretation of A's own choice of the a_A-form, the reverse ranking makes sense.[24] Equation (2.10) is here understood in the sense that, since the assertion forms g_A, b_A, s_A, r_A and k_A cover a decreasing number of possibilities (i.e., sequentially assigning fewer diagnoses to a medical case), the uncertainty levels assigned to the corresponding assertions should increase.[25]

In view of the above considerations, Table 2.1 lists possible combinations of the three uncertainty factors (A, CA and a_A) that may occur in medical decision-making. Let us study some examples of these combinations.

---

[23] I.e., here U is not A's own subjective uncertainty about the appropriateness of the chosen a_A form to represent the particular case.
[24] When Dr. Heller describes his state of mind about a case as r_A, the physician feels that he possesses a better understanding of the clinical case than when he describes it as b_A, and, as a consequence, Dr. Heller assigns smaller uncertainty U to r_A than to b_A.
[25] I.e., the more uncertain an assertion the more informative it is; naturally, the opposite is the direction of the probability levels (Section 3.6).

Table 2.1: Combinations of uncertainty factors.[26]

Case(s)	Medical expert(s)	Assertion form(s)
CA_1, CA_2	A	$a_{A,1}, a_{A,2}$
CA_1, CA_2	A, B	$a_{A,1} \equiv a_{B,2}$
CA_1, CA_2	A, B	$a_{A,1}, a_{B,2}$
CA	A, B	a_A, a_B
CA	A, B	$a_A \equiv a_B$

Example 2.24: To study an illustration of the first combination in Table 2.1, let us revisit Example 2.17. In that case expert A's sustained belief concerning $PM_{2.5}$ population exposure was

$$s_A(PM_{2.5} \in [5, 450\,\mu g/m^3]). \qquad (2.11)$$

Assume that in light of new evidence more possibilities need to be considered (say, move to a different cognitive level) so that A's mind state is reduced to the subjective belief form

$$b_A(PM_{2.5} \in [3.5, 485\,\mu g/m^3]).$$

Under these circumstances the following uncertainty inequality can be assumed to hold concerning the $PM_{2.5}$ distribution:

$$U[b_A(PM_{2.5} \in [3.5, 485\,\mu g/m^3])] \leq U[s_A(PM_{2.5} \in [5, 450\,\mu g/m^3])];$$

i.e., in technical terms it would be less uncertain (more probable) that b_A rather than s_A will turn out to be valid. Regarding the second combination in Table 2.1, assume that expert B's evaluation of the same case is expressed as the sustained belief

$$s_B(PM_{2.5} \in [3.5, 485\,\mu g/m^3]).$$

Since s_B permits more possibilities than s_A, technically one would expect the following uncertainty inequality to hold:

$$U[s_B(PM_{2.5} \in [3.5, 485\,\mu g/m^3])] \leq U[s_A(PM_{2.5} \in [5, 450\,\mu g/m^3])].$$

---

[26]The combination "A, CA, a_A" is trivial, whereas the combinations "$A, CA, (a_{A,1}, a_{A,2})$" and "$A, (CA_1, CA_2), (a_{A,1} \equiv a_{A,2})$" are counterintuitive, so are not included in Table 2.1.

In relation to the third combination in Table 2.1, Example 2.20 assumed that expert's belief

$$b_A(BT \in [37, 41°C])$$

allowed a larger set of possibilities than agent B's rationalization

$$r_B(BT \in [38.3, 39.1°C]).$$

Hence, the technical uncertainty of A's belief is lower, allowing more possibilities, than the uncertainty of B's rationalization that permits fewer possibilities. For the fourth combination in Table 2.1, A's belief $b_A(BT \in [37, 41°C])$ above allows the same set of possibilities as expert B's claim to knowledge

$$k_B(BT \in [37, 41°C]).$$

This is an interesting and by no means uncommon situation. The relevant uncertainties depend on each expert's qualifications and reputation. If A is a better expert than B, the b_A case form would be favored over k_B. In the opposite situation in which B is the better expert, the k_B form may be favored over b_A, although extra caution should be exercised. Lastly, the fifth combination is an ideal situation in which the two experts are in complete agreement and, accordingly, the corresponding uncertainties coincide. One may recall that belief is a concept that is like knowledge, except that it does not have the success element built into it; belief can be false. In this sense, B's claim to knowledge is considered to have zero uncertainty, whereas A's belief has a non-zero uncertainty.

2.4. SMR's View of Medical Connectives: Beyond Drug Digestion

Generally, logical connectives (Section 2.3.4) seek and systematize patterns of valid medical inferences in the linguistic recordings of decision-making activities. The aim is to develop reasoning rules operating on assertions a_A (statements, beliefs, claims) of the decision-making process such that, if the original a_A conveys a scientific truth, so will the assertions obtained by applying these rules. Choosing among logics essentially amounts to choosing among ways to give meaning to connectives in terms of the appropriate states of mind and the actual case conditions. In other words, sound reasoning in public and private practice depends critically on the health care professional's conception of the relations between connectives, mental states, case characteristics and language. Whether or not decision-making turns out to be sterile for all practical purposes, and generate results that are logically and biophysically incompatible, depends fundamentally on this conception. The goal is to set up an explanatory and predictive, not solely descriptive, framework in which to place *SMR* terms, notions and methods.

Health care professionals should keep in mind that two essential features of stochastic medical reasoning are substantiveness and uncertainty, meaning that *SMR* does not necessarily interpret logical connectives in exactly the same way as standard logic. It rather disallows the idealistic interpretations upon which mainstream quantitative analysis depends, by interpreting the connectives in a substantive way. As the readers might have guessed, the latter has several implications for medical thinking.

2.4.1. *Conversational (dialogical) connective interpretation*

A practicing physician is aware of the distinct meanings of the same logical operator in different cases, for an assertion may be linked to substantially different logical argumentations. Assume that the condition of truth functionality is valid, i.e., the *truth-value* (*TV*) of a connective,

$$\text{true } T \text{ (equivalently, 1) or } \textit{false } F \text{ (0)},$$

is determined by the *TV*s of its constituents. Then a *classical truth table* (*CTT*) can be introduced in medical decision-making.

Example 2.25: Consider two different diagnoses $DD^{(1)}$ (say, migraine) and $DD^{(2)}$ (say, tension-type headache). One can implement a *CTT* (see Table 2.2) *from left to right*, if diagnosis $DD^{(1)}$ is true, i.e. $TV(DD^{(1)}) = 1$ (or T), then the validity of the combined diagnosis $DD^{(1)} \vee DD^{(2)}$ is confirmed in the classical manner, i.e. $TV(DD^{(1)} \vee DD^{(2)}) = 1$ (or T).

Table 2.2: CTT of disjunctive medical diagnosis.

$DD^{(1)}$	$DD^{(2)}$	$DD^{(1)} \vee DD^{(2)}$
$T(1)$	$T(1)$	$T(1)$
$T(1)$	$F(0)$	$T(1)$
$F(0)$	$T(1)$	$T(1)$
$F(0)$	$F(0)$	$F(0)$

But, a health care provider may also look at the *CTT from right to left*, given that $DD^{(1)} \vee DD^{(2)}$ is true, one cannot confirm that the diagnosis $DD^{(1)}$ is valid or that diagnosis $DD^{(2)}$ is the case. Instead, there are three possibilities to consider (*TT*, *TF*, and *FT*).

Although this *CTT* does not demonstrate the individual truths of the two diagnoses, $DD^{(1)}$ and $DD^{(2)}$, it imposes certain restrictions on these individual truths. The *CTT* of the disjunction of the two diagnoses, e.g., excludes the possibility that both diagnoses are false. In other words, if the combined diagnosis $DD^{(1)} \vee DD^{(2)}$ is assumed to be valid in clinical practice, then it is not possible that the individual diagnoses $DD^{(1)}$ and $DD^{(2)}$ are both false in the actual case considered.

As the readers may have suspected, *SMR*'s interpretation of connectives needs to go beyond the classical interpretation of connectives as depicted in the *CTT* of Table 2.2. As a matter of fact, health care providers may find it interesting that in the preceding cases the original Eq. (2.7) may be seen in the light of *fallibilism* (or *Pyrrhonism*), which admits the possibility of knowledge that is incomplete, to a certain degree, but not of absolute certainty.[27] *SMR* is concerned with establishing criteria and justifications for assertion validity (especially knowledge claiming). Loosely speaking, these criteria and justifications act as a kind of thermostat, deterring and constraining excessive assertions made by investigators (knowledge claims, strong beliefs, unmistaken diagnoses, definite treatments). Without this kind of constraint, medical assertions might easily descend into exaggerated statements and dogmatic belief systems. These are factors that have an essential effect on clinical practice, which extends beyond drug digestion to an intimate interaction with patients. Relative to the latter is that in many cases the formal interpretation of a connective (assumed in std guidelines, medical software libraries etc.) and its *conversational* or *commonsensical* interpretation (used by health care providers and patients in their communication with doctors) can differ substantially. We will deal with these interpretational matters in more depth in Section 2.6. At the moment, we should keep in mind that in everyday reasoning the physicians' conversational conditionals are inherently uncertain rather

---

[27] According to this philosophical perspective, taken for granted in natural sciences, humans could be wrong about their assertions (claims, beliefs, expectations), or their understanding of the world. Accordingly, one is open to new evidence, experience and argumentation that would disprove some previously held assertions (Kompridis, 2006).

Table 2.3: Truth table for two interpretations of disjunction (classical and conversational).

RT_p	$LF_{p'}$	$RT_p \vee LF_{p'}$ (Classical)	$RT_p \vee LF_{p'}$ (Conversational)
$T(1)$	$T(1)$	$T(1)$	$F(0)$
$T(1)$	$F(0)$	$T(1)$	$T(1)$
$F(0)$	$T(1)$	$T(1)$	$T(1)$
$F(0)$	$F(0)$	$F(0)$	$F(0)$

than strictly formal, e.g., they hold only "usually", "probably" or "under normal conditions". This kind of difference, if unnoticed by the health care professionals, can have serious consequences.

Example 2.26: In an analogous manner, the formal consideration of the "disjunction" connective is different from its everyday conversational interpretation, since the latter does not usually include the case that both components of disjunction are true (Table 2.3). To illustrate this situation consider the assertion that

> "on April 30, 2014 Mr. Pininfarina will undergo retinal detachment (RT_p) surgery in Los Angeles",

where, $p = (s, t)$, s denoting "the geographical location of Los Angeles" and t denoting the date "April 30, 2014"; and

> "on May 1, 2014 Mr. Pininfarina will take a long flight ($LF_{p'}$) to New York",

$p' = (s', t')$, with s' denoting "New York" and t' denoting "May 1, 2014". In formal or classical logic, when both RT_p and $LF_{p'}$ are assumed true, the disjunction $RT_p \vee LF_{p'}$ is also true (see, third column of Table 2.3). Yet, this is not valid in everyday conversational argumentations (fourth column of Table 2.3). I.e., unless certain special circumstances apply in the real world, to a rational individual it appears rather nonsensical that Mr. Pininfarina would undergo RT surgery when, in fact, he plans to take a long flight the next day (his doctors would never had permitted him to do so). Such important substantive features of commonly used connectives must be kept firmly in mind when health care providers communicate with their patients or use medical software libraries.

2.4.2. Content-dependent vs. content-independent connectives

A direct consequence of the preceding analysis is that a health care professional should not view the implementation of the logical arguments of Tables 2.2 and 2.3 as a mechanistic process, but one that depends on the content and context of the actual case, the relevant uncertainty and the required medical and other KBs. Accordingly, the validity or existence of the relevant assertions involving these connectives is CD, based on scientific knowledge and empirical evidence, rather than CI, which is the case of formal logic. Let us be more specific. Consider the assertions concerning

"human exposure (HE_p) at the city with location–time coordinates p",

and

"disease incidence (DI_p) at the city p",

denoted as triadics, $a_A H E_p$ and $a_A D I_p$, respectively. The assertion $a_A H E_p$ is substantively valid if there is a CD method (involving biomedical law, scientific theory, strong evidential support) the investigator can use to show that $a_A H E_p$ is valid in reality. Similar is the case with the assertion $a_A D I_p$. Also, the conjunction

$$a_A H E_p \wedge a_A D I_p$$

holds if there is a CD method to show that both assertions about HE_p, DI_p occur *in situ*. On the other hand,

$$a_A H E_p \vee a_A D I_p$$

is substantively valid if there is a CD method to show which one of the assertions about HE_p, DI_p, is the case in clinical or medical practice. Lastly, the entailment

$$a_A H E_p \therefore a_A D I_p$$

makes sense if the assertion about HE_p is a reality that leads to the assertion concerning DI_p through a CD process, including logical connectives.

Example 2.27: The assertion of the Sir Run Run Shaw hospital staff member A that

"Dr. Messerschmitt's influenza seminar is given in Room H ($a_A RH$) of the hospital", is in disagreement with the assertion of the hospital staff member B that

"Dr. Messerschmitt's seminar is given in Room L ($a_B RL$) of the same hospital".

When $a_A RH$ is observed, the statement "either $a_A RH$ or $a_B RL$" ($a_A RH \lor a_B RL$) is formally justified (in a manner similar to that of Table 2.2). Clearly, no expert medical knowledge is needed for this justification. On the contrary, for A's dialogical assertions

"Ms Conejo Enamorado properly takes steroids that cure the deadly kind of poison ivy ($a_A LS$)",

and

"Ms Conejo Enamorado died from poison ivy ($a_A LD$)",

$a_A(LS) \lor a_A(LD)$ is not substantively justified merely because a valid $a_A(LS)$ has been actually observed in the real world. Instead, the justification requires scientific knowledge about the course of the disease, its symptoms, about the fact that the harmful effects of poison ivy are cured with steroids etc.

As noted earlier, situations such as the above are handled by assigning a probability to each TV associated with a medical case, which is why in real professional practice one rather talks about stochastic rather than deterministic truth tables. Which brings us to the other essential feature of the physician's stochastic reasoning, that is, the rigorous incorporation of health status uncertainty in one's argumentation. Interestingly, medical statements that may be logically impossible in the context of formal certainty could be valid in the uncertainty setting of Eq. (2.7). This realistic flexibility makes it possible for health care providers to build into their medical information processing better ways to link to *in situ* data that add new value in their argumentation.

Example 2.28: Let DS_p denote a specified disease symptom that can make its appearance at location–time p. In standard logic, the $DS_p \land (\neg DS_p)$ of object language is definitely impossible in a clinician's everyday routine. However, the assertion $b_A DS_p \land b_A(\neg DS_p)$ of metalanguage could be valid in the sense that, due to uncertainty, expert A cannot exclude one of the two possibilities with certainty; i.e.,

"A believes that DS_p would occur with probability say, 0.75, but also believes that $\neg DS_p$ would be the case with probability 0.25".

More about this matter will come after we introduce the notion of probability in Chapter 3.

Lastly, there is also a *sociological* dimension in the context- and content-dependent analysis of connectives. For example, differing institutional contexts in medicine are likely to draw different conclusions. Which institutional structure is best for the case at hand is likely to depend on the essence of the medical problem at hand. What is considered a true judgment in one institutional context is likely not to be acceptable in another (Klein, 1996; Palmer, 2001; Lele and Norgaard, 2005).

2.5. Natural Laws and Scientific Models

It is important to recall that the substantiveness and semantics of *SMR* are intimately linked to the notions of natural law, scientific model and empirical evidence. When individuals interact, various kinds of relations tied to their internal structure emerge as a consequence of a causal association between internal structure and the way the individual interacts with others (e.g., infected individuals and their contacts with other people). Because *SMR* takes place in a space–time domain, these relationships can link either different case attributes (say, $CA_{1,p}$ and $CA_{2,p}$), or the same attribute at different location–time points ($CA_{1,p}$ and $CA_{1,p'}$), or different attributes at different points ($CA_{1,p}$ and $CA_{2,p'}$).

Undoubtedly, increasing patient acuity and health care complexity demand that health care providers fuse diverse bodies of knowledge. In this spirit, physicians can take advantage of the fact that case attributes are often linked by means of laws of spatiotemporal change and empirical models. Processing this kind of information would improve considerably the physicians' critical thinking skills, clinical judgment and multiple perspectives. As a matter of fact, during the last three decades considerable effort has been devoted to the *quantitative modeling* of important diseases leading to noticeable developments in mathematically rigorous and biophysically meaningful health modeling (Anderson and May, 1991; Crawford-Brown, 1997; West and Thompson, 1997; Roberts, 2004; Kane, 2009). *Stochastics*, in particular, offers considerably more information to health care providers as well as a more complete understanding of disease than mainstream deterministic health models (Christakos and Hristopulos, 1998; Allen and Burgin, 2000; Tan, 2000; Kolovos *et al.*, 2011; Angulo *et al.*, 2012). This information is semantically rich, and includes realistic representations of the *in situ* trend and variation of disease attributes (e.g., a hand–foot–mouth disease model connecting susceptible, infected and recovered population fractions; Angulo *et al.*, 2013), their temporal dependency and spatial heterogeneity (e.g., space–time covariances and higher cumulants of dengue fever; Yu *et al.*, 2011) and probability of attribute occurrences (e.g., extinction probability of $AIDS^{28}$ infecteds with time; see also Example 2.29 below). In any case, continuous model testing is necessary because one never understands a health model until one knows precisely its boundaries and applicability conditions — when and why it fails to provide good case assessment and disease prediction.

2.5.1. *Infectious disease and human exposure modeling*

Generally, scientific models result from the merging of understanding (biological, epidemiological and clinical aspects of disease spread) with rigorous calculation that is consistent with experiment and observation. Mathematical models of disease evolution and spread are useful for a number of reasons. First, these models can allow in-depth understanding of some basic features and principles of disease spread. Second, they can be used to assess the impact of risk factors and to screen for important risk variables for epidemic prevention and control purposes. Third, health care professionals may use mathematical models of an epidemic to evaluate and compare different prevention and control strategies. Fourth, mathematical epidemic models can be used to project future cases and disease prevalence. Physicians should not hesitate in doing the mathematical analysis and calculations required by quantitative decision-making. For illustration purposes, the readers may appreciate a few examples of laws and models that can support professional practice.

As noted earlier, a disease distribution is a fundamentally spatiotemporal phenomenon, its purely temporal and purely spatial models being merely convenient simplifications. The *generalized SIR (Susceptible–Infected–Recovered)* model that describes the distributions of the infected I_p, susceptible S_p and recovered R_p population fractions during time t within a spatial area around location s was proposed by Angulo *et al.* (2012):

$$I_{s,t+dt} = q_{s,t}(1 - a_{s,t} + b_{s,t}S_{s,t})I_{s,t} + \Lambda[(1 - a_{u,t} + b_{u,t}S_{u,t})I_{u,t}]$$
$$S_{s,t+dt} = q_{s,t}(1 - b_{s,t}I_{s,t})S_{s,t} + \Lambda[S_{u,t} - b_{u,t}I_{u,t}S_{u,t}]$$
$$R_{s,t+dt} = q_{s,t}(R_{s,t} + a_{s,t}I_{s,t}) + \Lambda[R_{u,t} + a_{u,t}I_{u,t}], \qquad (2.12\text{a--c})$$

---

[28] Acquired immune deficiency syndrome.

where $\Lambda[\cdot] = \int du \kappa_{s-u,t}[\cdot]$; $q_{s,t}$ is the population fraction that resides at (i.e. does not displace from) the location–time domain, $\boldsymbol{p} = (\boldsymbol{s},t)$, $1 - q_{s,t}$ is the fraction that migrates during the specified time period, $\kappa_{s-s',t}$ is a spatially homogeneous kernel (e.g., Gaussian kernel with finite variance) that controls population movement across space, its spatial integral being equal to $1 - q_{s,t}$, $a_{s,t} \in (0,1]$, is the probability that an infected individual at $\boldsymbol{p}$ recovers and becomes immune and $b_{s,t} \in (0,1]$ is the probability of infection transmission during an encounter of one infected and one susceptible individual ($a_{s,t}$ and $b_{s,t}$ can include information about regional topography and local climatic conditions).[29] Equations (2.12a–c) are stochastic constructions. Generally, the mathematical structure of an epidemic model varies considerably, depending on vital disease features.

Example 2.29: In the case of *AIDS*, in particular, mathematical models have been developed for the general or target populations, e.g., homosexuals or drug users (Tan, 2000). *HIV*[30] attributes, including the number of susceptibles, infecteds and clinical *AIDS* cases, mixing patterns (an individual's number of sexual partners per unit of time), infection times and transmission speeds, are represented in stochastic terms. The reasons for this modeling choice include that many *AIDS* attributes are subjected to considerable random variation and *in situ* fluctuations (clinical, social, cultural). Lastly, as a visualization illustration, Fig. 2.3 displays the geographical distribution of *bubonic plague* mortality in India at different times (Yu and Christakos, 2006). Modeling accounted for disease records, climate data, topography and transportation information.

Laws of change and empirical models have been developed for human exposure and population dynamics situations with important clinical and epidemiological applications.

Example 2.30: Under certain *in situ* circumstances, the pollutant concentration distribution, $PC_{\boldsymbol{p}}$, follows the chemical law (Christakos and Hristopulos, 1998),

$$PC_{\boldsymbol{p}'} = PC_{\boldsymbol{p}} e^{\beta|\boldsymbol{p}-\boldsymbol{p}'|}, \qquad (2.13)$$

---

[29] Under certain *in situ* conditions of disease control, such as *quarantine* and *vaccination*, the transmission and recovery rates are indeed time-varying.
[30] Human immunodeficiency virus.

where β is an empirical parameter. Also, one may recall that the local (s) species population at time t, SP_p, obeys the classical growth law (Anderson and May, 1991)

$$\frac{d}{dt}SP_p - aSP_p = 0, \qquad (2.14)$$

where a is an empirical coefficient. Furthermore, an exposure model of environmental tobacco smoke is as follows (Crawford-Brown, 1997):

$$IR_p = PC_p BR_p, \qquad (2.15)$$

where IR_p is the intake rate of the pollutant, PC_p is the concentration of pollutant particles in the air and BR_p denotes a person's normal breathing rate for air (this model is also studied below in Example 2.31).

Linked to the above considerations are a number of additional issues concerning the methodological assumptions and metalanguage of a physician's assertions. Recall that the SMR metalanguage may be linked to both a natural language and an abstract notation (as in the symbolic form of models (2.12)–(2.15)). Some health modelers cite the view that in the mathematical terms of Eqs. (2.12)–(2.15) it is not the interacting entities themselves (humans, real or abstract objects) that participate in the formal constitution of the system of equations, but rather their quantitative attributes and couplings (location, time, contaminant concentration, intake rate, disease distribution). As a consequence, these modelers argue, interaction may be conceived as the space–time change in the numerical values of these attributes. In other words, in models like those of Eqs. (2.12)–(2.15) physical objects and humans "disappear", so to speak, into groups of formal variables confined to holding numerical values that quantify attributes, such as the concentration of a chemical pollutant, the frequency of a human gene, the disease mortality of a population, the spread of an epidemic, the density of an electromagnetic field or the position and velocity of a radioactive cloud.

Example 2.31: Agents exposed to tobacco smoke are not represented as biophysical entities in Eq. (2.15). Rather, the agents are linked to a set of quantities specifying pollutant distributions in their space–time domain, their breathing and pollutant intake rates and pre-existing medical conditions. These numerical values are an appropriate abstraction, but only as long as the entities involved do not have an internal structure that changes and affects the internal structure of others.

It is worth reminding our readers that, in addition to its dominant scientific character, medical thinking has strong *social* and *dialogical* features, and it is not strictly monological and individual. These noticeable features have several useful consequences in quality care and patient safety practice. For example, clinicians can obtain early warnings of problematic or perilous cases by comparing their observations and insight to those of other health care providers.

2.5.2. *Medical syllogism and the justification of professional assertions*

As we will see in more detail in subsequent sections, medical reasoning relies on syllogisms[31] that link medical case premises (or antecedents) with the health care provider's valid conclusion (or consequent). The general form of a *medical syllogism* used in *SMR* is as follows:

$$\begin{array}{ll} \text{Medical case premises} & \begin{array}{l} \text{Premise}_1 \\ \text{Premise}_2 \\ \vdots \\ \text{Premise}_n \end{array} \\ \text{Care provider's conclusion} & \therefore\ \Phi[\text{Premise}_1 \ldots \text{Premise}_n] \end{array} \quad (2.16)$$

where the premises involve case attributes concerning disease symptoms, physical exams and lab test results, and Φ denotes a logical, causal or probabilistic function connecting the premises in terms of logical connectives, biophysical laws, scientific theories and probability calculus. In many clinical cases, the physician arrives at conclusion Φ by using published results together with personal insight and experience (the largest part of which comes from having seen similar patients and recalling their diagnoses). The premises themselves may be established by empirical observation, the conclusions of previous medical arguments or by definition. Since in practice the premises involve different assertions about case attributes, so should the conclusion, i.e. Φ. But, formally the care provider's premises must not depend on the conclusion, otherwise the syllogism (2.16) will be circular. Most relevant in medical applications are medical syllogisms that provide an underlying diagnosis (e.g., presence of benign or malignant histology in a residual mass after cancer treatment), an outcome occurring within

---

[31] The term was introduced by Aristotle in his works on logic.

a relatively short time after making the prediction (e.g., 30-day mortality rate), or a long-term outcome (e.g., ten-year incidence rate of coronary artery disease, with censored follow-up of some patients). Accordingly, a main objective of medical syllogisms of type (2.16) is to inform patients about their diagnosis properly (e.g., breast cancer) or their prognosis (e.g., after a breast cancer diagnosis has been made). The premises as well as the conclusion of the syllogism (2.16) involve triadic case formulas (Section 2.3.7) of the relevant case attributes. Depending on the a_A-forms (k_A, r_A, s_A, b_A or g_A), different justification degrees, quantitatively expressed in terms of probabilities of occurrence, are associated with the conclusion Φ (we will say more about the matter after the notion of probability has been introduced in Chapter 3).

At the intellectual level, to adequately diagnose and treat disease, clinicians generally need to: (a) exercise forethought and an ongoing grasp of a patient's health status, (b) rigorously assess their understanding of the case at hand (including how much they do not know due to incomplete information and limited cognitive ability) and (c) justify the assertions, syllogisms and inferences they make on the basis of (a) and (b) to ensure positive patient outcome. In real practice, a health care professional would probably follow one of three approaches or a combination of them:

(1) *fit* the medical case at hand to a well-established inference or a set of syllogisms (like those, e.g., listed in various tables throughout the book);
(2) *test* the applicability in the case at hand of *ad hoc* or circumstantial syllogisms and inferences (which happened to work in some clinical cases but their validity has not yet been rigorously proven in a general context) or
(3) *construct* new and valid syllogisms or inferences that are suitable for the case at hand and also logically and biomedically sound.

Example 2.32: Let us start with a simple case in which the Φ-function of Eq. (2.16) is purely logical.

"If it is asserted that the biopsy of Mr. Wu's tumor does not show it to be malignant ($\neg BM$), then it is asserted that Mr. Wu's tumor is not malignant ($\neg TM$)",

and

"the biopsy of Mr. Wu's tumor indeed does not show it to be malignant ($\neg BM$)", it is valid that

"it can be asserted that Mr. Wu's tumor is not malignant $(\neg TM)$".

In the terms of Eq. (2.16),

$$\begin{array}{c} \text{Case premises} \\ \\ \text{Care provider's} \\ \text{conclusion} \end{array} \quad \dfrac{a_A(\neg BM) \to a_A(\neg TM)}{\neg BM} \\ \therefore a_A \neg TM \qquad (2.17)$$

The inference of Eq. (2.17) is known as "contraposition" (see Table 2.5). The probability of occurrence associated with $a_A(\neg TM)$ will depend on the form of a_A. So, different probabilities may be assigned to the forms $a_A \equiv r_A$ (state of rational justification) and form $a_A \equiv b_A$ (state of mere belief).

Depending on the real circumstances of a clinical or medical study, the validity of a syllogism or the justification of an assertion can be made in different ways, two of which are next described to some extent.

Approach a: In justifying an assertion concerning the health effect of an exposure or the etiology of a disease, a care provider explains why the assertion was made, which places it in a frame that one makes sense out of it. In general, the more favorable the impact of an assertion on one's cognitive representation of the case of interest, the more justified the assertion. If the assertion is an intentional one, it should be understood in the context of goal pursuit and the assertion must further a goal. For mind states, the goal can be either to maximize true medical claims or to minimize false ones.

Example 2.33: Consider the case of *environmental tobacco smoke*, which is a rather common pollutant in a number of countries (Example 2.30). Expert A's assertion that for a person at location–time coordinates p,

"the pollutant intake rate (IR_p) is over $1800\,\mu^3/d$"

may follow from the physical fact that

"the concentration of pollutant particles in the air (PC_p) was estimated $110\,\mu/m^3$"

and the empirical assertion that

"from past experience, when PC_p is more than $95\,\mu/m^3$ the IR_p of an individual (with normal breathing rate for air, $BR_p = 20\,m^3/d$) exceeds the critical threshold of $1850\,\mu^3/d$".

In the symbolic terms of Eq. (2.16) the above syllogism can be expressed as:

Case premises
$$PC_p = 110\,\mu g/m^3$$
$$a_A(PC_p > 95\,\mu g/m^3) \to a_A(IR_p > 1850\,\mu^3/d)$$
$$a_A(BR_p = 20\,m^3/d),\ \text{normal breathing}$$

Care provider's conclusion
$$\therefore\ a_A(IR_p > 1800\,\mu^3/d)$$

(2.18)

where PC_p is measured by the appropriate physical instruments, and BR_p is based on biophysical evidence that the expert A finds reliable. The above process locates its conclusion, $a_A(IR_p > 1800\,\mu^3/d)$, in a context with the other premises about PC_p and BR_p. The Φ-function in inference (2.18) is a logical function (the reader may recall that Φ was introduced in Section 2.37). If these other premises happen to be justified, then $a_A IR_p$ will profit from this association. Concerning a_A, it may assume different forms in the various reasoning stages shown in inference (2.18). In some cases the belief state b_A may be chosen, in some others the knowledge state k_A may be appropriate etc. In light of the analysis in Section 2.3.8, one may comment that in technical terms the probability of the conclusion $a_A(IR_p > 1800\,\mu^3/day)$ could be larger than of $a_A(IR_p > 1850\,\mu^3/day)$ in the second premise.

Approach b: Alternatively, the case conclusion $a_A IR_p$ may be justified by its origin (biomedical, logical, causal, statistical), i.e., the $a_A IR_p$ came about in a way that assertions that are likely to be true come about (e.g., it follows rigorously from a natural law or on the basis of an exposure causal chain). In other words, here the justification process aims to show that $a_A IR_p$ fits into the real world, relative to the laws constraining the medical decision process that gave rise to the case conclusion.

Example 2.34: In the case of the environmental tobacco smoke considered earlier, the following inference is also valid:

Case premises
$$PC_p = 110\,\mu g/m^3$$
$$a_A(BR_p = 20\,m^3/d),\ \text{normal breathing}$$
$$M_{ICB}(IR_p = PC_p BR_p)$$

Care provider's conclusion
$$\therefore\ a_A(IR_p = 2200\,\mu^3/d)$$

(2.19)

i.e., expert A concludes that the

"pollutant intake rate (IR_p) at location–time coordinates p is $2200 \, \mu^3/d$",

where M_{ICB} is an exposure model that is part of A's core knowledge about the phenomenon. In other words, the conclusion about IR_p in inference (2.19), i.e. $a_A(IR_p = 2200 \, \mu^3/d > 1800 \, \mu^3/d)$, was derived by means of a scientific process, i.e., the Φ-function has a physical component too. Another facet of this process is that if one assumes that $a_A \equiv k_A$ in premise 2, then, in light of the exposure model (M_{ICB} is objectively valid at current time), it holds that $a_A \equiv k_A$ in the conclusion, as well.[32]

Care providers should pay attention to the fact that the difference between *approaches a* and *b* is subtle, with potentially serious consequences in medical studies. For one thing, the study standards may change depending on whether a physician uses the thinking mode of *approach a* or that of *approach b*. Yet, it seems that the existing health decision-making literature has paid insufficient attention to this issue. We call the readers' attention to the fact that *approach a* operates consistently within the health care provider's epistemic system (a_A's in Eq. (2.18)), whereas *approach b* moves outside of it by means of a logical process involving biophysical knowledge (M_{ICB} in Eq. (2.19)) that exists independently of A (some health care providers might even be unaware of it). Approach a functions in terms of reasons and justifies the result $a_A(IR_p > 1800 \, \mu^3/d)$ by placing it in a specified belief system (i.e., arguing that the coherence of the combination of assertions about PC_p, BR_p and IR_p is evidence for the likely validity of the result). *Approach b* derives $a_A(IR_p = 2200 \, \mu^3/d)$ by means of a network of causes, logical proofs and biophysical laws, arguing that the result is obtained in the way this combination works in similar cases. In Eq. (2.19), the intake rate $IR_p = 2200 \, \mu^3/d$ is asserted as valid because the scientific process that gave rise to it generally gives rise to correct intake values when it functions properly in similar environmental health circumstances. In other words,

"investigator A asserts IR_p because it also asserts PC_p and BR_p, to which IR_p is related" (*approach a*)

vs.

"IR_p is valid because of its content-based connections with scientific assertions about PC_p and BR_p" (*approach b*).

---

[32] Note that $k_A IR_p \therefore IR_p$, although the inverse, $IR_p \therefore k_A IR_p$, is not necessarily so.

While *approach a* concludes that the intake rate is merely above the threshold of 1800 μ^3/d, *approach b* calculates exactly how much above the threshold the actual intake rate is. In this quantitative sense, *approach b* is more adequate than *approach b*. In sum, *approaches a* and *b* seek to justify, in their own way, the calculated pollutant intake rate. In turn, both approaches leading to specified intake rate values should be justified themselves. This is a remarkable situation of *justifying the justification* that emerges rather frequently in medical decision-making. We will revisit the issue in later sections.

2.5.3. Reconstructing Chinese arguments in terms of Greek syllogisms

At this point, a historical diversion is intriguing. In particular, it would be interesting to study the reconstruction of argumentation styles appearing in ancient China in terms of Greek logic-type syllogisms, like that of inference (2.16). In doing so one should keep in mind that in many cases the ancient Chinese thinkers favored an implicit kind of logic rather than an explicit (formal) one. Generally, the reconstruction may involve four steps:

(1) identify typical forms of reasoning in early Chinese medicine writings,
(2) present them in terms of formal logic,
(3) identify general logical rules underlying them and
(4) compare them with ancient Greek logic, when this is possible.

When applying the procedure (1)–(4) one should keep in mind that many ancient Chinese inferences involve statements that cannot be considered as premises. Also, a usual situation in ancient argumentation is that the conclusions are implicit. An example helps to clarify matters considerably.

Example 2.35: The following type of argument, which can be found in ancient Chinese texts, involves six sentences:

[1] To "live longer" (LL) there is a way:
[2] If "one has a good health" (GH), "one lives longer" (LL).
[3] To have a good health (GH) there is a way:
[4] If "one has a healthy lifestyle" (HL), "one has good health" (GH).
[5] To have a healthy lifestyle (HL) there is a way:
[6] If "one practices self-care" (SC),[33] {one has a healthy lifestyle (HL)}.

---

[33] Maintain physical fitness and good mental health, practice good hygiene and avoids health hazards (e.g., smoking, drinking).

The first thing to notice is that the logical structure of this argument is transparent — it actually involves just three premises: sentences [2], [4] and [6]. And the argument's conclusion [6] is implicit (denoted by the brackets "{ }"), which is a rather common situation in ancient Chinese argumentation. The argumentation [1]–[6] can be reconstructed in terms similar to those in Eq. (2.16):

$$
\begin{array}{lll}
& GH \to LL & SC \to HL \\
\textit{Case premises} & HL \to GH \quad \text{or} & HL \to GH \\
& \underline{SC\{\to HL\}} & \underline{GH \to LL} \\
\textit{Care provider's} & & \\
\textit{conclusion} & \therefore SC \to LL & \therefore SC \to LL
\end{array}
\qquad (2.20\text{a–b})
$$

In its second formulation (2.20b) the Chinese logic argument takes the form of the sequential conditional mode of Greek logic. Clearly, when reconstructing an argument, certain pragmatic factors need to be taken into account, since its linguistic shape is often not sufficient for its logical reconstruction.

In reconstructions such as the above it is more appropriate, from a Chinese logic viewpoint, to refer to the admissibility of the premises or the conclusion than to their truth or falsity, as in ancient Greek logic (Chmielewski, 1962). Sometimes an argument contains *pleonastic* premises that should not be included in the logical reconstruction. Similar is the case with *elliptic* arguments, in which a conclusion is often omitted in an argument as being obvious. Moreover, in ancient Chinese texts some arguments were often based on *metaphors* (i.e., it was not the literal meaning of the words in the argument that determined its adequate understanding). Lastly, ancient Chinese thinkers had correctly recognized certain propositional logic principles as valid (conjunction, commutation), but at the same time, they lacked the linguistic tools to express some others.

Example 2.36: Chmielewski (1962) observed that in archaic Chinese there was no direct way to express the connective expressing disjunction. Instead of the direct "alternative" $CA_1 \lor CA_2$ in Chinese one finds the indirect one, $\neg CA_1 \to CA_2$. For illustration, suppose that an ancient Chinese physician has to decide between two traditional remedies,

$$TR_1(\textit{Hange Koboku-To}) \quad \text{or} \quad TR_2(\textit{Sanchitongshu}),$$

for someone who recently had a stroke. Then the Chinese physician's interpretation of Greek logic disjunction "either TR_1 or TR_2" is

"if not TR_1 then TR_2".

Also, Chinese logicians defined equivalence, $CA_1 \equiv CA_2$, as

$$(CA_1 \to CA_2) \wedge (\neg CA_1 \to \neg CA_2),$$

rather than as the mainstream definition, $CA_1 \leftrightarrow CA_2$, of Aristotelian logic.

2.5.4. Revisiting content-dependent and content-independent assertions

It was noted earlier that assertions that rely on natural laws are CD (content-dependent) assertions, i.e., they assume selective kinds of relationships as they are genuinely linked to the unique biophysical meanings of the attributes involved. On the other hand, CI (content-independent) assertions do not depend on the meanings of the attributes and their substantive interrelations. Clearly, CD assertions about a health or exposure case establish a more significant relationship between the case attributes than CI assertions, in that the former offer a more intimate connection between the case attributes than the latter. To illustrate some preliminary aspects of CD and CI assertions, let us study some rather simple examples.

Example 2.37: During a five-day flu treatment the patient is given a daily medicine dose of $MD = 2\,mg$, the total medicine amount during treatment being $TM = 5MD = 5 \times 2 = 10\,mg$. The following *modus ponens* inference is valid:

$$\begin{array}{ll} \text{Case premises} & \begin{array}{l} MD = 2\,mg \xrightarrow{TM=5MD} TM = 10\,mg \\ MD = 2\,mg \end{array} \\ \text{Physician's conclusion} & \therefore TM = 10\,mg \end{array} \quad (2.21)$$

One observes that inference (2.21) is a CD implication, symbolically

$$MD I_{CD} TM,$$

since MD and TM are uniquely linked via the physical dose-total amount relationship. If one replaces $TM = 10\,mg$ with, say, $TM = 7.5\,mg$ then inference (2.21) is not valid, since only for $TM = 10\,mg$, one gets $MD = 2\,mg$.

Next, suppose that several pharmaceutical companies produce the same popular medicine, let MB denote the specified medicine brand, and consider the following *modus ponens*:

$$\text{Case premises} \quad \frac{\begin{array}{l} MD = 2\,mg \to MB = P\!fizer \\ MD = 2\,mg \end{array}}{} . \quad (2.22)$$

$$\text{Physician's conclusion} \quad \therefore MB = P\!fizer$$

(That is to say, if the recommended dose is $2\,mg$ then *Pfizer* is the medicine's brand.) In this case, the logical implication established in Eq. (2.22) is *CI*, viz.

$$MD I_{CI} MB,$$

since MD and MB are not linked in any biomedically substantive manner. If one replaces $MB = P\!fizer$ with, say, $MB = Roche$, then Eq. (2.22) is still valid (since *Roche* also produces the same medicine). Actually, we observe that Eq. (2.22) may also be valid for any $MB = \neg P\!fizer$ (this includes, of course, the case $MB = Roche$), which is to say that in formal logic terms the I_{CI} can yield both MB and its negation.

When uncritically generalized in professional routine, the *CI* notion described earlier could potentially lead to some additional problematic results. This is not a small matter, since the phenomenon characterizes a significant number of cases in clinical and medical practice.

Example 2.38: Suppose that a diagnosis involves the attributes BT and HR (*body temperature* and *heart rate*). Suppose that a physician wants to test a model of the general form (object language)

$$BT I_{CI} HR$$

that relates BT and HR.[34] The *CI* would be considered biomedically useful if it predicted HR values that satisfy the medical condition

$$HR \in \Psi$$

(where Ψ may denote, say, certain physiologic conditions), which means that any prognosis $HR \in \neg\Psi$ disproves the model. However, because of the *CI* feature of the model, in principle both $HR \in \Psi$ and $HR \in \neg\Psi$ are

---

[34] Such a *CI* model can be a regression-based model; see Example 2.39 below.

possible, which shows that a CI model could, under certain circumstances, produce meaningless prognoses. This logical conclusion would be a serious concern in regression modeling of health statistics and biostatistics. Indeed, the readers may recall that the CI notion captures the essence of many techniques of health statistics. As usual, insight is gained with the help of an example.

Example 2.39: Space–time case attributes $CA_{1,p}$ and $CA_{2,p}$ are physically linked through a natural law $M_{CA_1CA_2}$.[35] Due to the presence of $M_{CA_1CA_2}$, the $CA_{1,p}$ implies $CA_{2,p}$ in a CD manner, i.e.,

$$CA_{1,p} I_{CD} CA_{2,p}$$
$$I_{CD}: M_{CA_1CA_2}.$$

This means that the implication does not hold if one replaces $CA_{1,p}$ in $M_{CA_1CA_2}$ with a different attribute $CA_{3,p}$. On the other hand, let us consider a different situation in which the $CA_{1,p}$ and $CA_{2,p}$ datasets do not obey any known scientific law but merely exhibit a high statistical correlation, $Corr^{\gg}_{CA_1CA_2}$, say, via a regression scheme. Not surprisingly, in this case $CA_{1,p}$ is associated to $CA_{2,p}$ merely in a CI manner, i.e.,

$$CA_{1,p} I_{CI} CA_{2,p}$$
$$I_{CD}: Corr^{\gg}_{CA_1CA_2},$$

which is semantically considerably weaker than $CA_{1,p} I_{CD} CA_{2,p}$. This happens because attribute $CA_{1,p}$ can be easily replaced with a substantively different attribute $CA_{3,p}$ that generates sets of data for which it is also valid that $Corr^{\gg}_{CA_3CA_2}$.[36] Therefore, the I_{CD} notion could be used to establish *causation*, whereas the I_{CI} clearly could not (more examples are discussed in Chapter 4).

Let us tahe stock. The relations that CD assertions establish between case attributes exist only by virtue of the particular meanings of the attributes and the substantive biophysical and logical links to each other via appropriate laws. In other words, while CI-based logic is done strictly inside a system, CD reasoning is done outside a system by drawing on medical

---

[35] Say, the two attributes are PC_p and IR_p satisfying the exposure law (2.15), above.
[36] As we will see later (Example 4.16) stock market prices in one place may be highly correlated with disease incidence rate in another far away place, yet there is no substantive link between the two.

insight and clinical experience, constructing analogies and metaphors, on occasion skipping procedural steps and working backwards, resorting to thought experiments and inventing informal tools (shortcuts, diagrams, images) and seeing what happens to the diagnostic or treatment process if a physician changes the formal rules of the case.

2.6. Substantive Conditionals in Medical Thinking

Medical decision-making is a *feedback process*, the diagnosis and treatment changing as the physician gradually collects new information about the patient, or as factors are discovered that can affect judgment regarding prognosis. Consequently, conditionals, in their various forms, play a major role in medical thinking. We already discussed the notion of conversational connectives in Section 2.3.4. In the natural language used in everyday medical practice, a *conversational* or *dialogical* conditional is the logical operation expressed by physicians' statements of the form "if $CA_{1,p}$ then $CA_{2,p}$". Unlike formal logic conditionals (e.g., material conditional), a conversational conditional does not have a stipulated definition.

Formal logic conditionals do not always function in accordance with everyday "if–then" reasoning. Indeed, there are well understood drawbacks to using formal logic conditionals to represent conversational "if–then" statements.

Example 2.40: When, during his clinical practice, Dr. Kavafis speaks of the conditional

"if influenza is a virus" then "1 pill +1 pill = 2 pills"

as being false, he means neither that this assertion is about the statement "influenza is a virus", which is already true, nor that the assertion is about the statement "1 pill +1 pill = 2 pills", whose truth is assured by certain mathematical conditions. Instead, Dr. Kavafis thinks of this conditional as an implication, i.e., in its conversational use the statement

"influenza is a virus" implies that "1 pill + 1 pill = 2 pills",

and this is apparently false, although the implication is valid in formal logic terms.[37]

---

[37] As usual, the term "implies" has degrees of epistemic strength (it may or may not involve causality etc.).

Example 2.41: Although in many cases a perfect correlation could exist between the premise and the conclusion, assuming substantive causality between them turns out to be invalid. The conditional

"if 20 is an odd number" then "20 cannot be divided by two" is true,

whereas the conditional

"if influenza is a virus" then "1 pill + 1 pill = 2 pills"

is false in dialogical terms. There cannot be any causality between "influenza is a virus" and "1 pill + 1 pill = 2 pills". Otherwise said, the correlation between "influenza is a virus" and "1 pill + 1 pill = 2 pills" is 100%, yet the causality between them is false.

To avoid these drawbacks,[38] several alternatives have been used in the real world. *Inter alia*, causal medical conditionals rely on logic connectives, pragmatic conditionals require a theory of context, and counterfactual (subjunctive) conditionals are based on possible worlds semantics. Most behavioral studies of the psychology of reasoning have been carried out with counterfactual conditionals, causal conditionals and non-monotonic conditionals. In this section we start by establishing a *prima facie* case for the appropriateness of introducing the notion of a *substantive conditional*: that in medical decision-making care providers need a *CD* conditional, the content being assigned by substantive factors like the available *KB* for the case (core and case-specific), situational aspects (*in situ* uncertainty, space–time domain) and study needs (prevention, treatment, cure).

2.6.1. *The notion of content-dependent conditional*

Like other *CD* syllogisms, substantive conditionals address the concerns of everyday reasoning in a manner that includes transforming the *CI* conditionals of classical logic into suitable *CD* formulations. In this way, the mathematical rigor of the former is combined with the biomedical soundness of the latter to adequately represent conversational "if–then" assertions of health care practice. Of equal importance, substantive logic avoids the paradoxes of formal logic that are responsible for various nonsensical results in clinical diagnosis and real decision-making. A physician's reasoning can

---

[38] For completeness of presentation, some of these drawbacks are discussed in more detail in Section 2.6 and elsewhere.

form explicit models for structuring one's understanding of clinical cases in terms of substantive conditionals generally represented symbolically as

$$a_A(CA_{2,p} \backslash CA_{1,p})$$

or (2.23a–b)

$$CA_{2,p} \backslash CA_{1,p} \vdots_{KB}$$

The second representation is usually more convenient when KBs are directly recognized in the argumentation considered. Formulation (2.23a–b) is a semantically rich conditional, the content being assigned by the relevant KB. This formulation is the metalanguage counterpart of $CA_{2,p} \backslash CA_{1,p}$ (object language), in which the symbol "$\cdot \backslash \cdot$" can assume different forms (Christakos et al., 2002). As in disjunction, there are a few different senses in which a substantive conditional can be interpreted, including the following:

Conditional title	Formulation	Explanation
CD material	$CA_{1,p} \rightarrow CA_{2,p} \vdots_{KB}$	If $CA_{1,p}$ then $CA_{2,p}$ in light of KB
CD equivalence (biconditional)	$CA_{1,p} \leftrightarrow CA_{2,p} \vdots_{KB}$	$CA_{2,p}$ iff $CA_{1,p}$ in light of KB
CD counterfactual (subjunctive)	$CA_{1,p} \Rightarrow CA_{2,p} \vdots_{KB}$	If $CA_{1,p}$ were valid then $CA_{2,p}$ would be the case in light of KB
CD causal	$CA_{1,p} \mapsto CA_{2,p} \vdots_{KB}$	If $CA_{1,p}$ is the cause then $CA_{2,p}$ is the effect in light of KB
CD statistical	$CA_{2,p} \mid CA_{1,p} \vdots_{KB}$	$CA_{2,p}$ given $CA_{1,p}$ in light of KB

In other words, substantive conditionals account for the objective features of a case and the physician's own understanding of it based on the available core and case-specific knowledge bases, denoted by the subscript KB (including scientific theories, disease models, empirical evidence and clinical data). The conditional can be used to aid the physician's goal of selecting and interpreting diagnostic tests, and assessing their effect on prescribing between different treatments. As we will see later, $CA_{2,p} \backslash CA_{1,p} \vdots_{KB}$ can be expressed in terms of laws of nature or biophysical models that link antecedent and consequent of the medical case. Several other issues are worth investigating, such as for which $\cdot \backslash \cdot$-forms the *transposition relationship* $(\neg CA_{1,p}) \backslash (\neg CA_{2,p}) \therefore (CA_{2,p} \backslash CA_{1,p}) \vdots_{KB}$ holds.

Let us concentrate on some of the most commonly used conditionals in medical decision-making: the causal, the material and the statistical conditionals. Causal conditional reasoning essentially means drawing inferences from a conditional statement that refers to causal content. Statements of medical causality require the premise to precede or coincide with the conclusion in time, whereas the other conditionals do not require this temporal order. Causal conditional reasoning interprets the premise content specifically as a statement referring to a cause and its effect. In addition, it even considers what physicians know about alternative causes for this effect, or about inhibitory factors that need to be absent for the effect to occur.

The truth-value (TV) of the CD material conditional is determined for all four possible combinations of TVs of the premise (antecedent) and the conclusion (consequent), which is why it is termed a truth functional (Table 2.4). This is not the case of the CD statistical conditional, which is indeterminate when the premise is false (also, Table 2.4). The statistical conditional expresses that there is a (symmetric or asymmetric) correlation between its two components, premise $CA_{1,p}$ and conclusion $CA_{2,p}$. Moreover, according to Table 2.4, $CA_{1,p} \to CA_{2,p}\vdots_{KB}$ is logically equivalent to $-(CA_{1,p} \wedge \neg CA_{2,p})\vdots_{KB}$.

Hence, the material conditional can be constructed in terms of logic connective, like $-$ and $\wedge$, whereas the statistical conditional cannot. Interestingly, by viewing $CA_{2,p} \mid CA_{1,p}\vdots_{KB}$ as a mathematical entity with basically three values — T (when both $CA_{1,p}$ and $CA_{2,p}$ are valid), F (when $CA_{1,p}$ is valid and $CA_{2,p}$ is invalid) and IN (indeterminate, when $CA_{1,p}$ is invalid) — several investigators point out a certain correspondence between the TVs of $CA_{2,p} \mid CA_{1,p}\vdots_{KB}$ and $(CA_{2,p} \wedge CA_{1,p})\vdots_{KB}$. Generally, depending on the particular medical case, one conditional may be favored over the other, which is a decision to be made by the competent health care professional.

Table 2.4: Truth table of material and statistical conditionals (IN denotes indeterminate).

$CA_{1,p}$	$CA_{2,p}$	$CA_{1,p} \to CA_{2,p}\vdots_{KB}$	$CA_{2,p} \mid CA_{1,p}\vdots_{KB}$
$T(1)$	$T(1)$	$T(1)$	$T(1)$
$T(1)$	$F(0)$	$F(0)$	$F(0)$
$F(0)$	$T(1)$	$T(1)$	IN
$F(0)$	$F(0)$	$T(1)$	IN

Example 2.42: Let RS and CA denote "Mr. Radon is a smoker" and "Mr. Radon died of cardiac arrhythmia", respectively. The $CA|RS\dot{:}_{KB}$ measures the correlation between RS and CA in a statistical manner, if one wants for the conditional to relate

"Mr. Radon is a smoker and he died of cardiac arrhythmia" ($RS \wedge CA\dot{:}_{KB}$)

to

"Mr. Radon is a smoker and he did not die of cardiac arrhythmia" ($RS \wedge \neg CA\dot{:}_{KB}$).

And there is not necessarily any causality assumed between RS and CA. In a stronger milieu, one may have reasons to assume that there is a causality relation between RS (say, smoking) and LC (say, lung cancer). In this setting, one is not speaking of the (conditional) truth of LC, but of the truth of an implication, since there is abundant evidence concerning the matter.

In Chapter 3, the substantive conditional of SMR inference will be enhanced with the notion of probability, which will allow it to account for the real uncertainties of a medical case in a rigorous and effective way (the 1 and 0 values of the conditionals in Table 2.4 will be replaced by the corresponding probability values).

While formal logic is descriptive, SMR inference is creative. In the former case one cannot obtain new scientific truths that are not obtainable by means of tautologies whereas in the latter, causal patterns may be obtained that are not logical tautologies. By employing the notion of the substantive conditional above, medical reasoning can describe an assertion about the relationship of two entities (cases, attributes, events), i.e., a CD conditional has meaning in terms of the assertion that a certain logical, factual or biophysical relation between entities exists. As such, the substantive conditional captures the notion of relevance (physical, factual or logical) between premise and conclusion of a true implication, which is ignored by formal conditionals of deterministic truth-functional logic. Perhaps health care providers might be able to ignore the use of CD conditional reasoning in medical cases in which they are absolutely certain that the premise is false (something that rarely happens in clinical practice, as the readers may admit). But they cannot ignore the use of conditionals whose premises they consider *likely* to be false.

Clearly, in the *SMR* setting substantive conditionals are not limited by the unnecessary restrictions of classical conditionals. Consider the material conditional (object language) $CA_1 \rightarrow CA_2$ for the case attributes CA_1 and CA_2. In the standard sense, the conditional is true if CA_1 is false or if CA_2 is true, which basically assumes that the *TV*s of CA_1 and CA_2 are known before one can settle the status of $CA_1 \rightarrow CA_2$. In clinical reality, this is a rather unrealistic scenario. Due to *in situ* uncertainty characterizing many health models of object language, the models are expressed in terms of assertions that belong to metalanguage. As usual, the form of the conditional assertions of metalanguage ($a_A \equiv k_A, r_A, s_A, b_A, g_A$) depends on the relevant *KB* and associated uncertainty level. *Perfect* knowledge of the case is a rare condition that characterizes only deterministic derivations and the assertion takes the form of rigorous proof — one seeks a definition of a proof of the implication $CA_1 \rightarrow CA_2$ in terms of possible proofs of CA_1 and CA_2. In other words, the implication is correct if the correctness of the conclusion CA_2 can be shown as soon as the correctness of the premise CA_1 has been established, which, as noted earlier, shares some similarity with intuitionistic logic that also involves the notion of proof.

The substantive conditional also differs from the *indicative conditional* in a crucial sense. The indicative conditional is often described as a logical or factual statement about the relationship between case premise and conclusion, denoting that the "meaning" to be preserved is the assertion that the relationship between entities holds. For an indicative conditional to be true it suffices that the conclusion (diagnosis) really follows from the premise (symptom, sign), even though the two components thus linked are false; whereas to be false it suffices that the sequence from premise to conclusion is invalid, even though the two components are true.[39] For a substantive conditional to hold, on the other hand, the conclusion must really follow from the case premise, and the two components thus linked must be valid.

In sum, experienced health care providers are aware of the difference between substantive conditional and formal logic conditional — the latter holds if the sequence it signifies is valid (invalid), even though the two

---

[39]This is because in the indicative conditional the judgment bears only upon the conjunction of the two components (a necessary conjunction that we declare to be or not to be). The conditional, e.g., "if ten is an odd number" then "ten cannot be divided by two" is true; and the conditional proposition, "if Crete is an island" then "$9 + 1 = 10$", is false.

entities are false (true). For a substantive conditional to be valid the sequence it signifies must be valid, i.e., the conclusion really follows from the premise in terms of laws of nature and/or laws of *CD* logic. In other words, in substantive conditionals the physician's judgment bears upon the conjunction of the two entities — a conjunction that the physician declares that does or does not exist on the basis of scientific knowledge or *CD* logic. In this way, substantive logic avoids many of the paradoxes of mainstream logic.

2.6.2. *Paradoxes of mainstream logic*

Table 2.5 presents a list of the most common relationships and rules of mainstream (standard) logic. It is important that the readers realize that not all of these formal relationships and rules apply in substantive reasoning of everyday routine. As a matter of fact the application of several of them can lead to several *paradoxes*, which are responsible for nonsensical conclusions in medical research and clinical practice.

Paradoxes in logic have a very long history, and indeed in ancient times it was realized that formal logic could lead to paradoxes. In 4th century BC, *Euboulides* posed the now famous question:

Is the assertion "this statement is false" true?

If it is (true), then it is not; and if it is not, then it is. The above is linked to the problem of *self-reference* one finds in certain medical statements (self-reference is literally the quality of a statement's referring to itself). Cases of what is referred *to* are also found in the Mohist *Canons* of Chinese logic (see later in this section). In medical reasoning it is used to characterize statements that include themselves within their scope of reference. For example, let us recall that the cause of certain infections spread in hospitals is poor hand washing. Consider the assertion,

"all medical staff of the St. Judas hospital must either wash their hands
or their hands are washed by the hospital's chief nurse".

This statement is self-referential, as the chief nurse, apart from being "the chief nurse" referred to, is also one of the "medical staff of the hospital".

In previous sections we remarked that, if used in a "black box" manner, mainstream (classical or formal) logic can be the executioner's handmaiden or a fool's ideal accomplice, metaphorically speaking. Indeed, we can find in medical decision-making several statements and inferences using conditionals whose meaning and validity are severely distorted when the involved

Table 2.5: Standard logical relations between case attributes (object language).

L_1 _Identity relation_ $CA \equiv CA$	L_2 _Transposition relation_ $CA_1 \to CA_2 \equiv \neg CA_2 \to \neg CA_1$
L_3 _Double negation relation_ $CA \equiv \neg\neg CA$	L_4 _Exportation relation_ $(CA_1 \wedge CA_2) \to CA_3 \equiv CA_1 \to (CA_2 \to CA_3)$
L_5 _Excluded middle relation_ $CA \vee \neg CA$	$L_6 - L_7$ _De Morgan's relations_ $\neg(CA_1 \vee CA_2) \equiv \neg CA_1 \wedge \neg CA_2$ $\neg(CA_1 \wedge CA_2) \equiv \neg CA_1 \vee \neg CA_2$
$L_8 - L_9$ _Tautology relations_ $CA \underset{\leftrightarrow}{\equiv} \begin{cases} CA \wedge CA \\ CA \vee CA \end{cases}$	$L_{10} - L_{11}$ _Association relations_ $CA_1 \vee (CA_2 \vee CA_3) \equiv (CA_1 \vee CA_2) \vee CA_3$ $CA_1 \wedge (CA_2 \wedge CA_3) \equiv (CA_1 \wedge CA_2) \wedge CA_3$
$L_{12} - L_{13}$ _Commutation relations_ $CA_1 \vee CA_2 \equiv CA_2 \vee CA_1$ $CA_1 \wedge CA_2 \equiv CA_2 \wedge CA_1$	$L_{14} - L_{15}$ _Distribution relations_ $CA_1 \wedge (CA_2 \vee CA_3) \equiv (CA_1 \wedge CA_2) \vee (CA_1 \wedge CA_3)$ $CA_1 \vee (CA_2 \wedge CA_3) \equiv (CA_1 \vee CA_2) \wedge (CA_1 \vee CA_3)$
L_{16} _Noncontradiction relation_ $\neg(CA \wedge \neg CA)$	L_{17} _Mutual exclusiveness relation_ $CA_1 \wedge CA_2 = \varnothing$
L_{18} _Addition relation_ $CA_1 \to (CA_1 \vee CA_2)$	L_{19} _Collective exhaustiveness relation_ $CA_1 \vee CA_2 = \Omega$
$L_{20} - L_{21}$ _Absorbtion relations_ CA_1 $\equiv \begin{cases} CA_1 \wedge (CA_1 \vee CA_2) \\ CA_1 \vee (CA_1 \wedge CA_2) \end{cases}$	L_{22} _Resolution relation_ $(CA_1 \vee CA_2) \wedge (\neg CA_1 \vee CA_3) \to (CA_2 \vee CA_3)$
$L_{23} - L_{24}$ _Complementation relations_ $CA \wedge (\neg CA) \equiv \varnothing(F, 0)$ $CA \vee (\neg CA) \equiv \Omega(T, 1)$	$L_{25} - L_{26}$ _Material equivalence relations_ $CA_1 \leftrightarrow CA_2 \equiv \begin{cases} (CA_1 \to CA_2) \wedge (CA_2 \to CA_1) \\ (CA_1 \wedge CA_2) \vee (\neg CA_1 \wedge \neg CA_2) \end{cases}$
$L_{27} - L_{28}$ _Material implication relations_ $CA_1 \to CA_2$ $\equiv \begin{cases} \neg(CA_1 \wedge \neg CA_2) \\ (\neg CA_1 \vee CA_2) \end{cases}$	L_{29} _Simplification relation_ $(CA_1 \wedge CA_2) \to CA_1$ L_{30} _Disjunction relation_ $((CA_1 \vee CA_2) \wedge \neg CA_1) \to CA_2$

conditionals are understood as formal material conditionals. As a matter of fact, if they look carefully into the matter, health care providers will find out that classical logic grants absurdities and oddities such as the following:

(1) *Self-contradiction.* The inference form

$$CA \to \neg CA$$

is consistent in formal logic terms, yet it is clearly self-contradictory in the dialogical or substantive terms of clinical practice, i.e., $CA \not\to \neg CA\dot{:}_{KB}$ (see Example 2.43 below).

(2) *Falsehood implies any assertion.* This absurdity yields the *negated antecedent to conditional* paradox. More precisely, assume that a physician's everyday use of conditionals employs the formally valid inference[40]

Case premises	$\neg CA_1$
Care provider's conclusion	$\therefore CA_1 \to CA_2$

In conversational terms, however, the above inference would mean that the falsity of $CA_1\dot{:}_{KB}$ is logically sufficient for the truth of $CA_1 \to CA_2\dot{:}_{KB}$ for any $CA_2\dot{:}_{KB}$, which is obviously counterintuitive (also, Example 2.43).

(3) *Affirmed consequent to conditional.* If a physician's everyday use of conditionals was logically equivalent to a material conditional, the formally valid inference

Case premises	CA_2
Care provider's conclusion	$\therefore CA_1 \to CA_2$

should also be considered valid in dialogal terms, i.e. the truth of $CA_2\dot{:}_{KB}$ should be logically sufficient for the truth of $CA_1 \to CA_2\dot{:}_{KB}$.[41] This is, again, counterintuitive in real clinical practice (Example 2.44 below).

---

[40] Recall that the $CA_1 \to CA_2$ is valid for every false CA_1 and for every true CA_2.
[41] In the following, on occasion the subscript "KB" used in substantive logic will be dropped, when implied by the context. So, $CA_1 \vee CA_2\dot{:}_{KB}$, $CA_1 \to CA_2\dot{:}_{KB}$, $CA_2 \mid CA_1\dot{:}_{KB}$ etc. will be simply denoted as $CA_1 \vee CA_2$, $CA_1 \to CA_2$, $CA_2 \mid CA_1$ etc. Arguably, the ensuing gain in simplicity and clarity more than offsets any potential

(4) *Premise strengthening* or *strengthening of the antecedent*. If a physician's everyday use of conditionals was assumed in formal logic terms, the inference

$$\begin{array}{ll} \textit{Case premises} & \underline{CA_1 \rightarrow CA_2} \\ \textit{Care provider's conclusion} & \therefore (CA_1 \wedge CA_3) \rightarrow CA_2 \end{array}$$

should be valid in conversational terms.[42] Yet, it is usually supposed that premise strengthening is invalid for indicative conditionals as well as for counterfactual (subjunctive) conditionals (Example 2.45, below), which can lead to meaningless argumentation. Premise strengthening is closely related to the affirmed consequence to conditional (*paradox 3* above), because the latter can be thought of as the strengthening of a null antecedent. Also, this situation is related to the so-called *monotonicity*. A conditional is monotonic when the addition of new information to its premise cannot undermine the support of the conclusion already tendered by the original premise.[43] Material conditional is a monotonic conditional, i.e., adding premises (disease symptoms, tests, hypotheses) to a valid medical assertion it does not allow the withdrawal of its conclusion.[44] In their medical practice, however, physicians frequently retract their earlier conclusions when new evidence strongly suggests so.

(5) *Conditional excluded middle*. If a physician's everyday use of conditionals was logically equivalent to the formally valid inference

$$\begin{array}{ll} \textit{Case premises} & \underline{CA_1 \rightarrow (CA_2 \wedge CA_3)} \\ \textit{Care provider's conclusion} & \therefore (CA_1 \rightarrow CA_2) \vee (CA_1 \rightarrow CA_3) \end{array},$$

then this should lead to absurdities in dialogical reasoning (an illustration can be found in Example 2.46).

---

confusion. This makes sense since the book is concerned mainly about substantive rather than formal logic.

[42] This is because it cannot be that "the conclusion is F if the premise is T".

[43] That is, a conditional is monotonic if whenever $CA_1 \rightarrow CA_2$ holds $(CA_1 \wedge CA_3) \rightarrow CA_2$ must also hold.

[44] It may only increase the number of possible conclusions (disease diagnoses, prognoses).

(6) *Conditional disjunctivity*. If a physician's everyday use of conditionals was logically equivalent to a material conditional, the inference

$$\begin{array}{ll} \text{Case premises} & \underline{CA_2 \vee CA_3)} \\ \text{Care provider's conclusion} & \therefore (CA_1 \to CA_2) \vee (CA_1 \to CA_3) \end{array}$$

should be valid in conversational terms, which is usually senseless (Example 2.47).

(7) *Tautology*. The inference form

$$(CA_1 \to CA_2) \vee (CA_2 \to CA_1)$$

is a tautology in formal logic (one could infer $CA_1 \to CA_2$ from the mere falsity of $CA_2 \to CA_1$ if they are understood as material conditionals). But in conversational terms the above claim implies that for any two assertions there must be one that implies the other, which is absurd (e.g., either smoking implies influenza or influenza implies smoking).

In sum, although the rules in *paradoxes* 1–7 above are valid in formal logic, they are not necessarily valid in other kinds of logic, including conversational or substantive (CD) logic of clinical routine. Potential absurdities can emerge in various health contexts, a few of which are discussed in the following examples.

Example 2.43: Let us start with *paradox* 1. The inference that

"Mr. Baker suffers from Parkinson's disease" implies that "Mr. Baker does not suffer from Parkinson's disease"

is formally valid but it is silly in conversational terms. Next, two cases of *paradox* 2. The argument,

"if Dr. Fumagalli is a virus" then "$5 + 4 = 9$",

is true when translated as a formal material implication. Yet, as far as the substantive conditional is concerned, the argumentation is intuitively false since a true implication must tie the premise and conclusion together by some notion of relevance (whether or not "Dr. Fumagalli is a virus" seems in no way relevant to whether "$5 + 4 = 9$"). Furthermore, the mere fact that

"donated blood does not belong to blood group Alpha" ($\neg BA$), entails the truth of

"if donated blood belongs to group Alpha" then "it is Rhesus-positive" $(BA \to BR^+)$.

Moreover, we could also infer from the mere fact $\neg BA$, that

"if donated blood belongs to group Alpha" then "it is not Rhesus-positive" $(BA \to (\neg BR^+))$.

But a care provider will certainly find it ridiculous to maintain both that $BA \to BR^+$ and that $BA \to (\neg BR^+)$.[45]

Example 2.44: Now, let us concentrate on *paradox* 3. In formal logic terms the mere fact that

"Ms Jefaisesbulles' meeting with her physical therapist will not be cancelled" (JT)

entails the truth of

"if Ms Jefaisesbulles dies the night before" then "Ms Jefaisesbulles' meeting with her physical therapist will not be cancelled" $(JD \to JT)$.

In this case, if it is a fact that JT, then it remains a fact even if JD, which sounds nonsensical in substantive terms (although formally valid).

Example 2.45: Concerning *paradox* 4. Again, in formal logic

"if Mr. Sun has a heart attack (HA) then a physician will perform a triple bypass on him" $HA \to B3$,

entails that

"if Mr. Sun has a heart attack and he dies (SD) then a physician will perform a triple bypass on him" $(HA \wedge SD) \to B3$,

which does not make sense in substantive terms.

---

[45] Remarkably, some supporters of the use of formal logic in medical decision-making responded that this paradox can be avoided by means of the intuitive test that

$$\frac{\text{if one knows for sure that } \neg CA_1}{\therefore \text{ one does not have use of } CA_1 \to CA_2}$$

That is, conditionals have no role to play in such cases, and the physician has no practice in assessing them. Say, when a physician knows for sure that "Mr. Haramis did not die of Ebola", it does not make sense to go in for "if Mr. Haramis died of Ebola then".... Needless to say that this response remains controversial.

Example 2.46: Concerning *paradox* 5, the following inference is valid in formal logic: that

"if Cialis and Viagra are erectile dysfunction medications (ED) then either Cialis is made by Pfizer (CP) or Viagra is made by Eli Lilly & Co. (VE)" ($ED \to (CP \lor VE)$)

entails that

"either if Cialis and Viagra are ED then Cialis is made by CP or if Cialis and Viagra are ED then Viagra is made by VE". (($ED \to CP$) $\lor$ ($ED \to VE$)).

Yet, while in formal logic ED is sufficient to ensure that either CP or VE is valid, this same fact is sufficient to ensure that neither entailment above holds in reality.

Example 2.47: Concerning *paradox* 6. In formal logic terms, the inference

"either physician will prescribe Delsym or will prescribe Robitussin ($DL \lor RO$)"

entails that

"either if physician diagnoses liver cancer then will prescribe Delsym ($LC \to DL$), or if physician diagnoses liver cancer then will prescribe Robitussin ($LC \to RO$)".

But, from the fact that $DL \lor RO$, it does not follow in the physician's substantive language that either $LC \to DL$ or that $LC \to RO$, which shows that the above formal inference does not hold in clinical practice.

The following example presents a health case in which the statistical conditional seems to make more sense than the formal material conditional, the latter displaying some rather paradoxical features.

Example 2.48: A kidney transplant is a surgical procedure in which a healthy kidney from one person is placed into another whose kidneys have stopped working. An organ bank provides the doctors of Hammersmith Hospital with the opportunity to perform this procedure and see if

"the transplanted organ is a success for Ms Li who suffers from a serious kidney disease" (KD)

on the condition that

"Ms Li takes the laboratory test (LT)",

Table 2.6: Truth table for material and statistical conditionals (IN denotes indeterminate).

LT	KD	$LT \to KD$	$KD \mid LT$
T	T	T	T
T	F	F	F
F	T	T	IN
F	F	T	IN

which can detect the disease with certainty, if it is the case. The organ offer is withdrawn if testing is cancelled, i.e., $\neg LT$ ("Ms Li does not take the required laboratory test"). As we can see in Table 2.6, the material and the statistical conditionals display some noticeable differences. If the conditioning event LT does not happen, the conditional event $LT \to KD$ still has a $TV(T)$ in classical logic. On the other hand, when considered in substantive logic, the conditional event $KD \mid LT$ cannot realistically hold if LT does not happen (i.e., one cannot say whether the cadaver donor kidney could or could not save Ms Li's life, only that the organ offer is cancelled). The conditional event being indeterminate if the conditioning is false, the $KD \mid LT$ is not truth functional. Here, the $KD \mid LT$ interpretation seems to make more sense in everyday health practice than that of $LT \to KD$.

Serious damage to decision-making may be caused by the improper inclusion of formal logic rules in medical software when the associated paradoxes are more difficult to detect as they often constitute part of the intrinsic process. Before leaving this section on paradoxes, let us consider *Mohist reasoning* (Section 1.1.5) and some apparently paradoxical inferences linked to early cases of *medical ethics*. Remarkably, the later Mohists were fully aware of the self-reflexive paradoxes concerning truth and falsity. One of the *Canons* states that,

"to reject denial is inconsistent".

This was justified by the Mohists on the grounds that, no matter whether the rejection is to be rejected or not, this amounts to not rejecting the denial (Harbsmeier, 1998). The Canon stating that,

"non-existence does not necessarily (*pi*) presuppose existence"

is a situation of "what is referred to". In the case of non-existence of an entity, the entity has to exist before it is in this way non-existent (e.g., in the case of the non-existence of the case of "aspirin curing breast cancer", the case is considered non-existent without ever having occurred).

In order to reconcile the execution of thieves with love for all men, some Mohists maintained that although even a murderous thief is a man, executing this kind of criminals is not like killing men, in general. Not surprisingly, several contemporary thinkers rejected this suggestion as *sophistry*, based on the inference that "thief is a man" implies that "killing thieves is killing men". The Mohists replied that there are similar sentences that do not belong to the above inference form, but rather belong to another form, such as "Ms Li's brother is a handsome man" may imply that "loving her brother is not loving handsome men, in general " (it is not valid that "loving her brother is loving handsome men").

Specifically, the Mohists used a four-stage process to establish that the "thief–men" argument belongs to the latter kind of inference:

(a) *illustrating* the topic (thief) with the entity (man) of which formally similar statements may be made;
(b) *matching* parallel sentences about the illustrations and the topic;
(c) *adducing* supporting arguments for the most relevant parallels by expanding them and showing that the parallelism still holds and
(d) *inferring*, i.e. using the topic's similarity to what the person being argued with (Mohist) accepts in order to propose what the person does not accept.

The following example analyzes the Mohists' line of thought in a possible medical ethics context.

Example 2.49: Let iA, eA and eM denote, respectively, "is an assassin", "executing an assassin" and "executing a man" ("man" is here understood as a decent human being). For a Mohist physician supervising the execution, the ethical dilemma may be handled in terms of the four-stage inference that is represented in symbolic terms as

$$\begin{array}{ll} \textit{Illustrating} & A, M \\ \textit{Matching/Adducing} & \neg(m_k A \to m_k M) \quad (k = 1, \ldots, n) \\ \textit{Inferring} & \therefore \neg(m_{n+1} T \to m_{n+1} M) \end{array},$$

where m_k means matching parallel statements about a thief and a man. For example, for $k = 1, 2, 3$ the m_k may denote, respectively, *abundance in*, *being without* and *loving*; more precisely,

$\neg(m_1 T \to m_1 M)$: it is not valid that "abounding in assassins" implies "abounding in men";

¬($m_2T \to m_2M$): it is not valid that "being without assassins" implies "being without men";

¬($m_3T \to m_3M$): it is not valid that "loving assassins" implies "loving men".

Lastly, by analogy with m_1, m_2, m_3 the Mohist physician infers that for $k = 4$, $m_4 = e$, and

¬($eT \to eM$): it is not valid that "executing assassins" implies "executing men".[46]

The Mohist's line of thought rather belongs to a kind of *analogical argumentation* that assigns a certain content to the term "man" by means of the associated humanity level. For a Mohist an assassin is not as human a man as is, say, a virtuous monk. This line of argumentation can be further explored in terms of *stochastic inferences* (Example 3.35). On the other hand, in the eyes of some readers, the above example might reinforce the belief that when trying to find logic in early Chinese philosophy, one often has to come to grips with the complementarity of Greek logic and Chinese semantics (e.g., this is what the ancient school known as *Ming-Chia* mostly discussed).

This being the situation in real health care practice, one needs to proceed with due caution, perceiving the suitability or unsuitability of the theoretical context, recognizing what is salient and what is insignificant, and appreciating the unique characteristics of the particular case. We now turn our attention to another important feature of substantive conditionals, namely, associations between conditionals and metalanguage. These associations are broader and deeper than statistics hypotheses, and they frequently contain rich semantics.

2.6.3. *Conditionals and metalanguage*

An important conclusion of the preceding analysis is that useful conditionals can be defined *in situ* with the help of the metalanguage used in substantive reasoning. Let us focus on the equivalence conditional that expresses "sameness in meaning". As noted earlier, in a metalanguage setting medical

---

[46] As a Mohist would say, "although an assassin is a man, not loving assassins is not loving men, loving assassins is not loving men, and executing assassins is not executing men".

thinking is concerned with conditional operators that provide a substantive method by means of which the validity of the case conclusion can be deduced from that of the case premise. Not surprisingly, natural laws provide a powerful method for such a purpose.

Adequate metalanguage allows a physician to rigorously define reasoning operators that cannot be defined by themselves. This is the case of the CD statistical conditional operator "$\cdot|\cdot$" linked to Bayesian inference, the latter viewed as a theory of consistency between prior (old) medical information and posterior (new) data, physical examinations and laboratory tests.

Example 2.50: By itself the operator "$\cdot|\cdot$" cannot be defined formally (Example 2.48). Yet it obtains a definite meaning in the context of SMR inference, which has significant implications in decision-making. First, note that the physician's assertion

"if $RE = 16.5$ and $RU = 10.9$ (per million people) are the observed rates of a certain cancer in exposed and unexposed people, respectively" then "the relative risk is about $RR = 1.51$",

can be written in terms of the triadic case formula (Section 2.3.7) as

$$a_A(RR = 1.51 | RE = 16.5 \wedge RU = 10.9).$$

Moreover, the operator "$\cdot|\cdot$" also obtains a rigorous meaning in the probability setting (see Chapter 3).

The preceding analysis is likely to include a variety of beliefs and statements about the evolution of a health state in the space–time domain of interest. Health care providers who practice SMR in their daily professional activities need to investigate several possibilities about the clinical or medical case at hand. In quantitative terms, this investigation is carried out in the rigorous S/TRF metalanguage setting. To each possible world (realization) $CA_1^{(i)}, i = 1, 2, \ldots, N$, of $CA_{1,p}$ the metalanguage can ascribe a counterfactual with a possible world $CA_2^{(i)}$ of some other S/TRF $CA_{2,p}$ linked to the first case attribute via some biomedical law or other scientific mechanism. Let us look at a few examples.

Example 2.51: Given the medical knowledge available, physicians argue that ultraviolet radiation, UV_p, at location–time p would cause skin cancer, SC_p, to an individual or a population. In the S/TRF setting, if a particular population was exposed to sun with *ultraviolet radiation index* $UV^{(4)} = 8$

(possible world, $i = 4$), then the population *skin cancer rate* $SC^{(4)} = 0.005\%$ would be the case. Accordingly, the physician's reasoning is that

$$r_A(UV^{(i)} \Rightarrow SC^{(i)}), \quad i = 1, 2, \ldots, N$$

i.e., the physician argues rationally (that is, on the basis of sound science and credible evidence) that ultraviolet radiation would cause skin cancer with a different probability assigned to each possible world (realization).

Other possibilities exist that involve different combinations of S/TRF worlds, depending on the decision-making circumstances.

Example 2.52: Let PM_p represent the population mortality distribution in a location–time p due to influenza. The $b_A(PM_p \leq 0.01)$ denotes that

"physician A believes that mortality at p is less or equal to 1%"

(probably based on incomplete or subjective data). If $k_A(M_{PM})$ denotes that

"physician A knows that the mortality distribution of an infectious disease obeys the epidemic model M_{PM}",

then the triadic case formula $r_A(M_{PM} \to PM_p \leq 1\%)$ means that,

"using model M_{PM} physician A rationally derives that $PM_p \leq 1\%$".

Since PM_p is mathematically represented as a S/TRF, the triadic case formula $k_A(PM_{p_1} = 1\% \wedge PM_{p_2} \leq 1.5\%)$ may be interpreted as,

"in all realizations of PM at p_1 and p_2 that are consistent with A's knowledge, the mortality is equal to 1% and less than or equal to 1.5%, respectively",

which could be a highly justified assertion.

Many scholars have cited the view that physicians' training emphasizes learning how to look at a case, which is much simpler than learning how to *see*, often leading to apparent conflicts in case descriptions and medical literature. The examples discussed so far, as well as others that follow, aim to improve the physicians' argumentation style and allow them to see deeper into a case by using *SMR* concepts and inference tools in various research and clinical settings.

2.6.4. *Conditionals and natural laws*

The readers may have already anticipated some intriguing connections between natural laws and logical connectives of medical decision-making. These associations could offer additional insight into the notion of causality (etiology) and its application in real health systems.

Example 2.53: In an *environmental health* study PC_p and DI_p represent, respectively, pollutant exposure (concentration values in the range 0–10 ppm) at location–time p (geographical coordinates of the city of Patras during April 4, 2003) and population disease incidence (values in the range 0–35%), which are linked via the simple empirical law (object language)

$$DI_p - 3.5 PC_p = 0. \qquad (2.23)$$

Law (2.23) can become part of an exposure expert's argumentation by expressing it in terms of logical truth tables. In particular, Table 2.7 displays the truth table corresponding to, say, $PC_p = 2\,ppm$ and $DI_p = 7\%$.

As one can see, the TVs of law (2.23) are the same as those of the equivalence conditional $PC_p \leftrightarrow DI_p$. In fact, Eq. (2.23) establishes a one-to-one correspondence between events, e.g., for $DI_p = 7\%$ the law (2.23) is valid only for $PC_p = 2\,ppm$. The readers may recall that in substantive (CD) terms one could write

$$PC_p \leftrightarrow DI_p \dot{:}_{(2.23)},$$

where the subscript "(2.23)" denotes that in this case the KB is the law of Eq. (2.23). Next, consider another empirical law

$$DI_p = 3TE_p^2, \qquad (2.24)$$

physically linking temperature exposure (TE_p) and population disease incidence (DI_p) at location–time p. The TVs of law (2.24) are the same as those of the material conditional $TE_p \to DI_p$, see columns three and four

Table 2.7: Truth table for exposure-incidence law (2.23).

$PC_p = 2$	$DI_p = 7$	Law (2.23)	$PC_p \leftrightarrow DI_p$
T	T	T	T
T	F	F	F
F	T	F	F
F	F	T	T

Table 2.8: Truth table for law (2.24).

$TE_p = 2°C$	$DI_p = 12$	Law (2.24)	$TE_p \to DI_p$
T	T	T	T
T	F	F	F
F	T	T	T
F	F	T	T

of Table 2.8. Accordingly, in substantive terms one could write

$$TE_p \to DI_p \dot{:}_{(2.24)},$$

where, as before, the subscript denotes the law of Eq. (2.24). One notices that Eq. (2.24) does not establish a one-to-one correspondence and, hence, the equivalence conditional may be not appropriate here. For $DI_p = 12\%$, say, the law of Eq. (2.24) is valid for both values $TE_p = 2°C$ and $TE_p = -2°C$. Lastly, the readers may have observed that the statistical conditional $DI_p|TE_p$ by itself cannot offer any logical representation of law (2.23) or law (2.24).

Example 2.54: In the case of Eq. (2.13), the material conditional

$$PC_p \to PC_{p'} \dot{:}_{(2.13)}$$

is substantively valid since $PC_{p'}$ is derived from PC_p by means of the chemical law. Similarly, in the case of Eq. (2.15),

$$(PC_p \land BR_p) \to IR_p \dot{:}_{(2.15)}$$

is substantially valid because IR_p is derived from PC_p and BR_p via a valid exposure law.

Interestingly, as we will see in Chapter 3, during their daily routines researchers and clinicians often realize that TVs such as those of Tables 2.7 and 2.8 above cannot be necessarily assumed to be "certain", which is why many of them consider it more realistic to assign a numerical probability value to each one of these TVs. Probability helps care professionals create a case characterization that distinguishes between two medical case formulas that are syntactically equivalent but semantically distinct.

2.6.5. Over-extending and extrapolating

In various parts of the book we discuss assertions (associated with symptoms, diagnoses, therapies) that, although they initially seemed plausible, upon closer inspection in the light of science-based reasoning turned out to be invalid. This includes paradoxes, oddities and anomalies appearing in the process of *in situ* medical decision-making, which need to be better understood rather than neglected. By expressing professional assertions in the substantive reasoning framework discussed here and in Chapter 3 that follows, one may find that several diagnoses and expert treatments that appear to be highly intuitive could be, in fact, inadequate or even invalid in a rigorous sense. A rather common situation is *over-extending* (i.e., wrongly extending an inference scheme that is valid in a certain group of clinical cases to any other case). This real possibility is illustrated with the help of the following examples.

Example 2.55: Consider the case of a medical treatment adapted by physician A as follows: that

> "physician A administers drug D_1 to Ms Lagrange to cure a disease given that she suffers from it",

and

> "it is not the case that A does not administer drug D_2 to counteract the harmful effects of D_1 given that Ms Lagrange suffers from the disease",

entails that

> "A administers D_1 to cure the disease given that Ms Lagrange suffers from the disease and A administers D_2".

The above medical statements can be expressed in terms of rational triadic case formulas,

$$\begin{array}{ll} \text{Case premises} & r_A(Drug\ D_1|Disease) \\ & \neg r_A(\neg Drug\ D_2|Disease) \\ \text{Physician's} & \\ \text{conclusion} & \therefore\ r_A(Drug\ D_1|Disease \wedge Drug\ D_2) \end{array} \quad (2.25)$$

which is a potentially useful inference in certain clinical cases.

Encouraged by successful representations, such as the one above, it is often assumed that the inference (2.25) is generally valid: for any three case attributes $CA_{1,p}$, $CA_{2,p}$ and $CA_{3,p}$ that may vary in space–time,

"if physician A rationally believes $CA_{2,p}$ that happens given $CA_{1,p}$",

and also

"A does not find it rational that $CA_{3,p}$ does not happen given $CA_{1,p}$",

then it makes sense that

"A rationalizes that $CA_{2,p}$ occurs given $CA_{1,p}$ and $CA_{3,p}$".

In triadic terms,

$$\begin{array}{c} \textit{Case premises} \\ \\ \textit{Physician's} \\ \textit{conclusion} \end{array} \quad \dfrac{\begin{array}{c} r_A(CA_{2,p}|CA_{1,p}) \\ \neg r_A(\neg CA_{3,p}|CA_{1,p}) \end{array}}{\therefore\ r_A(CA_{2,p}|CA_{1,p} \wedge CA_{3,p})}. \qquad (2.26)$$

To the surprise of some professionals, a careful analysis shows that inference (2.26) can be invalid in several cases in health practice. Perhaps this situation is easier understood with the help of a counter example from environmental health.

Example 2.56: In an environmental health study in the city of Constanta, experts argue that

> "the ambient pollutant distribution over the city of Constanta during the specified time (p) obeys a rigorous atmospheric law (PD_p)",

and

> "Mr. Carsteanu — who lives in Constanta — gets sick when experiencing high pollutant levels (PS_p)"

(say, due to Mr. Carsteanu's pre-existing medical condition). Next, let us suppose that

> "Mr. Carsteanu routinely leaves the city (LC_p) during high-level pollution times".

If the inference (2.26) is used in this exposure study, it would lead to the wrong conclusion. Indeed, within an expert's decision process, the state of mind (rationalization) $\neg r_A(\neg LC_p|PD_p)$ is a valid triadic case formula since the objectively established atmospheric law, PD_p, is not relevant to whether or not the decision, LC_p, is true. But, as one can see, given LC_p, the case $\neg PS_p$ is, in fact, valid. Which implies that $r_A(PS_p|PD_p \wedge LC_p)$ can be an invalid triadic formula, and the formulation (2.26) is thus violated (the same result can be reached by assigning to (2.26) a suitable probability formulation, as in Example 3.48).[47]

Another potential reasoning trap is physicians' frequent *anchoring* (i.e., over-emphasizing the initial data of a clinical case over subsequent developments), which often leads to misdiagnoses. Many physicians routinely start with the patient's response to their preliminary questions concerning the case, and then they run with it, ignoring new pieces of valuable information (due to time pressure, multitasking, weak synthesis skills). A relevant reasoning problem is due to the physician wrongly *extrapolating* the past onto the present.

Example 2.57: In the middle of a *flu epidemic*, physicians often conclude that almost every patient they see has flu, even when this is not the case. Similarly, many misdiagnoses are linked to the tendency of several physicians to base their diagnoses on *stereotypes* they have in their minds.

In light of the above concerns, when confronted with intuitive medical assessments or habitual arguments, health care providers could test their validity in a rigorous manner by translating the assessments or the arguments in *SMR* terms, and then see whether the resulting formulations are valid or not. In many cases, this may require stripping the physician's argument from all its "clothes". What is left is the underlying logic of the physician's thinking mode that can be tested in rigorous medical reasoning terms.

2.7. The Object Language–Metalanguage Connection

As noted earlier, whether a_A (metalanguage) is valid or not is determined by objective factors like well-established core and case-specific knowledge, biophysical and logical links between knowledge and the case of interest,

---

[47] Table 2.9 that follows gives a list of inferences with validity depending on the assertion form.

as well as by the health care provider's acumen and state of mind to realize the previous factors. In this setting, the readers may recall that medical reasoning contains a modality k_A for knowledge, so that $k_A CA$ means that CA is "known" (in the sense that CA is in principle accessible to an observing care provider A). The provider's reasoning goes beyond the domain of epistemic logic[48] by incorporating several other modalities (g_A, b_A, s_A, r_A) in addition to k_A; and by integrating his/her thinking mode with laws of natural change as well as case-specific evidence in the content- and context-dependent environment of spatiotemporal disease variation and case uncertainty.

2.7.1. Relations between states

In practice, it is possible that SMR inference combines different kinds of assertions. For example, physician A knows that disease symptom DS_1 is present ($k_A DS_1$), but physician B believes that symptom DS_2 occurs instead ($b_B DS_2$). Moreover, DS_1 generally causes health effect HE by means of a biomedical law. Accordingly, useful connections can be developed between actual states of nature and possible states of mind (recall that the former are the focus of object language and the latter are the domain of metalanguage), as well as between states of mind themselves.

Let us continue our discussion by repeating that different kinds of *in situ* relationships between general medical assertions a_A can be potentially[49] associated with standard logical schemes, some of which were studied in previous sections of the book. Undoubtedly, a physician's cognitive situation has a significant effect on clinical practice and medical research, which means that the physician must be fully aware which particular assertion form (k_A, r_A, s_A, b_A or g_A) applies in the case of interest.

Example 2.58: Consider the CD syllogism of the form

$$\text{Case premises} \quad \begin{array}{l} a_A(CA_1 \to CA_2) \\ \underline{a_A CA_1} \end{array}.$$

$$\text{Care provider's conclusion} \quad \therefore a_A CA_2$$

---
[48] Recall that epistemic logic is a logic of believing and knowing (see Section 2.3.4).
[49] "Potentially" here means that the validity of these inferences depends on the a_A-form assumed by the care provider. In turn, the choice of the appropriate a_A-form for the case of interest can be made in a rigorous manner using probabilistic methods (see Chapter 3).

In the CD setting, whether this is a valid reasoning scheme or not depends on the form of a_A and the associated cognitive condition of the health care provider. For instance, the reasoning

Case premises $\quad \dfrac{b_A(CA_1 \rightarrow CA_2)}{b_A CA_1}$

Care provider's conclusion $\quad \therefore \; b_A CA_2$

is *modus ponens* at the justification degree associated with the belief mind state (b_A). Yet the inference of the combined belief–knowledge states,

Case premises $\quad \dfrac{CA_1 \rightarrow CA_2}{b_A CA_1}$

Care provider's conclusion $\quad \therefore \; k_A CA_2$

is not generally valid. For both the above inference schemes, rigorous proofs are obtained using the probabilistic methods of Chapter 3. In the above case, e.g., Table 3.11 shows that $P_{KB}[CA_2] \in [p_2, 1]$, which confirms that $k_A[CA_2]$ is generally invalid.

Example 2.59: That

"the bacteria clostridium botulinum (CB) is actually the cause of botulism disease (BD)"

does not necessarily mean that physician A knows about it. That is, it is not necessarily valid that $(CB \rightarrow BD) \; \therefore \; k_A(CB \rightarrow BD))$.[50] However, if we include the metalanguage premise $k_A CB$, the object language premise $CB \rightarrow BD$ may yield $k_A CB \rightarrow BD$. And since in the course of A's own reasoning the physician reached the last result, A is entitled to state that A knows about it, $k_A(k_A CB \rightarrow BD)$, or $k_A k_A CB \rightarrow k_A BD$, i.e. A knows that the BD occurs in this case.

Our discussion so far makes it clear that the successful implementation of an inference scheme in practice depends on the individual health care professional (reflecting the fact that any individual may have more experience with some types of clinical cases than with others). Nevertheless,

---

[50]On the other hand, that A objectively knows that $CB \rightarrow BD$ entails that $CB \rightarrow BD$ is actually valid in the real world, i.e., $k_A(CB \rightarrow BD) \; \therefore \; (CB \rightarrow BD)$.

Table 2.9: Medical inferences linking object language and metalanguage.

	I_1:	I_2:	I_3:
Case premises	$CA_1 \therefore CA_2$ $b_A CA_1$	$CA_1 \therefore CA_2$ $s_A CA_1$	$CA_1 \therefore CA_2$ $k_A CA_1$
Care provider's conclusion	$\therefore b_A CA_2$	$\therefore r_A CA_2$	$\therefore r_A CA_2$
	I_4:	I_5:	
Case premises	$CA_1 \therefore CA_2$ $\neg r_A CA_2$	$CA_1 \therefore CA_2$ $\therefore (\neg r_A \neg CA_1 \therefore \neg r_A \neg CA_2)$	
Care provider's conclusion	$\therefore \neg k_A CA_1$		

a care professional should be aware that *the occasional is not necessarily general*.

The above examples bring to mind another interesting category of reasoning scheme that allow valid medical inferences to be made from actual to mental states. Some more representative inferences of this sort are listed in Table 2.9. As in previous cases, despite their occasional usefulness, not all inferences proposed in Table 2.9 are necessarily valid, in general. Indeed, the CA_1 in the first line of the schemes I_1 and I_2 can be false even if physician A believes or is justified to believe that CA_1 is the case.[51]

On the other hand, the inferences I_3 and I_4 are always valid, since in I_3 the CA_1 cannot be false given that the physician knows that CA_1; and I_4 is equivalent to I_3 since if $\neg r_A A_2 \to \neg k_A A_1$ then $k_A A_1 \to r_A A_2$ by *modus tollens*. To gain further insight, consider the case of Ms Spooner who visits her physician A with symptoms that make A diagnose either pneumonia or pulmonary embolism. The physician knows that if a sputum culture test is positive (SC) then Ms Spooner has pneumonia (SP). It is interesting to consider two possible syllogisms. If A knows that she was tested and the result was positive $(k_A SC)$, then A knows that SP $(k_A SP)$; if, however, Ms Spooner had indeed a positive SC test but A was not aware of the test result, then it is not necessarily valid that $k_A(SP)$; i.e.,

$$\frac{k_A(SC \to SP)}{k_A SC} \quad \text{and} \quad \frac{k_A(SC \to SP)}{SC}$$
$$\therefore k_A SP \qquad \qquad \therefore k_A SP$$

---

[51] That is, a mere belief or a sustained belief state does not guarantee certain knowledge.

where $\not{.\!\!\!.\!\!\!.}$ means that "it does not entail". Indeed, the first syllogism is *modus ponens*, but the second one means that it is not necessary that $k_A SP$; it is possible that $r_A(SP)$ is valid etc. This points to another methodological issue that is worth mentioning: if one is not certain about the validity of an inference, one can always try to show its equivalence with another medical inference that is known to be valid.

Furthermore, some rather obvious links exist between mental states (or assertion forms) about a case CA (see Eq. (2.9)) as well as relationships from the mental to the actual concerning a case CA, like

$$k_A CA \therefore CA, \qquad (2.27)$$

i.e., that A knows that CA occurs entails that CA actually exists (although the reverse is not necessarily true). As far as medical thinking is concerned, the exact meaning of the notion of *knowledge* remains a controversial matter (see Section 3.1). The "mind–nature" analysis above could throw some light on the matter, since the knowledge state k_A offers one way to define the notion: physician A is in a definite state of knowledge if Eq. (2.27) is the case, i.e., "when A's state of knowledge implies actuality".[52] Interestingly, such a knowledge definition might be useful in clinical judgment cases in which definite statements are made without realizing that they are merely assumptions.

Example 2.60: In the epidemiologic situation examined earlier (Example 2.53) PM_p denotes population mortality, and the inference $k_A(PM_p = 1\%) \therefore (PM_p = 1\%)$ is considered valid if $PM_p = 1\%$ is true in all possible realizations $PM^{(i)}$ of the S/TRF model representing mortality.

Additional k_A inferences between case attributes that produce *certain* knowledge are listed in Table 2.10. For comparison, Table 2.11 presents a list of medical syllogisms in which while the premises contain certain knowledge, the derived conclusions do not represent certain knowledge. Indeed, in all cases of Table 2.11, to the a_A-forms of the conclusions the care providers can assign a certain level of confidence (probability) but not certainty, i.e., generally $a_A \not\equiv k_A$ (with $P[a_A] \in [0,1]$ and $P[k_A] = 1$).[53]

Rules of sound reasoning like those in the above tables are one thing, but knowledge of these rules is another. The rules exist, but if the care

---

[52] This entailment is not generally valid for other assertion forms; e.g., $b_A CA \therefore CA$ does not hold, in general.
[53] $P[\cdot]$ denotes probability; see, also, Chapter 3.

Table 2.10: Inferences that produce certain knowledge.

	I_1		I_2	
Case premises	$\neg k_A(\neg k_A CA)$		$\neg k_A CA$	
Care provider's conclusion	$\therefore k_A(\neg k_A \neg CA)$		$\therefore k_A(\neg k_A CA)$	

$I_3 - I_4$: (Necessitation)

Case premises	$k_A CA$	$\neg CA$	$k_A CA$	
Care provider's conclusion	$\therefore CA$	$\therefore \neg k_A CA$	$\therefore k_A(k_A CA)$	

I_6 ($\setminus, \equiv\mid, \rightarrow$)

	I_7	
Case premises	$k_A CA_2$	$k_A CA$
Care provider's conclusion	$\therefore k_A(CA_2 \setminus CA_1)$	$\therefore \neg k_A \neg CA$

	I_8		I_9	
Case premises	$k_A(CA_1 \therefore CA_2)$		$k_A CA$	
Care provider's conclusion	$\therefore (k_A CA_1 \therefore k_A CA_2)$		$\therefore (\neg k_A \neg k_A CA \therefore k_A CA)$	

I_{10} (Resolution)

	I_{11} (Sequence)	
Case premises	$k_A(CA_1 \vee CA_2)$	$k_A CA_1$
	$k_A(\neg CA_1 \vee CA_3)$	$\ldots$
		$k_A CA_n$
Care provider's conclusion	$\therefore k_A(CA_2 \vee CA_3)$	$\therefore k_A \vee_{i=1}^{n} CA_i, k_A \wedge_{i=1}^{n} CA_i$

(Continued)

Table 2.10: *(Continued)*

I_{12} *(Expansion)*		$I_{15} - I_{17}$ *(Reduction syllogisms)*	
Case premises	$k_A CA_1$	Case premises	$k_A(CA_1 \vee CA_2)$ $k_A \neg CA_2$
Care provider's conclusion	$\therefore k_A(CA_1 \vee CA_2)$	Care provider's conclusion	$\therefore k_A CA_1$
I_{13} *(Simplification syllogism)*		Case premises	$k_A CA_1$ $k_A(CA_2 \mid CA_1)$
Case premises	$k_A(CA_1 \wedge CA_2)$	Care provider's conclusion	$\therefore k_A CA_2$
Care provider's conclusion	$\therefore k_A CA_1$	Case premises	$k_A \neg(CA_1 \rightarrow CA_2) \equiv k_A(CA_1 \wedge \neg CA_2)$ $k_A \neg CA_2$
I_{14} *(Statistical conditional)*		Care provider's conclusion	$\therefore k_A CA_1$
Case premises	$k_A CA_1$ $k_A CA_2$	I_{19} *(Exchange)*	
Care provider's conclusion	$\therefore k_A(CA_2 \mid CA_1)$	Case premises	$k_A(\neg CA_1 \mid (CA_1 \vee CA_2))$ $k_A(\neg CA_2 \mid \neg CA_1)$
I_{18} *(Excluded middle)*		Care provider's conclusion	$\therefore k_A(CA_2 \mid (CA_1 \vee CA_2))$
Case premises	$k_A(\neg CA_1 \rightarrow CA_2)$ $k_A(\neg CA_1 \rightarrow \neg CA_2)$		
Care provider's conclusion	$\therefore k_A CA_1$		

(Continued)

	$k_A(CA_1 \vee CA_2)$ $k_A(CA_1 \rightarrow CA_2)$
	$\therefore k_A CA_2$

	$k_A(CA_1 \wedge CA_2)$ $k_A CA_2$
	$\therefore k_A CA_1$

Table 2.10: (Continued)

I_{20}			I_{21} (*Strengthening*)	
Case premises	$k_A(CA_1 \leftrightarrow CA_2)$		*Case premises*	$k_A(CA_3 \mid CA_2)$
	$k_A(\neg CA_1 \vee \neg CA_2)$			
Care provider's conclusion	$\therefore k_A(\neg CA_1 \wedge \neg CA_2)$		*Care provider's conclusion*	$\therefore k_A(CA_3 \mid (CA_1 \wedge CA_2))$
$I_{22} - I_{23}$ (*Equivalence syllogisms*)			I_{24} (*Absorption*)	
Case premises	$k_A(CA_1 \to CA_2)$		*Case premises*	$k_A(CA_1 \to CA_2)$
	$k_A(CA_2 \to CA_1)$			
Care provider's conclusion	$\therefore k_A(CA_1 \leftrightarrow CA_2)$		*Care provider's conclusion*	$\therefore k_A(CA_1 \to (CA_1 \wedge CA_2))$
Case premises	$k_A(CA_1 \leftrightarrow CA_2)$		I_{25}: (*Conditional transfer*)	
	$k_A(CA_1 \vee CA_2)$		*Case premises*	$k_A(CA_2 \mid CA_1)$
Care provider's conclusion	$\therefore k_A(CA_1 \wedge CA_2)$		*Care provider's conclusion*	$\therefore k_A(CA_1 \to CA_2)$
I_{26} (*Modus tollens*)			I_{27} (*Modus ponens*)	
Case premises	$k_A(CA_1 \to CA_2)$		*Case premises*	$k_A(CA_2 \backslash CA_1)$ ($\backslash \equiv \mid, \to, \therefore$)
	$k_A \neg CA_2$			$k_A CA_1$
Care provider's conclusion	$\therefore k_A \neg CA_1$		*Care provider's conclusion*	$\therefore k_A CA_2$

(Continued)

Table 2.10: (Continued)

I_{28} (Chain syllogism)	
Case premises	$k_A(CA_1 \to CA_2)$
	$k_A(CA_2 \to CA_3)$
Care provider's conclusion	$\therefore k_A(CA_1 \to CA_3)$

I_{29}:	
Case premises	$k_A(CA_2 \mid CA_1)$
	$k_A(CA_3 \mid (CA_1 \land CA_2))$
Care provider's conclusion	$\therefore k_A(CA_3 \mid CA_1)$

$I_{30} - I_{32}$ (Contraposition)	
Case premises	$k_A CA_1$
	$k_A(CA_1 \to CA_2)$
Care provider's conclusion	$\therefore k_A CA_2$
Case premises	$k_A(CA_1 \to CA_2)$
Care provider's conclusion	$\therefore k_A(\neg CA_2 \to \neg CA_1)$
Case premises	$CA_1 \therefore k_A CA_2$
Care provider's conclusion	$\therefore (\neg k_A CA_2 \therefore \neg CA_1)$

$I_{33} - I_{34}$ (Equivalence elimination)	
Case premises	$k_A(CA_1 \leftrightarrow CA_2)$
	$k_A CA_1$
Care provider's conclusion	$\therefore k_A CA_2$
Case premises	$k_A(CA_1 \leftrightarrow CA_2)$
	$k_A \neg CA_1$
Care provider's conclusion	$\therefore k_A \neg CA_2$

I_{35} (Constructive dilemma)	
Case premises	$k_A(CA_1 \to CA_2)$
	$k_A(CA_3 \to CA_4)$
	$k_A(CA_1 \lor CA_3)$
Care provider's conclusion	$\therefore k_A(CA_2 \lor CA_4)$

I_{36} (Contradiction)	
Case premises	$k_A(CA_1 \to CA_2)$
	$k_A(CA_1 \to \neg CA_2)$
Care provider's conclusion	$\therefore k_A \neg CA_1$

Table 2.11: Syllogisms that do not produce certain knowledge.

$k_A(CA_2 \mid CA_1)$ $k_A \neg CA_1$	$k_A(CA_2 \mid CA_1)$ $k_A CA_2$	$k_A \neg (CA_2 \mid CA_1)$ $k_A \neg CA_2$
$\therefore a_A CA_2$	$\therefore a_A CA_1$	$\therefore a_A CA_1$
$k_A(CA_1 \to CA_2)$ $k_A CA_2$	$k_A(CA_1 \to CA_2)$ $k_A \neg CA_1$	$k_A(CA_1 \to CA_2)$
$\therefore a_A CA_1$	$\therefore a_A CA_2$	$\therefore a_A(CA_2 \to CA_1)$

professional is either unaware of them or is aware of them but is unable to follow them properly, a crucial diagnosis may be impossible to make, and a physician may fail to realize that case knowledge is impossible under the given circumstances.

Example 2.61: Dr. Sun needs to use a medical test device to help her decide the right dose of the drug Sunitinib for her patient's *cancer*. Dr. Sun considers the test device to be defective (although it is, in fact, quite accurate). This test device is the only one at her disposal. Let DO denote "drug dose appropriate for the patient's case", and DT demote "defective test". Given the circumstances, the rigorous inference rules of Table 2.10 show that knowledge is impossible in Dr. Sun's case, i.e., she cannot know that she actually knows that the drug dose (as she calculated with the help of the test) is appropriate; in symbolic terms,

$$\neg k_{Sun} k_{Sun} DO.$$

Specifically, the logical reasoning process leading to this conclusion is explained step-by-step in Table 2.12. Note that different premises and inference rules are used at each step of the process. The two premises at *step* 1 are that "Dr. Sun does not know that the test device is not actually defective", and "if the test device is defective then Dr. Sun does not know the correct drug dose". The meaning of *step* 2 is that "if Dr. Sun knows that the calculated dose is correct, DO, then the test device is not defective, $\neg DT$". In *step* 3, "Dr. Sun knows that 'if she knows DO then $\neg DT$'". In *step* 4, "if Dr. Sun knows that she knows DO then she knows that $\neg DT$". In *step* 5, "if Dr. Sun does not know that the test device is accurate then she does not know that she knows the correct calculated dosage". In *step* 6, then, the final conclusion is as above, i.e, "Dr. Sun cannot know that she actually knows that the drug dose is appropriate".

Table 2.12: Reasoning process showing the impossibility of knowledge in Dr. Sun's case.

Step	Inference	Explanation
1	$\neg k_{Sun}\neg DT$ $DT \therefore \neg k_{Sun}DO$	Premises 1–2
2	$k_{Sun}DO \therefore \neg DT$	From *premise* 2 of *step* 1 by "contraposition" rule I_{32} (Table 2.10)
3	$k_{Sun}(k_{Sun}DO \therefore \neg DT)$	From *step* 2 and awareness of one's own logical derivation
4	$k_{Sun}k_{Sun}DO \therefore k_{Sun}\neg DT$	From *step* 3 by rule I_8 (Table 2.10)
5	$\neg k_{Sun}\neg DT \therefore \neg k_{Sun}k_{Sun}DO$	From *step* 4 by "contraposition" rule.
6	$\neg k_{Sun}k_{Sun}DO$	From *premise* 1 of *step* 1 by "*modus ponens*"

Once again, the medical reasoning steps in Table 2.12 require awareness of the relevant rules of logic and inferences on behalf of the physician. If, e.g., Dr. Sun is unaware of the logical rules called "*modus ponens*" and "contraposition", she cannot proceed with the corresponding *steps* 2, 5 and 6. Also, in *step* 3, Dr. Sun needs to be aware of her own logical derivation.

Example 2.62: It is worth noticing that certain of the inferences of Table 2.10 underlie the implicit logic of ancient Chinese texts. For instance, the logical inference used in *Lao-Tsi*'s ethical arguments is an illustration of the chain syllogism (I_{28} of Table 2.10).

The readers should be convinced at this point that the proper implementation of *SMR* inferences or rules in medical research and clinical practice is by no means straightforward. Depending on the medical case of interest, their own level of *in situ* understanding (cognitive situation), as well as their awareness of basic rules of logic, physicians should be able to decide which ones of the *SMR* inferences listed in Tables 2.9–2.11 to implement (additional inferences will be presented in other parts of the book).

2.7.2. *Combinations of medical inferences and derivative assertions*

Combinations of the above inferences can be used too, each assuming a certain cognition level that generates a set of health assessments and medical judgments of specified reliability level.

Example 2.63: The combination of case inferences $C_1 = \{I_2, I_3, I_4, I_8\}$ of Table 2.10 may be assumed to be epistemically stronger than the combination $C_2 = \{I_2, I_3, I_4\}$ — in fact, the former combination implies the latter. If a researcher's or clinician's cognitive situation about the case of interest is not so strong, one may find it more realistic to use combination C_2 rather than C_1.

In sum, depending once more on the a_A form, some of the reasoning schemes used in medical decision-making may apply only under specific conditions, whereas others are of general applicability. Accordingly, several of the medical "facts" currently viewed as fully established should be continuously evaluated and, when necessary, replaced by new ones with the help of reasoning schemes and "mind–nature" relationships such as those discussed above. These are important matters that are revisited in various parts of the book and in different settings.

Additional states linked to a health care provider's decision-making process, the so-called *derivative* assertions (mind states), can also be defined on the basis of the fundamental states discussed in the preceding sections, a fact that contributes considerably to the flexibility of one's medical thinking. A few examples would illustrate sufficiently the use of derivative states.

Example 2.64: Consider the quantitative relationship $DI_p \leq 2\%$, where DI_p denotes *malaria* disease incidence (%) at a location–time p. Now suppose that a physician A feels more comfortable with the assertion

"A is convinced that", or in symbolic form, c_A.

In the *SMR* setting, the new assertion form c_A can be conveniently expressed in terms of k_A, as follows

$$c_A(DI_p \leq 2\%) \equiv \neg k_A[\neg k_A(DI_p \leq 2\%)], \qquad (2.28)$$

in which case what is known about operator k_A can be used, if the medical conditions so require, to produce useful results for c_A. In fact, this is the basic idea of derivative mind states (assertions).

Example 2.65: Consider the disease diagnosis of physician A properly described by the familiar guess state g_A, i.e.,

"A's guess is that the specified organisms are the cause of Mr. Bagratuni's *pneumonia*" ($g_A PD$).

In this case, $g_A PD$ may be conveniently defined as

$$g_A(PD) \equiv \neg r_A(\neg PD), \qquad (2.29)$$

which is a noticeable result for medical decision support purposes, since it connects two different assertion forms.

The following example introduces a real-time correspondence between a physician's diagnosis and an uncertain *in situ* situation.

Example 2.66: Here is an inference concerning the commonly encountered *gastroesophageal reflux disease (GERD)*. That

"it is possible that physician A is mistaken that Ms Ruvikova suffers from *GERD*"

entails that

"it is possible that *GERD* is not actually the case",

or in symbolic form,

$$\nabla m_A \, GERD \therefore \nabla[\neg GERD], \text{[54]} \qquad (2.30)$$

where symbol ∇ denotes that "it is possible", and m_A means that "physician A is mistaken that".

As we will see in Chapter 3, the above distinctions, as well as several logical and scientific relationships among clinical states, can be also consequential in the calculation of the relevant case probabilities.

2.7.3. Levels of justification and uncertainty

The justification status of a medical case is an issue that emerges when physicians ask questions like,

(1) What level of evidence is needed before treating it as settled that disease diagnosis, DD, happens in a care provider's inquiry?
(2) Will what amounts to acceptable justification vary with the intellectual capacity and knowledge of different care providers?

---
[54] The reverse is not necessarily true.

(3) Can one rationalize that a disease prognosis DP will occur, $r_A DP$, on one's belief that DD is valid, $b_A DD$, or does one need more evidence for it?

To put it in a slightly different way, the issue is what epistemic state (knowledge, true belief etc.) is associated with the a_A-form a physician must bear to a diagnosis DD for it to be warranted enough to be a reason for a disease prognosis DP. There are several possibilities concerning a *warrant* being a sufficient reason a physician has to assert something (e.g., to justify causation) such as, either such warrant requires *epistemic certainty* or it is subject to *pragmatic uncertainty*.

Example 2.67: An exposure expert A argues that for the assertion

"Mr. Li is exposed to high levels of environmental smoke (ES)",

to be considered justified, its appropriate form is

"medical expert A knows that ES is the case ($k_A ES$)",

so that it can be used as a reason for

"A's rational belief that Mr. Li will suffer from lung cancer within the next 20 years ($r_A LC$)",

i.e., to justify objectively lung cancer causation, $k_A ES \to r_A LC$.

Having said that, most physicians would probably agree that there is no such a thing as epistemic certainty or zero uncertainty in professional practice. Only very rarely are health care professionals in the state of certainty, which is why they frequently employ *safety guidelines* of the type,

"CA is valid, but better check just in the unlikely case that $\neg CA$ happens".

Here is an example of guideline implementation:

"it is the case that Ms Lancaster's left hip needs to be replaced (CA), but the surgeon had better check the patient's chart 5 minutes before operating just in case it is not ($\neg CA$)".

We will revisit the subject in Chapter 4 (section on disease etiology).

At this point, the readers are reminded of the assertion ordering in Eq. (2.9). This ordering is linked to some important inequalities in terms of justification, uncertainty and probability. The first one is discussed here, the second one was derived in Eq. (2.10), whereas the third one will have to wait until Eq. (3.40). Most health care providers would probably agree that the justification of assertions, such as the ones discussed in earlier sections, admits of *degrees*, J. The degree to which the assertion g_A about the CA case needs to be justified is lower than b_A, which, in turn, is (logically, physically, socially) lower than s_A, which is lower than r_A which, finally, is not as high as the degree for k_A; i.e.,

$$J[g_A CA] < J[b_A CA] < J[s_A CA] < J[r_A CA] < J[k_A CA]. \quad (2.31)$$

There are different substantive interpretations of (2.31).

Example 2.68: One interpretation of (2.31) may focus on the set of possible worlds of CA of the triadic case formula, $a_A CA$. Let us revisit Example 2.9, where $CA \equiv DD$ (disease diagnosis). Suppose that the physician A's belief regarding a case includes four of the possible diagnoses studied in Example 2.9, $DD^{(i)}, i = 1, 2, 3, 4$; and physician B's set of possible realizations concerning the same case includes a single realization $DD^{(1)}$. Clearly, it requires less effort to justify the former (larger) set of possibilities than the latter (single) possibility regarding the medical case at hand. Another interpretation focuses on the a_A form. Suppose that two care providers study the same case. Care provider A's assertion form is

g_A: "mere guessing that the patient is infected with H9N7 avian influenza virus",

which requires a lower level of justification than care provider B's knowledge claim,

k_A: "knowing that the patient is definitely infected with H9N7 virus".

Based on the ranking in Eq. (2.31), it is likely that in the context of many clinical practice and exposure studies one can look at the choice of the assertion

g_A to be based on sparse data or no data at all,
b_A is justified merely on the basis of a dataset (possibly uncertain),
s_A to be supported by an empirical association or phenomenological relationship,
r_A to rely on a science-based model or theory,[55]
k_A to require the consideration of fundamental natural laws.

---

[55] For example, toxicokinetics theory or carcinogenesis model (Crawford-Brown, 1997).

Quantifying the justification degree of a case assertion is a key component of health modeling. As we will see in Chapter 3, a useful quantification method is provided in terms of probability, which is another way of saying that the degrees of justification are linked to the concept of uncertainty.

Physicians' awareness of the sequence of Eqs. (2.31), or (2.10), which duly characterize the understanding of a medical case, enables them to make adequate case assessments in terms of the assertions forms. One often takes advantage of the fact that an uncertainty level associated with an assertion about a case attribute is transferred to any other attribute that is logically or physically related to it.

Example 2.69: For illustration, assume that the *pollutant concentration* at the location–time coordinates p, PC_p, is linked to the intake rate of an individual at p, IR_p, by means of the simple environmental health model,

$$M_{CR}: IR_p = c_m\, PC_p, \qquad (2.32)$$

where c_m is a function of the minute volume (volume of air entering the lungs each minute) and the filtering efficiency (pollutant fraction taken out of the air by, say, a face mask). Due to observation errors, the empirical evidence about PC_p is considered incomplete (Tamerius *et al.*, 2006). This means that investigator A's assertion concerning exposure PC_p (i.e., $a_A PC_p$) depends on A's level of understanding of the *in situ* phenomenon. As in previous cases, in view of Eq. (2.32) one may assume that $a_A \equiv g_A, b_A, s_A, r_A$ or k_A. Depending on whether the investigator considers that a triadic case formula, say $b_A PC_p$ or $k_A PC_p$, is appropriate for the present exposure case, the following reasoning schemes are possible (Table 2.9 above):

$$\begin{array}{lll} \text{Case premises} & \dfrac{PC_p \therefore IR_p}{b_A PC_p} & \dfrac{PC_p \therefore IR_p}{k_A PC_p}\, . \\[1em] \text{Care provider's conclusion} & \therefore b_A IR_p & \therefore k_A IR_p \end{array} \qquad (2.33\text{a–b})$$

Here, the $PC_p \therefore IR_p$ is a logical representation of Eq. (2.32). Obviously, in inferences (2.33a–b) the assertion form of the second premise has a direct effect on that of the conclusion. Also, readers may recall that although the entailment is valid, the occurrence of PC_p itself in the first line of Eq. (2.33a) could be false (even if the care provider A believes that PC_p occurs). On the other hand, the inference (2.33b) is always valid since in this case PC_p cannot be false, given that the care provider knows that PC_p (i.e., $k_A PC_p$). In view of the environmental health model M_{CR}, the uncertainty level of A's understanding of PC_p has a direct effect on the accuracy of intake rate predictions, IR_p, for an individual at p; i.e., the probability of the latter can be expressed in terms of the probability of the former via law (2.32).

Let us turn our attention to medical cases in which using the *false connection* between states of mind and states of nature could potentially lead to wrong or nonsensical conclusions. One should recall that for a case attribute CA_p the physician is justified to claim that $k_A CA_p \therefore b_A CA_p$, but not $b_A CA_p \therefore k_A CA_p$. Only the knowledge state $k_A CA_p$ may imply CA_p with certainty, whereas this is not generally true for other states ($r_A CA_p, b_A CA_p$ etc.). It would be helpful to have a realistic health situation to refer to.

Example 2.70: Let CI_p denote *chikungunya* disease incidence rate at location–time p. Risk management of a potential *chikungunya* epidemic is based on the investigators' assessment that the state $k_A(CI_p \leq 0.08\%)$ properly represents regional epidemic spread. Nevertheless, it turns out that in reality the weaker state $b_A(CI_p \leq 0.12\%)$ is the appropriate representation of the *chikungunya* situation. The consequence of this mind state error on health management can be significant. For one thing, k_A assumes that the available epidemiologic evidence about *chikungunya* allows a higher degree of justification than the real evidence actually does. Accordingly, based on k_A the investigator may wrongly conclude that $CI_p \leq 0.08\%$ etc.

Let us take stock: Rules of logic exist, but if the care professional is unaware of them (or is aware of them but is unable to follow them properly), a crucial diagnosis may be impossible to make. Readers are reminded that a key *SMR* feature is that medical inference does not depend strictly on its logical formulation, as happens in standard decision-making, but also on the epistemic form of the relevant assertions. So, for example, while *modus ponens* is always valid in formal logic, it is not valid when the realistic conditions of everyday clinical practice are considered. This is well understood by experienced physicians who use critical thinking skills such as neutral examination of beliefs, perspective switching, emotional detachment and assessment of the current context. The *SMR* theory aims to give an account of argument construction under conditions of *in situ* uncertainty that explains when medical judgments are acceptable, which clinical inferences (involving physician's assertions and logic connectives linking them) are sound and why this linguistic construction is so important in decision-making. The epistemic strength of the various assertion a_A-forms is currently ranked as in Eq. (2.9). This ranking had a direct effect on the justification degrees and uncertainty levels of the corresponding forms $(g_A, b_A, s_A, r_A, k_A)$ as in Eq. (2.31). More concisely, the more possibilities the triadic case formula forbids, the more informative (uncertain) it is, and the less probable it is.

2.7.4. Does postmodern decision analysis make sense?

As Sampson (2000) notices, postmodernism is an outgrowth of cultural relativism that opens the doors to differing concepts of medicine. Each medical system performs services required by the specific culture's needs, which may not be those of so-called Western standards (cure of diseases, increased longevity etc.), instead, through ritual, a postmodern medical system may serve to assure, e.g., the society's cohesiveness. This postmodern medical system is then viewed on an equal footing to scientific biomedicine developed in technologically advanced Western societies. Interestingly, the postmodern thinking mode allows the misuse of language in medical reasoning and the generation of absurd propositions in recreating clinical reality. Physicians and patients alike can construct their own reality through their altered views, describing it by ascribing new meanings to words along the way. When cultural relativism provides a basic operating system for the health care providers' thinking process, medical knowledge is relative and logically derived clinical opinions are biases; relativism gives equal weight to both rational and absurd conclusions; and language distortion (propaganda) reverses meanings. In this setting, postmodern medicine argues, phrases such as "according to the patient's worldview" create a shield protecting absurd case assertions against critique. It is characteristic that the comment "it depends what the meaning of 'is' is" is the topic of long discussions in postmodern literature.

The conceptual relationships between clinical assertion (metalanguage) and truth (object language) can have significant consequences in medical decision analysis. Taking their theme to the extreme, postmodernists believe that "truth" is an empty word, which rather implies that there is no distinction between language and metalanguage. For purposes of illustration, let A be the substantive physician and assume that B is a postmodern physician. For A there is a close link between the words "truth" and "justification," but this link is not that of identity. One cannot maintain that "true" and "justified" convey the same thing, since justified presupposes the very notion of truth. Postmodern physician B, on the other hand, argues that the words "truth" and "justification" mean the same thing (e.g., "B has reasons to believe that CA" and "CA is true" signify the same thing). Physician A adopts a rather realistic viewpoint according to which the clinical assertion $a_A CA$ expresses a certain representation of truth or reality. By "CA is justified but it isn't true" the physician is saying "CA is justified for A but it is not the way CA is in reality (i.e., the contrast is between the reasons A has to believe CA and the way things are in reality). On the contrary, for physician B the assertion $a_B CA$ is always "relative to a group of individuals". By "CA is justified but it isn't true" the postmodern B is saying "CA is justified for this group but not for that group" (i.e., the contrast is between one group vs. another). For B, people should not search for truth, instead, they should create truth based on their culture and individual viewpoints however diverse these may be.

In other words, physician A does not reject the view that one might never know with certainty the absolute truth concerning all the causes of a particular disease — quite the opposite, there is plenty of room in A's professional thinking that accounts for real world uncertainty and the possibility of incomplete knowledge, which is why the notion of probability plays a major role in A's clinical judgment and decision support. What physician A actually rejects is the postmodern notion of truth as an empty word. Indeed, it is hard for A to see how a sound decision support system can be developed and maintained in the context of postmodern medicine where logic is a relative term, truth is a word void of meaning, no objective standards exist of medical practice and quality care, and rational or scientific ways of knowing are no more reliable than intuitive or personal ways of knowing. Similarly, in the postmodern environment there are no universal medical ethics norms, since all are culturally determined.

It is difficult for a thoughtful physician to accept that a crucial decision concerning whether to perform a life-or-death surgery is not based on an optimal understanding of clinical reality, high-level medical skills and real experience, but is a matter of opinions that vary widely between communities (e.g., for a postmodernist the professional assessment of a top neurosurgical expert essentially carries the same weight as the opinion of an inexperienced community doctor, since personal experience and subjective intuition carry as much authority as high caliber expertise and valid research). It is equally difficult to give credence to the view that quality care can be assured when quality is defined individually. Postmodern health care providers strongly reject care regulations because, they argue, they prevent them from making their own decisions in health care (O'Mathuna and McCallum, 1996). However, the consequences of this thinking mode are serious, and may have dangerous long-term effects on people's health (e.g., diabetic patients have exchanged insulin for alternative medicines and suffered serious complications; Gill *et al.*, 1994). In sum, lack of scientific evidence carries little weight in a postmodern medicine environment, whereas alternative medicine should be accepted as a way to celebrate diversity. Yet, sound medical decision-making must be based on more than merely personal feelings.

Chapter 3

The Role of Probability

3.1. How Much Understanding is Sufficient in Medical Investigations?

The rigorous answer to this key question can exert a profound effect on medical decision-making, population risk assessment and public health management. As a matter of fact, in many countries the development of a nationwide *health information infrastructure* is a major undertaking that is confronted with enormous challenges closely linked to the understanding of both the medical cases themselves and their interoperation with electronic systems (record processing, information base feedback, patient-centered care delivery models, sustained improvements in performance reporting, public health surveillance and other broader initiatives).[1] This twofold understanding is very consequential indeed, because it enables transferring sound science and information technology into comprehensible and high-functioning electronic medical records, and it also allows clinical datasets to flow efficiently and reliably between health care organizations and institutions.

3.1.1. *Medical assertions and partial understanding*

The question concerning understanding applies in every aspect of human inquiry, and not only in health sciences. The answer is often more complex

---

[1] US Congress allocated nearly $30 billion in an ambitious effort to build a nationwide health care information infrastructure. The widespread adoption of health information technology aims at supporting, *inter alia*, new care delivery models, such as patient-centered medical homes, alongside broader initiatives, such as performance reporting and public health surveillance.

than one might have thought, which is why it has occupied the attention of scientists, logicians and philosophers for many years.

As we discussed in the previous chapter, stochastic medical reasoning (*SMR*) and inference may be broadly seen as the interplay between the three key reasoning elements of the triadic case formula (Section 2.3.7), $a_A CA$:

(e1) *Health care provider* A who studies a medical case CA
(e2) Care provider A makes an *assertion* about CA, the form a_A of which is based on
  (e2a) pure guessing, or is justified by
  (e2b) personalized means (based on A's acumen and personal views)
  (e2c) subjective means (strong evidence and inductive logic)
  (e2d) objective means (indisputable evidence and deductive logic)
(e3) CA actually *occurs*.

Depending on A's cognitive situation, the term "assertion" may denote a mental state labeled as:

Guessing, a state characterized by reasoning element $e1$, $e2a$

Personal belief, characterized by elements $e1$, $e2b$

Sustained belief, characterized by elements $e1$, $e2c$

Rationalization, characterized by elements $e1$, $e2d$

Knowledge, characterized by elements $e1$, $e2d$, $e3$.

Some interesting *in situ* relationships may hold between the different states. Under certain cognitive conditions the belief may be defined in terms of knowledge. For example, if a health care provider A admits that in the case of interest it is valid that $b_A CA \to b_A k_A CA$ (i.e., to believe that CA in this sense is to believe that A knows that CA), then it can be shown that

$$b_A CA \equiv \neg k_A \neg k_A CA$$

is also valid. The above may be considered a definition of belief in terms of knowledge, which is, though, based on the interpretation of belief as *subjectively* indistinguishable from knowledge.

The matter clearly concerns the intimate *connection* between clinical assertion and medical understanding in professional practice. To have any confidence in the assertion $a_A CA_p$, physician A needs to possess a certain level of comprehension of CA_p. In other words, one does not need to have a complete understanding of the case — which is often not possible due to a number of factors (incomplete data, case complexity, equipment inadequacy, human uncertainty, cognitive limitations) — but rather an understanding that is adequate for the study purposes. Chapter 2 also examined the process of assigning different degrees of justification (J) and levels of uncertainty (U) to an assertion a_A. The above imply that a partial (or incomplete) understanding of a medical case is usually what happens in the real world, whereas perfect understanding remains a remote possibility. Accordingly, the determination of partial understanding cannot exclude the possibility of error and misunderstanding. In mainstream decision-making and medical judgment, understanding is often a process that involves a circularity that refers to the inductive mode of thinking rather than the deductive one (which is considered the case of complete comprehension). This is not necessarily valid when a physician uses medical reasoning tools (SMR) that can incorporate stochastic deduction.

As it turns out, uncertainty and the impossibility of complete knowledge are key aspects of realistic health systems (Griffith and Christakos, 2007; Koch and Denike, 2007; Liao et al., 2011). In a strict logical setting, to claim that a complete and perfectly accurate model of a complex real world system can be constructed may lead the health care provider to possible contradictions, as follows:

(1) asserting that an *in situ* health system Σ is complex means that Σ is not perfectly intelligible to a health care professional;
(2) if the professional's intention was to construct a model M that is the exact representation of Σ, this M must not be perfectly intelligible, according to *step* 1;
(3) therefore, only an incomplete and uncertain model M of the system Σ is a real possibility.

Assessing the actual levels of knowledge incompleteness and human uncertainty is a key factor in estimating how much health care providers do not know, simultaneously giving rise to fundamental questions, such as:

(1) If a cancer mechanism exists, how confident should physicians be about their ability to comprehend it?
(2) How often is a costly research program directed at discovering disease structures that do not exist?
(3) Is a line of medical decision-making in accordance with the structure of the real case rather than a new branch of pure mathematics?

3.1.2. On rationality and belief

Let us reflect on the notions of *rationality* and *belief*, which frequently emerge in medical thinking. An issue that immediately arises is what one means by "rationality", i.e., what makes a medical argument rational and what makes it irrational. *In situ* rationality may be associated with the partial or incomplete understanding of a clinical case when the physician is aware of some truth conditions of the case for the right reasons (e.g., for the same reasons that someone who recognizes all truth conditions does so). It may come as no surprise to our readers that the meaning of rationality is not the same in all fields of human inquiry. In biology, e.g., a "rational" argument is one generated by the scientific method, whereas in public health finances a "rational" argument is frequently one that maximizes expected utility. Also, in *evidence-based medicine* (*EBM*) "rational" is what is supported by existing evidence, whereas in conventional medicine the term refers to the outcomes of insight-based reasoning. Hence, it is possible that two health care providers working on the same case reach different conclusions, which could both be characterized as rational conclusions, although not necessarily in the same sense.

Example 3.1: It is not unusual in decision-making that one is confronted with a case in which $r_A(DS)$ is the rational mind state of care provider A regarding a particular disease symptom DS, whereas $\neg s_B(DS)$ is the sustained belief of care provider B about DS.

In this context, a simple description of belief is as an "assumed truth" about a medical case. A belief is not an eternal truth, but it can change in light of new findings about the medical case. Health care providers create beliefs of all sorts to anchor their comprehension of real phenomena at specified times and places. A belief system is a set of precepts from which care providers live their daily routine, those that govern their thoughts, words and actions, and without which they could not function. That belief may be treated as a psychological state (in which a physician holds a set of propositions or premises about the case to be valid) does not necessarily imply that a_A denotes a purely psychological concept (e.g., about how brain functions generate beliefs), although it may be a meaningful cognitive representation linked to this concept. Beliefs have different sources, like physicians' own experience and reflections of reality and their active participation within it; or published results and the acceptance of other professionals' views concerning reality, sometimes without argument and debate. In other words, the essence here seems to be focusing on the role and not solely the person (physician).

3.1.3. Knowledge theory revisited: Platonism, context and continuity in medical thinking

In addition to real cases, inference dynamics can play a useful role in *knowledge theory* development. Physician A understands a case in so far as A recognizes its truth conditions, which rather implies that understanding comes in degrees (e.g., linked to the different a_A-forms, the associated justification degrees J and uncertainty levels U, as discussed in previous sections).

Historically, *Plato* provided the classic definition of knowledge as the "justified true belief".[2] In order to know a medical case:

(1) the physician must believe it,
(2) the physician must have a justification for believing it and
(3) the case must be true.

Although formally imperfect, this definition has been very useful in the real world. Platonic knowledge (*conditions* 1–3) may be written in terms of the notions introduced earlier as,

$$k_A \equiv a_A[e1, e2d, e3], \qquad (3.1)$$

where elements $e1$, $e2d$ and $e3$ are linked to the Platonic *conditions* 1, 2 and 3, respectively. In view of earlier relationships, Eq. (3.1) would be also written as $k_A \equiv r_A \wedge e3$, where r_A expresses *conditions* 1–2 (health care professional A adequately understands the essence of the situation, this understanding being highly credible due to instrumental accuracy, cognitive adequacy), and $e3$ obviously expresses *condition* 3.

Quite often the *context* in which health care professionals use the term "knows" determines the standards relative to which "knows" is being attributed. For instance, in an everyday context k_A can be valid, and in

---

[2] Certain issues are worth the readers' attention, such as whether Plato's term "justified" should be viewed in a subjective or an objective sense.

a case context with higher standards for knowledge, the same k_A may be questionable. The readers should notice the difference between Eq. (3.1) and the strong reasoning $k_A CA \therefore CA$ for a case attribute CA. The former is about what knowledge "is", whereas the latter about what it "implies".

The implementation of Eq. (3.1) in medical investigations and public health studies is by no means a trivial matter. A practicing physician, e.g., is sometimes led to mistake true belief for knowledge, whereas, in fact, knowledge is epistemically richer than that. One reason as to why physicians focus on true beliefs, or on whether the content of knowledge is true, is that the true belief aspects of medical knowledge are often more salient and overshadow the justificatory or basing conditions of knowledge and, as a result, in many contexts the physicians care only about knowledge's factivity (i.e., that knowledge entails truth). Several examples of knowledge relationships were given in Table 2.10.

Example 3.2: In the exposure case of Example 2.34 the health scientist appreciated the physical meaning of the case attribute

$$\text{pollutant concentration } PC_p,$$

and considered the value $PC_p = 110 \, \mu g/m^3$ as highly credible to be expressed in object language terms (due to the accuracy of the observation instruments, favorable *in situ* conditions, rigorous theoretical support). Hence, the scientist can assume that $k_A PC_p \therefore PC_p$, in this case.

We will resume this section by noticing that another essential question of medical understanding addressed by *SMR* inference is related to *knowledge continuity*. Given that $k_A G$ (as usual, G denotes core or general knowledge), and assuming that new case-specific data S emerge, will health care provider A always be able to gain this new knowledge and, at the same time, to keep the previous one?

In symbolic terms, a physician may need to examine the validity of

$$\begin{array}{rl} \text{Case premises} & \dfrac{\begin{array}{c} k_A G \\ S \end{array}}{\therefore \neg(k_A S \Rightarrow \neg k_A G)} \\ \text{Care provider's conclusion} & \end{array} ; \qquad (3.2)$$

i.e.,

"if A knows G"

and

"new information S emerges",

it is entailed that it is not valid that

"if A were to know S, then A would not continue to know G".

In other words, given that A knows G, the care provider will be able to preserve G in the face of emerging S. The readers should keep in mind that knowledge continuity is closely related to the notion of the *conditional* (i.e., updating one's knowledge in light of new data; Section 3.2).

3.1.4. Concerning medical expertise

Relevant to the preceding analysis is the issue of medical *expertise*. The first thing to notice is that expertise in a particular medical field or human exposure discipline is about having an unusual accumulation of knowledge and facts regarding a case, but it is not necessarily about having a reliable means of translating facts into accurate diagnoses and prognoses. That is to say, disciplinary experts are by no means required to be wise judges, and they may not have any special competence or higher qualifications to predict the future of a medical case.

Example 3.3: Although it is true that different kinds of exposure data support scientists' careful evaluation of environmental conditions influencing health events, decisions and predictions, an expert specialized in data collection alone should know that this data alone rarely forms a satisfactory basis for a realistic disease explanation and prognosis.

Yet, a serious obstacle to real progress in environmental health research has been the fact that often expert conclusions are drawn based on data collected in large numbers, but without testing whether or not they satisfy elementary rules of sound reason. These conclusions frequently violate basic principles of sound reason, are self-contradictory and even paradoxical. It is well known to most experienced physicians that which facts or expert knowledge are germane to a medical case need to be determined by means of a synthesis of theory and prior evidence. The accumulated facts known to experts must be placed in the service of a sound theory, transparent method and replicable analysis of evidence if these experts are to translate facts to valid inferences. In health care, expertise is not necessarily a general skill that translates well across all kinds of clinical cases. Even though an expert in a particular kind of case acts on the basis of the best clinical and scientific knowledge available, it is entirely plausible that a medical professional could be an expert with respect to a certain kind of clinical case and patient population and, at the same time, be less skilled as regards other kinds of cases and populations.

In sum, clinical practice is an arena in which a blind eye is sometimes turned to sound theory, logical rigor and methodical analysis in favor of ill-conceived expert opinion. As a result, instead of using a systematic and critical evaluation of alternative options (concepts, notions, methods, thinking modes) to arrive at sound judgments, one is presented with an evaluation of the idiosyncratic and opaque means of views adopted by certain "influential" experts. This is a crucial matter that deserves the attention of medical decision-making in the light of probability theory.

3.2. Space–Time Probabilities of Medical Cases

Time for a reality check. Living in an uncertain world, physicians frequently interpret disease symptoms in different ways and they are not fully confident of the diagnoses they make. Similar is the situation with most health care providers. Accordingly, the introduction of the notion of *probability*[3] in medicine was made necessary due to significant uncertainties in the physician's thinking linked to changing environments, varying physiologic functions, incomplete case information, cognition limits, instrument inadequacy, laboratory discrepancies and biases of different kinds (Section 1.1.2, and Section 2.1).

3.2.1. *Common probability interpretations in health care practice*

In their daily routine, health care professionals experience multi-sourced uncertainties, which makes probability concept a basic component of the decision-making process. In accordance, care providers need to use different probability interpretations as appropriate (ontic or epistemic, technical or physical, logical or subjective). In most clinical practices phrases like "rule in", "rule out", "high likelihood", "doubt" and the like are used to describe

---

[3] If needed, the readers may consult the rich literature on probability theory and its various applications (e.g., Jeffreys, 1961; Hacking, 1975 and Kottegoda and Rosso, 1997).

the certainty level of a physician's assertions and conclusions about a case. It will be helpful to have some cases to refer to as in the example below.

Example 3.4: Consider some typical case assertions

"the combination of the lab test results rules in the diagnosis of pneumonia",

"the physician doubts that the patient has pulmonary embolism",

"the prostate-specific antigen (PSA) value is 8.2, thus the likelihood of prostate cancer increases" and

"the specified medication has a high likelihood of curing the disease".

Such phrases may be associated with a wide range of numerical probabilities, which must be assigned an appropriate interpretation. There are many cases in clinical practice that the cited probabilities were assigned an inappropriate interpretation, thus leading to the wrong case prognosis. Typically, when physicians argue that

"the specified medication has an 85% probability of curing the disease",

they mean that in a group of 100 people with the disease being given the same medication (with all other conditions being the same) approximately 85 people would be cured and 15 not cured. It is important to realize, however, that one cannot say beforehand that Mr. Volonte who is picked out from a group of people who take the medication will turn out to be one of the 85 people who would be cured. All the physician can say is that Mr. Volonte would have an 85% probability of being cured. Similarly, the assertion that

"45% of patients with severe septic shock will die within 28 days"

involves a certain confidence level, because previous medical research and clinical practice have found that this is the proportion of patients who died of the disease. However, it is much harder to translate this into a prognosis for an individual: additional information is needed to determine whether a patient belongs to the 45% of those individuals who will succumb, or to the 55% who will survive. That is to say, probability works well for groups of people, but not necessarily for a particular individual.

Probability in medicine may provide a quantitative characterization of the information a physician gathers during the interview and physical examination and express a clinician's opinion regarding the likelihood of a medical condition being present (a diagnosis that a patient has pneumonia) or expected to occur in the future (a prognosis that a patient will experience a myocardiac infraction within six months). Typically, a clinician derives a probability estimate for a disease hypothesis by using personal experience and the published literature (Sox *et al.*, 2013). Physicians' language expresses their uncertainty about the case and portrays their own understanding of this uncertainty. Confusion commonly arises when physicians use the same statement to express different judgments about a clinical case (e.g., physicians A and B both state that "Alzheimer's disease seems to occur in the case of Ms Pennelegion," but they come up with different interpretations of this statement). Similarly, physicians' language reflects their uncertainty when they use different statements to express the same judgment concerning disease occurrence (e.g., physician A says that "coronary artery disease cannot be ruled out in the case of Mr. Nightingale," whereas B says that "coronary artery disease is likely in the case of Mr. Nightingale"). A way out of the confusion is offered by the notion of probability that allows physicians to assign numbers to the medical statements. In Ms Pennelegion's case, A and B may assign different numbers (say, 0.7 vs. 0.9 probability) to the same words "seems to occur." In Mr. Nightingale's case, A assigns probability 0.3 to the words "cannot be ruled out," and B assigns the same probability to the words "is likely."

To a certain extent, probabilities are assigned by health care providers based on published literature, their past experience with similar cases, laboratory testing and experimentation, as well as their cognitive skills (analyzing critically, reasoning logically, applying standards, discriminating among possibilities, seeking new data, transforming knowledge and predicting).

Example 3.5: Based on published literature and past experience, a physician argues that about 33% of patients with possible deep venous thrombosis of the leg have physical signs of the disease, or that patients with ischemic pain have at least 90% probability of coronary artery disease. The calculation of many probability values is based on laboratory testing. This is clearly the case of a physician arguing that

> "the cytological examination has an 82% probability for the mammary tumor malignity diagnose at domestic carnivores".

Moreover, physicians should adequately *communicate* key medical decision notions to their patients (see also Section 2.3.3). Accordingly, Ms Pappas who suffers from OCD^4 understands that the physician's statement that

> "Ms Pappas has a 70% probability to respond to the anti-depressant Prozac"

does not mean that she would respond to Prozac 70% of the time, but rather the appropriate interpretation should be that she would either respond to the anti-depressant (with probability 70%) or not respond to it (with probability 30%).

As the readers are aware, a public health domain in which probability is routinely used by health care professionals is *survey data analysis*. They are also aware that probabilistic assertions, which are uncertain, should not be confused with deterministic statements, which are certain. This is one of the reasons why the implementation of the probability notion in unstructured domains of health care requires considerable concrete experience with real situations.

Example 3.6: When regional health care providers assert that survey data analysis concluded that

> "a 15% correlation exists between an observed population disease (PD) and the corresponding population behavior (PB)",

they practically mean that

> "PB has a 15% probability of causing PD".

This is often a misleading interpretation, since in many *in situ* cases the PB has no actual effect on the PD; instead, it has some correlation with the use of some unidentified product, which is the true cause (i.e., it correlates 100% with the PD).

The preceding considerations support the view that, despite the considerable experience gained by the profession, it is not uncommon that basic axioms and principles of probability theory are violated in the course of medical decision-making and health risk assessment. In addition to the ones discussed above, one can find numerous examples of such violations that are grounded in the relevant literature and, consequently, they can advance compelling arguments. Below we consider a typical case of

---

[4] Obsessive–compulsive disorder.

medical prognosis discussed by Redelmeier *et al.* (1995) where elementary probability rules have been clearly violated.

Example 3.7: A 67-year-old man was presented to the hospital emergency department with chest pain of four hours' duration. The initial diagnosis was *acute myocardial infarction*. Physical examination showed no evidence of pulmonary edema, hypotension or mental status changes. The electrocardiographic (*EKG*) showed *ST*-segment elevation in the anterior leads, but no dysrhythmia or heart block. Past medical history was unremarkable. The patient was admitted to the hospital and treated in the usual manner. Each physician was randomly assigned to evaluate one of four disease prognoses for this patient:

"dying during this admission" ($DP^{(1)}$),
"surviving this admission but dying within one year" ($DP^{(2)}$),
"living for more than one year but less than ten years" ($DP^{(3)}$) or
"surviving for more than ten years" ($DP^{(4)}$).

The average probabilities assigned to the above prognoses, $DP^{(1)}$, $DP^{(2)}$, $DP^{(3)}$ and $DP^{(4)}$, by the physicians were 0.14, 0.26, 0.55 and 0.69, respectively, which add up to 1.64. Profoundly, a basic rule of probability theory was violated, namely, that the probabilities of these four outcomes must sum up to one. As it was observed, the physicians apparently overweighted the possibility that was explicitly mentioned relative to the unspecified alternative.

3.2.2. *Probability of a case assertion (mind state)*

The justification of a physician's thinking in terms of degrees discussed earlier implies that in health practice the assertion forms (g_A, b_A, s_A, r_A, k_A) of the relevant triadic case formulas rather describe uncertain states of mind of varying uncertainty levels (from low to high). Since medical assertions (beliefs, symptoms, diagnoses etc.) express mind states, one is tempted to assume that the basic rules of a physician's inference engine are similar to the rules of probability logic. More precisely, the relevant metalanguage statements are not definite but rather uncertain, changing across space–time, and conditioned on the often incomplete information available, which is why physicians are using probabilistic notions with increasing frequency to support their assertions scientifically. Yet, this is not the complete story.

As a matter of fact, modern sciences generally recognize that the application of natural laws in the real world can never be expressed in certainty statements, but only in probabilistic statements. Functioning in an integrative manner, in addition to the logic of knowledge and belief, a physician who practices *SMR* inference essentially involves a rigorous space–time uncertainty assessment in terms of a twofold notion of medical probability: formal and substantive. As expected, the *formal* aspect assures that probability has a rigorous mathematical structure. The *substantive* aspect, on the other hand, links probability to actual medical practice. Epistemically, probability provides an inverse assessment of the notion of "uncertainty" discussed earlier. That is, probability gives a measure of "how much one knows", whereas uncertainty is rather an assessment of "how much one does not know".

In quantitative terms, when health care professionals apply *SMR*, in many cases they assume that the probability of a health event or medical case generally varies from place to place and from time to time, i.e., probability is a function of the location–time coordinates $\boldsymbol{p}$. Probability is evaluated through the acquisition of clinical data, the implementation of logical argumentation and the questioning of inconsistencies, whereas, at the same time, it allows for the revision of diagnoses, prognoses, actions and goals. Accordingly, Eq. (2.7) is completed by incorporating the probability of the case

$$P[a_A] = \eta,^5 \tag{3.3}$$

where η varies between 0 and 1. More precisely, the triadic case formula (consisting of A, $CA_{\boldsymbol{p}}$ and a_A) is assigned the mathematical probability function (3.3). For convenience, the $a_A CA_{\boldsymbol{p}}$ or "$CA_{\boldsymbol{p}}$ in light of KB" is sometimes represented simply by the subscript KB (denoting the knowledge base that substantiates the probability value η assigned to the physician's assertion about the case). Then, the probability in Eq. (3.3) is also written as

$$P[a_A CA_{\boldsymbol{p}}] = P_{KB}[CA_{\boldsymbol{p}}].$$

As in Eq. (2.7), KB is directly linked to A's cognitive condition, which frequently involves biases in inductive judgment, errors in deductive reasoning, fallacies and perceptual illusions. The probability function $P_{KB}[\cdot]$

---

[5]Care provider A's assertion about $CA_{\boldsymbol{p}}$ has probability η.

is assumed to account for all these uncertainty factors. In this setting, it makes sense that "the probability of a_A" is the same as "the probability that the assertion is true". More precisely, the probability of the assertion is the sum of the probabilities of the possible worlds or realizations (recall, S/TRF terminology in Section 2.2.2) in which the assertion can be considered valid.

Most emphatically, Eq. (3.3) does not focus on the case, but rather on the assertion a_A made by the competent physician about the case. That some assertions are cognitively stronger than others (in varying degrees, and even for the same conclusion) is represented by the different forms that a_A can take. In other words, a care professional may view Eq. (3.3) as a measure of how strong the a_A-form chosen is supported by the available KB. A critical finding of probability analysis of this kind is that rigorous testing can increase the certainty of a physician's diagnostic reasoning, thus selecting, say, the antibiotics that are most likely to treat an infection.

Example 3.8: Example 2.9 considered several diagnoses concerning a patient's health condition, to which physician A associates different a_A forms. Assume that A's cognitive situation regarding the disease diagnosis

$$DD^{(2)} : \{migraine,\ sinus\ disease,\ tension\text{-}type\ headaches\}$$

is expressed in terms of the form $b_A(DD^{(2)})$, which is subjective belief given the resources available. Specifically, a physician who understands the probabilistic method can argue that $b_A(DD^{(2)})$ holds with probability $P[b_A(DD^{(2)})]$.

The SMR's metalanguage has the advantage of viewing a probability value as the TV of the medical assertion in conditions of uncertainty, which is termed its *stochastic truth-value* (STV). This is a considerable generalization of the two-valued standard logic [T (true) or 1, and F (false) or 0] to a multi-valued stochastic logic possessing a continuous scale of values from 0 to 1. In symbolic terms, the $TV[CA_p] = 0, = 1$ is replaced by

$$STV(a_A CA_p) = P[a_A CA_p] = \eta, \qquad (3.4)$$

where $P_{KB}[TV(CA_p) = 1] = \eta \in [0,1]$. In the same setting, one may find it appropriate to assume that $STV(\neg a_A CA_p) = P[\neg a_A CA_p] = 1 - P[a_A CA_p] = 1 - \eta$, and $P_{KB}[TV(CA_p) = 0] = 1 - \eta$. Of some interest are the special probability values $\eta = 1$ and $\eta = 0$, in which case one may suggest that $a_A \equiv k_A$, with $k_A CA_p \therefore CA_p$ and $k_A(\neg CA_p) \therefore \neg CA_p$, respectively. Herein, the subscript $p = (s,t)$ will be included in CA_p only when it is necessary to indicate that the case attribute is a function of the location–time coordinates. Otherwise, it will be omitted for notational simplicity.

It is common knowledge that, due to the incomplete understanding of a case and the associated uncertainties, several of the statements made in everyday clinical practice can turn out to be rather meaningless and irrational — *absurd*.

Example 3.9: In clinical routine conversations, one hears statements like

"Ms Euler may survive (ES) if she has gastric bypass (GB) surgery"

is a reason her physician has to perform the surgery, and the serious risk that

"Ms Euler may not survive ($\neg ES$) if she has gastric bypass (GB) surgery"

is a reason not to perform the surgery. These sorts of statements are absurd because ES cannot be a reason Ms Euler has to GB, whereas $\neg ES$ is a reason she has to $\neg GB$. The truth of the "ES if GB" entails the falsity of the second conjunct "$\neg ES$ if GB". There are other similarly absurd assertions, even though the truth of their first part does not entail the falsity of the second part. Commonly encountered absurd assertions are conversationally expressed as,

"physician A cannot even guess whether ES" (ES but $\neg g_A(ES)$),
"A does not believe that ES" (ES but $\neg b_A(ES)$),
"A does not have good enough evidence to rationalize ES" (ES but $\neg r_A(ES)$),
"A does not know that ES" (ES but $\neg k_A(ES)$) and
"there is a significant chance that $\neg ES$" (ES but $P_{KB}[\neg ES] >$).

3.2.3. *Basic probability rules*

Probability is not an arbitrary quantity. Rather, it needs to satisfy certain rigorous mathematical conditions and follow well-defined rules. For the readers' benefit, Table 3.1 lists a set of probability rules that are customarily implemented in *SMR* and decision-making. Several probability relationships are directly derivable from logical relationships of a medical case, which is a significant tool of decision-making. In the process the determinism of the logical relationship is replaced by the indeterminism of the probabilistic relationship. The following is an example of such a derivation.

Table 3.1: A review of basic rules of probability.

Rule	Formulation
Normality[1]	$0 \leq P_{KB}[CA] \leq 1, \quad P_{KB}[CA] + P_{KB}[\neg CA] = 1$
Certainty (Ω or T: sure to occur case)	$P_{KB}[\Omega] = 1, \quad P_{KB}[\Omega\|CA] = 1$ $P_{KB}[\Omega \wedge CA] = P_{KB}[CA]$
Impossibility ($\varnothing$ or F: impossible to occur case)	$P_{KB}[\varnothing] = 0, \ P_{KB}[\varnothing\|CA] = 0$ $P_{KB}[\varnothing \wedge CA] = 0$
Multiplication	$P_{KB}[\wedge_{i=1}^{m} CA_i] = P_{KB}[CA_1] P_{KB}[CA_2\|CA_1]$ $\qquad P_{KB}[CA_3\|(CA_1 \wedge CA_2)]$ $\qquad \ldots P_{KB}[CA_m\|\wedge_{i=1}^{m-1} CA_i]$
Summation	$P_{KB}[CA_1\|CA_2] + P_{KB}[\neg CA_1\|CA_2] = 1$
Overlapping	$P_{KB}[CA_1 \vee CA_2] = P_{KB}[CA_1] + P_{KB}[CA_2]$ $\qquad - P_{KB}[CA_1 \wedge CA_2]$
Bayes	$P_{KB}[CA_1\|CA_2] P_{KB}[CA_2] = P_{KB}[CA_1] P_{KB}[CA_2\|CA_1]$
Total probability ($P_{KB}[CA] \in (0,1)$)	$P_{KB}[CA_2] = P_{KB}[CA_2\|CA_1] P_{KB}[CA_1]$ $\qquad + P_{KB}[CA_2\|\neg CA_1] P_{KB}[\neg CA_1]$
Conjugacy	$P_{KB}[CA] \in [p_1, p_2] \quad \text{iff} \quad P_{KB}[\neg CA] \in [1 - p_2, 1 - p_1]$
Mutual exclusiveness ($CA_1 \wedge CA_2 = \varnothing$)	$P_{KB}[CA_1 \vee CA_2] = P_{KB}[CA_1] + P_{KB}[CA_2]$ $P_{KB}[CA_1 \wedge CA_2] = 0$
Collective exhaustiveness ($CA_1 \vee CA_2 = \Omega$)	$P_{KB}[CA_1 \vee CA_2] = 1$ $P_{KB}[CA_1 \wedge CA_2] = P_{KB}[CA_1] + P_{KB}[CA_2] - 1$
Statistical independence	$P_{KB}[CA_1 \wedge CA_2] = P_{KB}[CA_1] P_{KB}[CA_2]$ $P_{KB}[CA_2\|CA_1] = P_{KB}[CA_2]$
Logical equivalence	$CA_1 \equiv CA_2$ entails $P_{KB}[CA_1] = P_{KB}[CA_2]$
Logical consequence	$CA_1 \therefore CA_2$ entails $P_{KB}[CA_1] \leq P_{KB}[CA_2]$[2]
Additiveness (a, b real numbers)	$P_{KB}[CA \geq a] = P_{KB}[CA + b \geq a + b]$ $\qquad = P_{KB}[(CA + b)^2 \geq (a + b)^2]$

[1] A measure is said to be "normalized" if it is put on a scale between 0 and 1.
[2] For the *equality*, $P_{KB}[\neg CA_1 \wedge CA_2] = 0$).

Example 3.10: Of particular interest in decision-making is the so-called *logical equivalence* rule. Starting from the valid logical identity

$$CA_1 \equiv (CA_1 \wedge CA_2) \vee (CA_1 \wedge \neg CA_2),$$

the care provider can derive the probabilistic relationship,

$$P_{KB}[CA_1] = P_{KB}[(CA_1 \wedge CA_2) \vee (CA_1 \wedge \neg CA_2)]$$
$$= P_{KB}[CA_1 \wedge CA_2] + P_{KB}[CA_1 \wedge (\neg CA_2)],$$

which is another expression of the total probability rule (Table 3.1). The rules of Table 3.1 hold if we replace $P_{KB}[CA_1]$, $P_{KB}[CA_2|CA_1]$ and so on, by $P[CA_1|KB]$, $P[CA_2|CA_1 \wedge KB]$ and so on.

That is, it is valid that,

$0 \leq P[CA_i|KB] \leq 1, \quad i = 1, 2$
$P[CA_1 \wedge CA_2|KB] = P[CA_1|CA_2 \wedge KB]P[CA_2|KB]$
$P[CA_1|CA_2 \wedge KB] + P[\neg CA_1|CA_2 \wedge KB] = 1$
$P[CA_1 \vee CA_2|KB] = P[CA_1|KB] + P[CA_2|KB] - P[CA_1 \wedge CA_2|KB]$
$P[CA_1|CA_2 \wedge KB]P[CA_2|KB] = P[CA_1|KB]P[CA_2|CA_1 \wedge KB]$
$P[CA_2|KB] = P[CA_2|CA_1 \wedge KB]P[CA_1|KB]$
$\qquad + P[CA_2|\neg CA_1 \wedge KB]P[\neg CA_1|KB].$

(3.5a–f)

The next example presents a KB-based application of these equations frequently used in this book.

Example 3.11: If we let $KB \equiv G$ (core knowledge), the probabilities in Eqs. (3.5a–f) become, $P_G[CA_i] = P[CA_i|G]$, $i = 1, 2$, $P_G[CA_1 \wedge CA_2] = P_G[CA_2|CA_1]P_G[CA_1]$, $P_G[CA_1|CA_2] = P_G[CA_2|CA_1]\frac{P_G[CA_1]}{P_G[CA_2]}$ etc.

3.2.4. *Probability interpretations in object language and metalanguage*

As noted earlier, there is a correspondence between the notions and principles employed in object language expressions and those employed in metalanguage expressions. Such a correspondence can also be established between the probability interpretation in the object language and metalanguage contexts. To illustrate the matter, consider an exposure case attribute CA_p distributed in space–time with N equiprobable realizations $CA^{(i)}$ ($i = 1, \ldots, N$) in the S/TRF sense of Section 2.2.2. We are interested about the probability that $CA_p \leq \zeta$, where ζ is a specified health threshold.

The frequentistic definition of this probability is (object language),

$$P[CA_p \leq \zeta] = \frac{1}{N} n_{i=1}^{N}(CA^{(i)} \leq \zeta) \qquad (3.6)$$

where $n_{i=1}^{N}(CA^{(i)} \leq \zeta)$ denotes the number of realizations that satisfy the condition $CA^{(i)} \leq \zeta$. With this in mind, for each realization the metalinguistic expression

physician A's assertion that "$CA^{(i)} \leq \zeta$" is T, or $(CA^{(i)} \leq \zeta)$ is T,

is equivalently expressed as $TV[CA^{(i)} \leq \zeta] = 1$.[6] In light of this equivalence, Eq. (3.6) can be written as

$$P[CA_p \leq \zeta] = \frac{1}{N} n_{i=1}^{N}(TV[CA^{(i)} \leq \zeta] = 1)$$
$$= \frac{1}{N} n_{i=1}^{N}(a_A(CA^{(i)} \leq \zeta) \text{ is } T), \qquad (3.7)$$

which shows that the frequencies of realizations $CA^{(i)} \leq \zeta$ employed in the object language probability interpretation may be substituted with frequencies of assertions about realizations, i.e. $a_A(CA^{(i)} \leq \zeta)$, in its metalanguage interpretation. Accordingly, all results of the object language probability interpretation can be transferred to the metalanguage probability interpretation, and be seen as its special case. Furthermore, consider the metalinguistic representation

$$P[a_A(CA_p \leq \zeta)] = \frac{1}{N} n_{i=1}^{N}(a_A(CA^{(i)} \leq \zeta)), \qquad (3.8)$$

where $n_{i=1}^{N}(a_A(CA^{(i)} \leq \zeta))$ is the number of realizations in which A asserts that $CA^{(i)} \leq \zeta$ holds. Representation (3.8) differs from that of Eq. (3.7) in that in (3.7) all assertions $a_A(CA^{(i)} \leq \zeta)$ are true;[7] whereas in Eq. (3.8) not all of them are necessarily true (some may be wrongly assumed to be true, and some others wrongly assumed to be untrue). In the rather rare case that A possesses a perfect knowledge of the phenomenon, all $a_A(CA^{(i)} \leq \zeta)$ in (3.8) are true (say, $a_A \equiv k_A$), in which case Eq. (3.8) reduces to Eq. (3.7), i.e., $P[a_A(CA_p \leq \zeta)] = P[CA_p \leq \zeta]$.

To gain further insight about probability interpretations, next we consider some examples from medical practice.

---

[6] Recall that since $\eta = 1$, $a_A(CA^{(i)} \leq \zeta) \therefore (CA^{(i)} \leq \zeta)$.
[7] Ibid.

Example 3.12: As a follow up to Eq. (2.7), which goes from events to assertions about events, Eq. (3.3) regards a probability, not as a property of the event, but as a property of the assertion about the event. This is because, once more, the readers are reminded that while object language focuses on attributes and events, metalanguage is concerned with assertions about attributes and events. For illustration, instead of saying that

"the probability that a tumor will be detected if a CT scan is performed is 0.75", $P[CT] = 0.75$

(object language interpretation of probability), one can say that

"the probability of the assertion that 'a tumor will be detected if a CT scan is performed' is 0.75", $P[a_A(CT)] = 0.75$

(metalanguage interpretation of medical probability). By means of this transition, the notion of probability is used to rate the relevant assertions, and the probability assertions belong to the physician's metalanguage rather than object language.

Another thing worth noticing is that the probability value $\eta \in [0,1]$ is linked to the cognitive clout of the assertion a_A in a direct manner, i.e., the larger the η, the cognitively stronger the a_A.

Example 3.13: Suppose that the attribute PC_p represents the space–time distribution of an individual's exposure rate to nicotine (in $\mu g/m^3$) at the location–time coordinates $p = (s,t)$, and consider

"the probability of the belief that the PC_p values at p vary within the interval $(39\,\mu g/m^3, 54\,\mu g/m^3)$,"

$$P_{KB}[PC_p \in (39,54)] = P[b_A(PC_p \in (39,54))] = \eta.$$

The substantiation of expert A's belief about exposure admits of degrees that are directly reflected on the value of $\eta \in [0,1]$. Since a belief is not necessarily true but rather probably true, the higher the value of η, the more probable the validity of $b_A(PC_p \in (39,54))$.

The application of modern physics in medicine has been very fruitful (Kane, 2009). Many results of modern physics (x-rays, lasers, magnetic resonance imaging) have been transformed into medical technology (imaging, surgery). Quantum physics is combined with molecular biology and genetics to obtain many of the findings of modern medicine. It then comes as no surprise that the metalanguage of these scientific fields is linked to noteworthy interpretations of the probability notion.

Example 3.14: In the metalanguage of quantum physics used in medicine, each atomic assertion carries along an assertion degree, a complex number in this case, which has been interpreted as *probability amplitude* (Bohm, 1989).

Let us take stock: a health care provider practicing *SMR* carefully distinguishes between the object language interpretation and the metalanguage interpretation of probability. For a case CA, the *in situ* probability $P_{KB}[CA]$ may generally be different than the actual (but unknown) probability $P[CA]$. The former probability is relativized to the current medical KB, whereas the latter probability is non-relativized to any epistemic situation but is a characteristic of nature. It is noteworthy that when the actual probability is unknowable, some professionals tend to assume that $P_{KB}[CA] = P[CA]$, which is not without consequences in clinical practice, disease risk assessment and public health management (see discussion and examples in Christakos (2010)).

As exemplified earlier, the space–time probabilistic analysis depends on the physician's cognitive situation in terms of different a_A-forms (k_A, r_A, s_A, b_A and g_A). Note that many care providers view the probability of their knowledge state, $P[k_A]$, as the "objective" probability of a medical diagnosis based on unassailable scientific theory and/or extensive empirical evidence obtained in similar situations. Let physician A's disease diagnosis, $DD^{(1)}$, be assigned the knowledge state $k_A DD^{(1)}$; and that of physician B's, $DD^{(2)}$, be assigned the state $k_B DD^{(2)}$. In light of Eq. (2.27), the diagnosis inference holds,

$$\begin{array}{ll} \text{Case premise} & \dfrac{k_A DD^{(1)} \therefore k_B DD^{(2)}}{} \\ \text{Care provider's} & \therefore P[DD^{(1)}] \leq P[DD^{(2)}] \\ \text{conclusion} & \end{array} \qquad (3.9)$$

where the inequality holds between actual probabilities (object language). It is cognitively valid that, $P[k_A DD^{(1)}] \leq P[k_B DD^{(2)}]$, $k_A DD^{(1)} \therefore DD^{(1)}$ and $k_B DD^{(2)} \therefore DD^{(2)}$. Note that Eq. (3.9) does not hold in terms of b_A, e.g., since it is not generally the case that $b_A DD^{(1)} \therefore DD^{(1)}$ or $b_A DD^{(2)} \therefore DD^{(2)}$.

Example 3.15: Suppose the diagnosis of physician A regarding Ms Chan's health status is

$$r_A(migraine),$$

whereas that of physician B is

$$r_B(migraine,\ sinus\ disease).^8$$

Arguably, the latter implies the former (r_A is a subset of r_B). As far as Ms Chan is concerned, the diagnosis of physician B is technically more probable than that of physician A,[9] since it includes more possibilities. Yet, if it turns out to be true (after the event), A's diagnosis was more informative than B's (before the event).

For a case CA, the assertion form $a_A CA$ allows in-depth encounters with uncertainty sources, say, in terms of laboratory tests, equipment limitations and observation errors. In a certain sense, this uncertainty is smaller if a physician can justifiably assume as appropriate the form b_A rather than r_A. Also, one must be aware of the different probabilities

$$P[CA|k_A CA]\ \text{vs.}\ P[CA|b_A CA]\ \text{vs.}\ P[CA|r_A CA].$$

Since $r_A CA \therefore b_A CA$,[10] a physician can claim that $P[r_A CA] \leq P[b_A CA]$ (this is an inequality between probabilities belonging to the metalanguage). The preceding and similar conclusions, further stress the existence of substantive differences between medical decisions that are based on k_A vs. s_A vs. r_A vs. b_A vs. g_A. We will have more to say about the subject later in the book.

3.2.5. *Body of evidence and medical interventions*

A professional's state of mind (and related assertions) may be based on various kinds of medical evidence and health status assessment together with a description of how they were obtained. A physician's state of mind, which guides the physician's choice of diagnostic tests, depends on information about the absolute and/or relative test accuracy, whether the test results could change the clinician's mind, the patient's feelings about the health situation, and the physician's careful estimate of the prior disease probability. Unlike previous health systems that were reliant upon *game*

---

[8] In more terms, $r_A(DD^{(1)})$ and $r_B(DD^{(2)})$, where "$DD^{(1)}$: *migraine*" and "$DD^{(2)}$: *migraine, sinus disease*".

[9] The statement is in agreement with the well-known inequality of probability theory, $P[DD^{(1)} \wedge DD^{(2)}] \leq P[DD^{(1)}]$, $P[DD^{(2)}] \leq P[DD^{(1)} \vee DD^{(2)}]$.

[10] Recall that $r_A CA \therefore b_A CA$ also implies $U[r_A CA] \geq U[b_A CA]$. As a numerical illustration, A may claim that $CA \in [0, 0.5]$, whereas B that $CA \in [0, 1]$; i.e., A allows fewer CA realizations than B.

theory (Binmore et al., 1993), *SMR* inference employs the more general and realistic perspective of stochastics theory. The health care provider is able to introduce more than one probability interpretation, such as logical probability (based on a scientific relation between the body of evidence and the medical hypothesis) and subjective probability (about the hypothesis before any connection is established between it and the existing evidence).

In general, clinicians distinguish between disease probability before a body of evidence was gathered by means of diagnostic tests (prior or pre-test probability) and disease probability after the test results have been interpreted (posterior or post-test probability). A general principle is that the interpretation of the test results depends on the pre-test disease probability. Based on a large study of patients with chest pain, Weiner et al. (1979) calculated that the pre-test probability of coronary artery disease among males depended on the patient's history. In particular, for histories classified as typical angina, atypical angina, and non-anginal chest pain, the corresponding probabilities were 0.89, 0.70 and 0.22. Following abnormal exercise testing, the post-test probabilities for the above history types were increased as follows: 0.96, 0.87 and 0.39.

Example 3.16: David Hume's relentless questioning of the validity of inductive inference opened the door to major developments, including the connection between the notions of probability and causality. Consider a case attribute CA_p (body temperature, blood pressure, cancer incidence, plague mortality). Assume that the core KB is that the space–time distribution of CA_p obeys the scientific law M_{CA} (object language) with boundary-initial conditions CA_0.[11] Then,

$$P_{KB}[CA_p] = Prob[a_A(CA_p) \; in \; light \; of \; M_{CA}, CA_0] = \eta, \quad (3.10)$$

where the subscript $p = (s, t)$ is included to indicate that the case attribute is generally a function of the location-time coordinates (through the law), and the η value is justified by appeal to core knowledge and case-specific conditions. The physician must make sure that the domain of the case of interest lies within these conditions. If it lies outside the domain, no matter how well the content of the conditions is described, the medical findings could be inadequate and even false.

---

[11] These conditions determine the focus of investigation and establish the space–time domain of concern.

At this point, it is worth reminding our readers that inductive medical arguments cannot be valid in the same way as deductive ones. The care provider's conclusion is never implicit in the provider's premises, which can only supply evidence in support of the probability of the conclusion being valid. Yet, most of the decisions and actions in real clinical practice are based on induction — as is in most of science.

It is common knowledge that physicians frequently disagree with one another about whether some facts are evidence that a certain diagnosis or hypothesis is true, or, if it is, about how strong that evidence is. In several other cases, the physicians do not disagree that a phenomenon has occurred, a fact has been observed or an experimental result was obtained. Rather they disagree whether what has been observed constitutes evidence (full or partial) for a diagnosis. Suppose that two groups of physicians have the same case-specific data S in front of them, but different core knowledge bases, say G_1 and G_2, so that the probabilities of a disease diagnosis DD, for one group of physicians is $P_{G_1}[DD|S]$ and for another $P_{G_2}[DD|S] \neq P_{G_1}[DD|S]$. This sort of reasoning is formally expressed in terms of total knowledge, $K_i = G_i \cup S$ ($i = 1, 2$), as

$$P_{G_i}[DD|S] = P_{K_i}[DD] = \eta_i, \qquad (3.11)$$

meaning that S is evidence for the G_i-based diagnosis DD with probability η_i. The higher η_i is, the stronger is the suggestion that S constitutes evidence for DD. The probability of a diagnosis is absolute with respect to K_i but is conditional with respect to "G_i given S". The preceding discussion emphasizes how essential the knowledge subscript (G, S, K) is which a physician assigns to the probability symbol P and this subscript is intimately linked to the assertion form. Often this subscript is neglected, which could cause misunderstandings in medical decision-making and health risk assessment (e.g., one may confuse metalanguage probability with object language probability). This is because, *inter alia*, whether S constitutes evidence for a case and to what extent it depends on G. If G in Eq. (3.11) is objective knowledge, one talks about S as an objective notion of evidence. If G is subjective knowledge, one assigns to S a subjective notion of evidence.

Interestingly, as we will see later, having more therapy options and processing more clinical data does not necessarily produce fewer violations of logic and probability theory or leads to better medical judgments. Equally or more important than data availability is the soundness of the physician's thinking mode and the theory of knowledge (also Section 3.1.3) he/she relies upon. The contribution of an adequate theory of knowledge is to assure the scientific and epistemic adequacy of assertions about a case (concerning symptoms, diagnoses, therapies) and, on occasion, to make sense of rather than ignore counterintuitive findings, paradoxes and anomalies. On the other hand, the quality of decision-making is improved when care professionals are confronted with challenging yet insightful and penetrating case studies. These studies illuminate basic concepts and precepts, on occasion reveal new avenues for research or theory development, and have the potential to broaden and deepen knowledge and understanding in a way that might not be available in terms of trivial cases.

Some readers may cite the view that the varying medical interventions (diagnostic, therapeutic) in terms of a_A and $P[a_A]$ that are suggested by different physicians about the same case have a lot to do with the physician's own developmental stage. The clinical behavior of novice physicians is rather limited and inflexible; they make decisions on a slower pace, putting reflection before acting, and are more error prone. As more experience is gained, the interventions of competent professionals are faster, more confident and usually more accurate. Proficient physicians have advanced the ability to recall what they have learned (based on previous encounters with similar patient populations), and to comprehend a case spontaneously and as a whole. Expert physicians apply forward reasoning and take advantage of their superior skills, experience and rich knowledge base to make fast, intuitive and fluent decisions.

3.3. Probabilities of Medical Conditionals

As we saw in Section 2.6.1, the "if–then" argument commonly encountered in professional practice can be interpreted in terms of the substantive conditional, $CA_2 \backslash CA_1 \vdots_{KB}$ (i.e., as a conditional event). Accordingly, it is interesting to derive the mathematical expressions of the probabilities of certain forms of the conditional introduced in Section 2.6.1. These expressions are very useful in decision-making pertaining to medical research, clinical practice and survey data analysis.

3.3.1. *Standard logical relations and inference rules*

For the benefit of the readers, we will start with a review of some rather elementary results of standard logic. In particular, Table 2.5 presented a list of logical relationships between case attributes (CA) that are generally valid in medical decision-making. The list is of considerable value in the development of a health care provider's thinking style. Recall that Table 2.10 presented some of the most common inference rules in current use. A link between Tables 2.5 and 2.10 is the basic knowledge condition ($k_A CA \therefore CA$). Also, concerning the statistical conditional $CA_2|CA_1$ the readers may recall that, for mainstream decision-making purposes, it can be viewed as a mathematical entity with basically three values: T (when both CA_1 and CA_2 are valid), F (when CA_1 is valid and CA_2 is invalid), and IN (indeterminate, when CA_1 is invalid). When used properly in real clinical practice and exposure conditions, the medical inference rules of Table 2.10 aided by the logical relations of Table 2.5 can offer critical analyses of the underlying issues, identify and challenge case assumptions, present synthetic discussions and reach interpretive conclusions. The example below may help the readers to gain some insight regarding the implementation of the above.

Example 3.17: At the Hospital Kuala Lumpur (HKL) a health care manager needs to assess the appropriateness of the decision-making process leading to Ms Yeoh's transfer to the intensive care unit (ICU). It is observed that Ms Yeoh's vital signs[12] are not within normal limits today ($\neg VS$), and she also feels more uncomfortable than usual (denoted as UC). Medical guidelines require that a HKL physician will perform the surgery (PS) only if Ms Yeoh's VS are within normal limits. If the surgery is not performed, Ms Yeoh needs close monitoring (CM) and support to maintain normal bodily functions, in which case she must be sent to the ICU today. The HKL care manager can check the validity of the above argumentation by means of an appropriate combination of the logical relationships and inference rules presented in Table 2.5 and 2.10. The validity of the HKL's argumentation mode, leading to Ms Yeoh's transfer to the ICU, is explained step-by-step in Table 3.2. Note that different premises are used at each distinct step of the decision-making process.

As a matter of fact, in a decision-making process such as that of Table 3.2, there is the desire to retain, as far as possible, substantive interpretations of clinical observations and health risk assessments. Ideally, a HKL care professional is trying to develop reasoning in such a way

---

[12] Clinical measurements, specifically pulse rate, temperature, respiration rate and blood pressure, that indicate the state of a patient.

Table 3.2: *HKL*'s argumentation process.

Step	Inference	Explanation
1	$(\neg VS) \wedge UC$	Observations
2	$\neg VS$	From *step* 1 by "simplification" relation (L_{29} of Table 2.5; or I_{13} of Table 2.10)
3	$PS \to VS$	Premise
4	$\neg PS$	From *steps* 2–3 by "*modus tollens*".
5	$(\neg PS) \to CM$	Premise
6	CM	From *steps* 4–5 by "*modus ponens*".
7	$CM \to ICU$	Premise
8	ICU	From *steps* 6–7 by "*modus ponens*".

that the sound justification of case assertions embodies content-dependent (*CD*) arguments (scientific principles, empirical evidence etc.) rather than content-independent (*CI*) ones.[13]

Typically, the validity of a medical inference or hypothesis is demonstrated by translating them into the formal language of mathematical logic and showing that the conclusion can be deduced from the premises in the corresponding proof system. As has become customary in this book, another word of warning is appropriate at this point. Rigorous medical decision-making requires that health care professionals formulate a series of hypotheses on the basis of the input they obtain from their patients (signs, interviews, examinations, symptoms, observations). The validity of these hypotheses should be regarded with skepticism as tests are chosen and performed to develop each one of them. The professionals must bear in mind the unavoidable test inaccuracies and errors, as well as the biases associated with the interpretation of the test results. Actually, it is not only errors due to lack of knowledge or omission that lead to diagnostic failure, but also errors in judgment or interpretation applied to the hypotheses the professionals have created.

3.3.2. Choosing a conditional probability form

In probability theory, inferences are frequently justified by developing a model of the medical case under investigation, and the relevant propositions are subsequently analyzed in the context of the model. Medical conditionals

---

[13] In Chapter 4 we will see that this line of argumentation constitutes a partial answer to the question: why would one want to reinterpret mainstream logic in a substantive way?

(Section 2.6.1) are essential components of this model. Specifically, *material* conditional, *equivalence* conditional (or biconditional) and *statistical* conditional probabilities are defined as, respectively,

$$P_{KB}[CA_1 \to CA_2] = P_{KB}[CA_2|CA_1]P_{KB}[CA_1] + P_{KB}[\neg CA_1] \quad (3.12)$$

$$P_{KB}[CA_1 \leftrightarrow CA_2] = 2P_{KB}[CA_2|CA_1]P_{KB}[CA_1]$$
$$+ P_{KB}[\neg CA_1] - P_{KB}[CA_2] \quad (3.13)$$

$$P_{KB}[CA_2|CA_1] = P_{KB}[CA_1 \wedge CA_2]P_{KB}^{-1}[CA_1]. \quad (3.14)$$

It is worth noticing that the above conditionals account for the fundamental principle of clinical medicine that the interpretation of new medical information depends on what the physician believed beforehand. One way to look at these equations is as expressing the strength of the case probabilities $P_{KB}[CA_1 \to CA_2]$, $P_{KB}[CA_1 \leftrightarrow CA_2]$ and $P_{KB}[CA_2|CA_1]$. Equation (3.12) is also expressed as

$$P_{KB}[CA_1 \to CA_2] = P_{KB}[\neg(CA_1 \wedge \neg CA_2)],$$

which is also a useful formula in clinical decision-making. The probability of the equivalence conditional, Eq. (3.13), provides an intuitive measure of relatedness, or logical correlation, between two case attributes. It measures the degree to which two case attributes are simultaneously true or simultaneously false. Equation (3.14) is linked to the well-known *Bayes* rule of clinical diagnosis on the basis of which one can calculate how much new information (say, a new diagnostic test) should change the prior (or pre-test) probability of the diagnosis and obtain its posterior (or post-test) probability. A Bayesian approach has certain advantages. Firstly, it seems to be the natural choice for decision-making and prediction, as it focuses on the probability that the case parameters have certain values given the data, rather than the probability of the data given particular values of case parameters. Secondly, the Bayesian approach allows one to include not only the information in the data (likelihood), but also other (pretest) information one may have about the parameters, derived from expert opinion or from previous studies of similar diseases. On the other hand, the Bayesian approach in its classical formulation has certain limitation as well, such as a probability may be difficult to define prior to any observation, probability raising is sometimes confused with confirmation, the conditional probability of a test result is independent of the prior test result as well as of the prior disease probability, overreliance on *a priori* assessment, the reversibility of the Bayesian rule may not be meaningful on biophysical grounds (Christakos, 2010).

In any case, an important thing to keep in mind is that the statistical conditional (3.14) is not the only choice clinicians have when they need to know how much to adjust an initial probability of disease diagnosis when additional information becomes available. As a matter of fact, there are cases in which approaches based on the conditionals (3.12)–(3.13) may be more appropriate choices providing better estimates of how much a physician's uncertainty about a patient's actual state could change. Just like Eq. (3.14), Eqs. (3.12)–(3.13) can also be used, for example to design clinical trials, to calculate the probability that a patient with chest pain has significant coronary artery narrowing given an abnormal exercise electrocardiogram, to judge whether to admit a patient suspected of having an acute myocardial infarction, to develop compiled strategies for diagnosing and treating pulmonary emboli, or to estimate how likely it is that a patient has cancer after a test is negative. For numerical illustration, as we shall see later in Example 3.61 Mr. O'Rorke has a pre-test probability of *lung cancer* (LC), $P_{KB}[LC] = 0.40$ (estimated in published studies). The chest radiogram of the patient shows a lung mass (LM) with probability $P_{KB}[LM] = 0.26$. Then, statistical conditional (post-test) probability of the patient having lung cancer (LC) was found to be $P_{KB}[LC|LM] = 0.909$. Care providers can go even further and, using Eqs. (3.12)–(3.13), they can also find that the material and equivalence conditional (post-test) probabilities are, respectively, $P_{KB}[LM \to LC] = 0.976$ and $P_{KB}[LM \leftrightarrow LC] = 0.816$.

> "Which one of the three post-test probabilities is the most appropriate interpretation in the particular case?"

the readers may wonder. Certain possibilities could be examined: the first probability (0.909) is a purely statistical result (the ratio of the number of cases where both LM and LC occur over the total number of LC cases), the second probability (0.976) refers to the presence of LM logically implying the occurrence of LC, and the third probability (0.816) refers to the presence of LM if and only if LC is the case, which has the smallest probability because it is the most restrictive case (i.e., it requires that both LM implies LC and LC implies LM). From stochastic theory it is valid that $P_{KB}[LM \to LC] \geq P_{KB}[LC|LM], P_{KB}[LM \leftrightarrow LC]$; but $P_{KB}[LC|LM]$ may be larger or smaller than $P_{KB}[LM \leftrightarrow LC]$, depending on the particular case conditions.

Despite their considerable value in decision-making, the nature of probability calculus is such that, if they are used uncritically, the probability conditionals may lead to some paradoxical results, which the care providers ought to be able to interpret adequately. Here is an example.

Example 3.18: Let DS denote a highly unlikely disease symptom and DD a disease diagnosis. In this case the care provider knows that $P_{KB}[\neg DS] \gg$, i.e. the probability of the lack of symptom is very high, and Eq. (3.12) yields,

$$P_{KB}[DS \to DD] = a + P_{KB}[\neg DS] \gg$$

with $a \geq 0$.[14] Consequently, from a very unlikely symptom one can draw any conclusion about DD with a high degree of probability. Not being aware of this possible paradox means a physician's reasoning can generate misleading case conclusions.

Complications may arise due to the fact that conversational statements are in many cases characterized by *lexical ambiguity*, which can arise when a word has more than one possible meaning. As a result, it is possible that different results are obtained when different probabilistic formulations are used for the same conversational statement.

Example 3.19: Let ED and ET denote the *Ebola disease* and an *Ebola test*, respectively. In clinical practice ambiguity can arise regarding questions like:

(1) Which one of the two following probabilities

$P_{KB}[ET|ED]$: "ET detects ED given that ED is present" or

$P_{KB}[ED|ET]$: "ED is present given that ET detects ED"

should be used to represent the accuracy of ET in detecting ED?
(2) Should the probability that ET reported a negative result although the patient does not have ED be denoted as

$P_{KB}[\neg ET|\neg ED]$ or as $P_{KB}[\neg ED|\neg ET]$?

---

[14]Here, we use $P_{KB}[DS \to DD]$ to specify the strength of the conditional "if DS then DD".

Note that the conditional probabilities $P_{KB}[ET|ED]$ and $P_{KB}[\neg ET|\neg ED]$ are also called the test *sensitivity* and *specificity*, respectively. In the case context, physicians ideally seek medical tests with the highest sensitivity and specificity values. However, this may not always be feasible, due to the high cost of the tests or their undesirable side effects. In practice, clinicians usually choose an expensive test if the clinical value of the post-test probability of the disease increases considerably compared to the pre-test one. Wrong answers to such questions concerning the appropriate association between a conversational statement and its probabilistic formulation can have profound consequences in medical decision-making. As we shall see in Example 3.22 below, the probability of the conditional $ET|ED$ is 99.99% but that of $ED|ET$ is only 0.02%. So, depending on the interpretation in question (1) above, very different case conclusions are reached.

Some relevant probability theorems and relations of significant value to health care professionals are listed in Table 3.3. Among other things, the readers may notice that statistical and material conditional probabilities share the same order relations. Many probability formulas can be directly related to Bayes theorem, because the latter is essentially a tautology. Indeed, one can start from the obvious equality $\frac{a}{a} = \frac{b}{b}$, with $a, b \neq 0$, which clearly holds for $a = P_{KB}[CA_1]$ and $b = P_{KB}[CA_2]$, i.e., $\frac{P_{KB}[CA_1]}{P_{KB}[CA_1]} = \frac{P_{KB}[CA_2]}{P_{KB}[CA_2]}$; this tautology implies

Table 3.3: Fundamental relations of probability conditionals of *SMR*.

Total probability theorem	$\begin{cases} P_{KB}[CA_2] = P_{KB}[CA_2|CA_1]P_{KB}[CA_1] + P_{KB}[CA_2|\neg CA_1]P_{KB}[\neg CA_1] \\ \quad = P_{KB}[CA_1 \to CA_2] + P_{KB}[\neg CA_1 \to CA_2] \\ \quad - P_{KB}[\neg CA_1] - P_{KB}[CA_1] \end{cases}$
Bayes theorem	$\begin{cases} P_{KB}[CA_2|CA_1] = P_{KB}[CA_1|CA_2]P_{KB}[CA_2]P_{KB}^{-1}[CA_1] \\ P_{KB}[CA_1 \to CA_2] = P_{KB}[CA_2 \to CA_1] + P_{KB}[CA_2] - P_{KB}[CA_1] \end{cases}$
Probability relations	$\begin{cases} P_{KB}[CA_1 \leftrightarrow CA_2] = 2P_{KB}[CA_1 \to CA_2] - P_{KB}[\neg CA_1] - P_{KB}[CA_2] \\ \quad = 2P_{KB}[CA_2|CA_1]P_{KB}[CA_1] \\ \quad\quad + P_{KB}[\neg CA_1] - P_{KB}[CA_2] \\ \quad = P_{KB}[(CA_1 \wedge CA_2) \vee (\neg CA_1 \wedge \neg CA_2)] \end{cases}$
Order relations	$\begin{cases} P_{KB}[CA_2|CA_1] > P_{KB}[CA_3|CA_1] \\ \textit{iff} \\ P_{KB}[CA_1 \to CA_2] > P_{KB}[CA_1 \to CA_3] \end{cases}$

Table 3.4: A comparison of probability conditionals.

$P_{KB}[CA \to CA] = 1$	$P_{KB}[CA\|CA] = 1$	$P_{KB}[CA \leftrightarrow CA] = 1$
$P_{KB}[CA \to \neg CA] = P_{KB}[\neg CA]$	$P_{KB}[\neg CA\|CA] = 0$	$P_{KB}[CA \leftrightarrow \neg CA] = 0$
$P_{KB}[\neg CA \to CA] = P_{KB}[CA]$	$P_{KB}[CA\|\neg CA] = 0$	$P_{KB}[\neg CA \leftrightarrow CA] = 0$
$P_{KB}[\varnothing \to CA] = 1$	$P_{KB}[CA\|\varnothing] = 0/0$	$P_{KB}[\varnothing \leftrightarrow CA] = P_{KB}[\neg CA]$
$P_{KB}[CA \to \varnothing] = P_{KB}[\neg CA]$	$P_{KB}[\varnothing\|CA] = 0$	$P_{KB}[CA \leftrightarrow \varnothing] = P_{KB}[\neg CA]$
$P_{KB}[\Omega \to CA] = P_{KB}[CA]$	$P_{KB}[CA\|\Omega] = P_{KB}[CA]$	$P_{KB}[\Omega \leftrightarrow CA] = P_{KB}[CA]$
$P_{KB}[CA \to \Omega] = 1$	$P_{KB}[\Omega\|CA] = 1$	$P_{KB}[CA \leftrightarrow \Omega] = P_{KB}[CA]$

that $P_{KB}[CA_1 \wedge CA_2]\frac{P_{KB}[CA_2]}{P_{KB}[CA_2]} = P_{KB}[CA_1 \wedge CA_2]\frac{P_{KB}[CA_1]}{P_{KB}[CA_1]}$; or that $P_{KB}[CA_1|CA_2]P_{KB}[CA_2] = P_{KB}[CA_2|CA_1]P_{KB}[CA_1]$, which is, of course, Bayes theorem. Moreover, Table 3.4 presents an illustrative comparison of the three main kinds of probability conditionals in certain special cases. There are cases in which all three probabilities lead to the same result, as well as cases in which the results differ substantially. Remarkably, the case probabilities of the material vs. equivalence conditionals yield similar results in a larger number of cases than do the probabilities of the statistical vs. material and the statistical vs. equivalence conditionals. Only in two cases do the probabilities of all three conditionals coincide.

Experienced care providers know that deciding the optimal probabilistic formulation of conversational statements routinely made in everyday practice is by no means a straightforward affair. One can find in the literature different perspectives concerning the appropriate type of conditional to be used in a case. For example, studies in classical logic, decision theory and artificial intelligence endorse the material conditional probability (Fulda, 1989; Tidman and Kahane, 1999; Nguyen et al., 2002; Layman, 2005; Hurley, 2012). On the other hand, certain psychological studies on "if–then" suggest that statistical conditional probability would offer the best means for human understanding of "if–then" (Liu et al., 1996; Evans et al., 2003; Over and Evans, 2003). Accordingly, there is not a single conditional form that applies in all medical cases in professional practice. Instead, professionals need to decide which one of the two better represents their understanding of "if–then" for the specified case. Furthermore, as the readers might have guessed, the numerical probabilities of the various conditional forms can differ substantially.

Example 3.20: Physicians A and B examine the same clinical case. Suppose that A makes the disease diagnosis $DD^{(1)}$, whereas B makes the diagnosis $DD^{(2)}$. Medically, the $DD^{(1)}$ and $DD^{(2)}$ are mutually exclusive ($DD^{(1)} \wedge DD^{(2)} = \emptyset$) in this case. Also, let $P_{KB}[DD^{(1)}] = a$ and $P_{KB}[DD^{(2)}] = b$, where $a, b \in [0, 1]$. The probability of the statistical conditional is

$$P_{KB}[DD^{(2)}|DD^{(1)}] = P_{KB}[DD^{(1)} \wedge DD^{(2)}]P_{KB}^{-1}[DD^{(1)}] = 0 \times a^{-1} = 0,$$

which is a sensical result.[15] On the other hand, the material conditional probability is

$$P_{KB}[DD^{(1)} \rightarrow DD^{(2)}] = P_{KB}[DD^{(2)}|DD^{(1)}]P_{KB}[DD^{(1)}]$$
$$+ P_{KB}[\neg DD^{(1)}] = 0 \times a + 1 - a = 1 - a,$$

which does not make sense in the present case (given the medical implausibility of the relation "if $DD^{(1)}$ then $DD^{(2)}$"), unless $a = 1$, i.e., $DD^{(1)}$ is the case for sure ($DD^{(2)}$ cannot happen). Note that both probability conditionals above are independent of b, the probability of diagnosis $DD^{(2)}$. Lastly, the equivalence conditional probability is

$$P_{KB}[DD^{(1)} \leftrightarrow DD^{(2)}] = 2P_{KB}[DD^{(2)}|DD^{(1)}]P_{KB}[DD^{(1)}]$$
$$+ P_{KB}[\neg DD^{(1)}] - P_{KB}[DD^{(2)}]$$
$$= 2 \times 0 \times a + 1 - a - b = 1 - a - b,$$

which is zero when $a + b = 1$ ($DD^{(1)}$ and $DD^{(2)}$ are the only two possibilities); or when $a = 1$ and $b = 0$ (see case of material conditional probability earlier), or $a = 0$ and $b = 1$ (i.e., $DD^{(1)}$ has zero chance, which is rather meaningless).

The following is another example in which the material and equivalence conditional probabilities seem to be more appropriate choices than the statistical conditional probability in the testing of medical hypotheses and theories that are *a priori* highly unlikely (they may contradict established viewpoints etc.).

Example 3.21: Recall the case presented in Example 2.11. Until the early 1980s the dominant view in gastroenterology was that

"peptic ulcer (stomach or duodenal) was due to excess acidity or stress".

---

[15] In other words, $P_{KB}[DD^{(2)}|DD^{(1)}]$ is determined only if $DD^{(1)}$ is valid. But if $DD^{(1)}$ is true, the $DD^{(2)}$ is false (since "if $DD^{(1)}$ then $DD^{(2)}$" is invalid); hence $PB_{KB}[DD^{(2)}|DD^{(1)}] = 0$.

As a result, peptic ulcer was seen as a permanent condition and the patients would spend many years on antacids. Then, in 1983 the physicians Marshall and Warren suggested a new medical hypothesis (Marshall and Warren, 1983) (let us denote it as MW):

"peptic ulcer is not a permanent condition but a bacterial infection",

which led to the prediction that

"a patient can be cured with the help of antibiotics"

(denote it as CA). In probability terms, according to the new hypothesis, $P_{KB}[CA|MW] \approx 1$. Most experts dismissed the MW hypothesis as ludicrous, assigning to it zero probability of being valid, $P_{KB}[MW] \approx 0$, and, similarly, they considered CA as highly unlikely *a priori*, $P_{KB}[CA] = p \ll$. Yet, as we also saw in Example 2.11 it was eventually shown that the MW theory is valid and its prediction CA is correct. Remarkably, the statistical conditional gives a zero chance to the validity of the MW theory, i.e.,

$$P_{KB}[MW|CA] = P_{KB}[CA|MW]P_{KB}[MW]P_{KB}^{-1}[CA]$$
$$= 1 \times 0 \times p^{-1} = 0.$$

The equivalence conditional probability, on the other hand, allows the possibility that the new theory is valid:

$$P_{KB}[CA \leftrightarrow MW] = 2P_{KB}[MW|CA]P[CA] + 1 - P_{KB}[MW]$$
$$- P_{KB}[CA] \approx 2 \times 0 \times p + 1 - 0 - p - 0 = 1 - p > 0$$

(as the readers notice, since $p \ll$, the probability $1-p$ is considerable.). Similar is the result obtained in terms of the material conditional probability,

$$P_{KB}[CA \to MW] = P_{KB}[MW|CA]P[CA] + P_{KB}[\neg CA]$$
$$\approx 0 \times p + 1 - p = 1 - p.$$

The material conditional may be appropriate in *pathophysiological* (causal) reasoning that establishes cause–effect relations between clinical attributes within anatomical, physiological, biochemical, and ultimately, genetics-based representations of reality (Miller and Geissbuhler, 2009; Sox *et al.*, 2013). A pathophysiological derangement, PD_1 causes the expected clinical manifestations, CM, as inferred by logical deduction ($\to$), which means that an appropriate interpretation of the posterior probability is

$P_{KB}[PD_1 \to CM]$ rather than the statistical conditional $P_{KB}[CM|PD_1]$. In practice, in between PD_1 and CM other pathophysiological derangements PD_2 etc. may be involved (i.e., PD_1 causes PD_2 as inferred by $\to$, which, in turn, yields another derangement etc., until eventually CM occurs). Reasoning backward along this logical chain may lead to the cause of the disease. On the other hand, the statistical conditional (|) is useful when the clinicians use the associative reasoning mode (pattern recognition; Kulikowski, 2000), which establishes empirical associations between certain symptoms, clinical findings and complications and a pattern of disease manifestation (see, also, Bayesian statistics). Lastly, let us examine the issue of *medical testing*. No disease test is perfect, which is why if a lab test concerning a particular disease is performed on a patient, there is a chance (hopefully small) that the test will return a positive result, termed a *false positive*, even if the patient does not have the disease. The problem lies, however, not just in the chance of a false positive prior to testing, but determining the chance that a positive result is in fact a false positive. When *Bayesian* inference is assumed as in the example below, if a health condition is very rare the majority of positive results may be false positives, even if the test for that condition is (otherwise) reasonably accurate.

Example 3.22: As above, let ED denote Ebola disease, and suppose that the hospital physicians have an accurate Ebola test (ET) to their possession. The premises of the Ebola case include the probabilities shown in Table 3.5, in particular, the conditional probabilities of the ET being accurate given the presence or absence of the disease. Suppose that the ET is administered to a patient in the US, where the base-rate chance $P_{KB}[ED]$ of someone having Ebola is very small (see, also, third premise of Table 3.5). The ET of Mr. Balourdos reports positive for Ebola and his physician wants to estimate the conditional probability that Mr. Balourdos actually has the disease, $P_{KB}[ED|ET]$. The inference leading to the calculation of this conditional probability involves the Bayes and total probability rules, and is also outlined in Table 3.5. The required probability is found to be $P_{KB}[ED|ET] = 0.02\%$, i.e., very small (compare it with $P_{KB}[ET|ED] = 99.99\%$). The conditional probability of Mr. Balourdos having the disease given that the ET reported negative is $P_{KB}[ED|\neg ET] = 0.20 \times 10^{-7}\%$, i.e., negligible. The probability of a false positive is

$$P_{KB}[\neg ED|ET] \approx 1 - 0.02\% \approx 99.98\%.$$

Table 3.5: Probabilistic medical inference in the ebola case.

Case premises	$P_{KB}[ET\|ED] = 99.99\%$ $P_{KB}[ET\|\neg ED] = 0.5\%$ $P_{KB}[ED] = 0.0001\%$
Physician's intermediate conclusion	$\therefore P_{KB}[ET]$ $\stackrel{Total\ rule}{=} P_{KB}[ET\|ED]P_{KB}[ED] + P_{KB}[ET\|\neg ED]P_{KB}[\neg ED]$ $\approx 0.5\%$
Physician's final conclusion	$\therefore P_{KB}[ED\|ET] \stackrel{Bayes\ rule}{=} P_{KB}[ET\|ED]\dfrac{P_{KB}[ED]}{P_{KB}[ET]}$ $\approx 0.02\%$

Table 3.6: Sensitivity analysis of the Bayesian ebola inference ($P_{KB}[ET|ED] = 99.99\%$).

$P_{KB}[ET\|\neg ED]$ (%)	$P_{KB}[ED]$ (%)	$P_{KB}[ET]$ (%)	$P_{KB}[ED\|ET]$ (%)	$P_{KB}[\neg ED\|ET]$ (%)
0.5	0.0001	0.5	0.02	99.98
0.1	0.0001	0.1	0.1	99.9
0.1	0.1	0.2	50	50
5	0.1	5	2	98

The percentage of Ebola patients who test positive ($P_{KB}[ED|ET] \approx 0.02\%$) is 200 times the percentage before we knew the outcome of the test, i.e., $0.02 = 200 \times 0.0001$, or

$$P_{KB}[ED|ET] = 200\, P_{KB}[ED].$$

Despite the apparent high accuracy of ET, the disease incidence ED is so low (1 in 100,000, $P_{KB}[ED] = 0.0001\%$) that the vast majority of patients who test positive do not have the disease ($P_{KB}[\neg ED|ET] \approx 99.98\%$). Yet, some investigators argue that the case sensitivity analysis presented in Table 3.6 shows that the test is not useless. Re-testing may on occasion improve the reliability of the result. In particular, a test must be very reliable in reporting a negative result when the patient does not have the disease (i.e., $P_{KB}[ET|\neg ED]$ must be very small), if it is to avoid the problem of false positives. The test did well here since all $P_{KB}[ET|\neg ED]$ values are very small (see Table 3.6). If it is assumed that the base-rate chance is $P_{KB}[ED] = 0.1\%$ (i.e., much higher than the base-rate chance $P_{KB}[ED] = 0.0001\%$ assumed earlier, also in Table 3.6), and the test reported a negative result in patients without the disease with probability $P_{KB}[\neg ET|\neg ED] = 99.9\%$, then, the calculation yields a probability of a false positive of roughly $P_{KB}[\neg ED|ET] \approx 50\%$ (i.e., the usefulness of the test has been improved considerably). Clearly, the probability of a false positive is affected predominately by the base-rate chance of someone having the disease.

3.3.3. Stochastic truth tables: A second look

In view of the probability concept, medical thinking replaces the classical truth table (CTT) of mainstream logic (Section 2.4.1) with a *stochastic truth table* (STT). As a consequence, the strictly two-valued CTT is replaced by the more realistic, continuum-valued STT ("true" and "false" of CTT are replaced by "probable" and "improbable" of STT).

As noted earlier, a key difference between a formal connective, e.g. $(\cdot \rightarrow \cdot)$, and its substantive formulation, $(\cdot \rightarrow \cdot)\!:_{KB}$, is that the formal is CI, whereas the substantive is CD. Moreover, with the introduction of probability, the substantive formulation also accounts for the realistic uncertainties characterizing a medical case. In Table 3.7, the TVs of the case attributes CA_1 and CA_2 are the *key arguments* on the basis of which the TVs of all other composite arguments are derived.

For quantitative analysis purposes, the TVs, T for truth and F for false in CTT, are usually represented with the numbers 1 and 0, respectively. The TV of $CA_2|CA_1$ is indeterminate (IN) when the case attribute CA_1 is invalid ($TV[CA_1] = 0$). In other words, the IN in Table 3.7 should not be interpreted as meaning that there exists no TV, but rather that $CA_2|CA_1$ is an expression that is both valid and invalid when $TV[CA_1] = 0$, i.e. it is not an assertion and should not be involved in medical reasoning. An example will allow us to comment on some interesting points of the CTT.

Example 3.23: Suppose that $CA_1 \equiv DS$ and $CA_2 \equiv DD$, denote disease symptom and diagnosis, respectively. In view of Table 3.7 some instances

Table 3.7: A classical truth table (CTT); "IN" denotes indeterminate.

CA_1	CA_2	$CA_1 \wedge CA_2$	$CA_1 \vee CA_2$	$CA_1 \rightarrow CA_2$	$CA_1 \leftrightarrow CA_2$	$CA_2|CA_1$
1	1	1	1	1	1	1
1	0	0	1	0	0	0
0	1	0	1	1	0	IN
0	0	0	0	1	1	IN

are of interest. Assume that the symptom is not present, i.e., $\neg DS$ with $TV[\neg DS] = 1$ but the diagnosis occurs, i.e., $TV[DD] = 1$ (3rd row of Table 3.7). In this case one finds that:

(a) $TV[DS \to DD] = 1$, i.e., DS absence (still) implies DD occurrence;
(b) $TV[DS \leftrightarrow DD] = 0$, i.e., DS absence does not imply DD occurrence;
(c) $TV[DD|DS] = IN$, i.e., DS absence does not allow the statistical conditional to be involved in the physician A's reasoning about the case.

In Table 3.8, on the other hand, the TVs are replaced by the STVs that are equal to the corresponding probabilities. That is, unlike CTT, which features two key TVs ($TV[DS]$ and $TV[DD]$), the STT requires three key STVs:

$$STV[DS] = P[DS] = p, \ STV[DD] = P[DD] = q \text{ and}$$
$$STV[DS \wedge DD] = P[DS \wedge DD] = r.$$

These probability values must be selected so that

$$p, q, r \in [0, 1], \text{ and } p + q - 1 \leq r \leq q, p.$$

An alternative situation is described in the lower part of Table 3.8, in which case the probabilities

$$STV[DS] = p, \quad STV[DD] = q, \quad STV[DD|DS] = P[DD|DS] = r$$

are selected so that

$$p, q, r \in [0, 1], \text{ and } \frac{p + q - 1}{p} \leq r \leq \frac{q}{p},$$

using probability theory. Interestingly, several formulas can be derived from the first three arguments of Table 3.8 that express various kinds of relations between case probabilities.

The readers should notice that Table 3.8 contains the most general, so far, forms associated with material, equivalent and statistical conditionals.

Table 3.8: A stochastic truth table (STT).

$CA_1 \vdots KB$	$CA_2 \vdots KB$	$CA_1 \wedge CA_2 \vdots KB$	$CA_1 \vee CA_2 \vdots KB$	$CA_1 \to CA_2 \vdots KB$	$CA_1 \leftrightarrow CA_2 \vdots KB$	$CA_2 | CA_1 \vdots KB$
p	q	r	$p + q - r$	$1 - p + r$	$1 - p - q + 2r$	r/p

$CA_1 \vdots KB$	$CA_2 \vdots KB$	$CA_2 | CA_1 \vdots KB$	$CA_1 \vee CA_2 \vdots KB$	$CA_1 \wedge CA_2 \vdots KB$	$CA_1 \leftrightarrow CA_2 \vdots KB$	$CA_1 \to CA_2 \vdots KB$
p	q	r	$p + q - pr$	pr	$2pr + 1 - p - q$	$pr + 1 - p$

For many decision-makers, an adequate STT should be in accord with the perspective that when physicians reason from uncertain clinical premises, naturally they want their reasoning process to lead to probable medical conclusions.

SMR inference frequently employs the above criterion at every stage of the process as a guiding light to assure that the desired conclusions are derived (e.g., it should not be possible that the SMR premises are probable but its conclusions are impossible).

In essence, Table 3.7 may be seen as a special case of Table 3.8, by restricting the STVs of Table 3.8 to the two values, 0 and 1; in symbolic terms,

$$\left. \begin{array}{c} \text{Table 3.8} \\ STV[\cdot] = P[\cdot] \in [0,1] \end{array} \right\} \xrightarrow{P[\cdot]=0,1} \begin{array}{c} \text{Table 3.7} \\ TV[\cdot] = 0,1 \end{array}. \quad (3.15)$$

For illustration, we will consider a few numerical examples. Let $p = 1$ in Table 3.8, for $q = 1$ the third column gives $r = 1$, and for $q = 0$ it gives $r = 0$, which are the corresponding TVs in Table 3.7. Next, let $p = 0$ in Table 3.8, for $q = 0$ or 1 the third column gives $r = 0$. The proofs for the fourth to sixth columns are similar. It is worth noticing the case of the seventh column. If $p = 1$ in Table 3.8, then $r/p = q$ (note that the first two rows of the seventh column in Table 3.7 are the same as those of the second column). But when $p = 0$ in Table 3.8, the r/p is indeterminate (see "IN" in last two rows of the seventh column of Table 3.7).

Example 3.24: Interestingly, not only the statistical conditional assumes a meaning in situations like in Table 3.8, but under certain conditions it can do the same in the context of other, more involved, logical expressions. For illustration, Table 3.9 shows that the probability of the conditional $CA_2|CA_1\dot{:}_{KB}$, which is meaningful individually and for $P_{KB}[CA_1] > 0$, remains meaningful in the context of the logical expression $(CA_1 \vee CA_2) \wedge (CA_2|CA_1)\dot{:}_{KB}$.

Table 3.9: STT of $(CA_1 \vee CA_2) \wedge (CA_2|CA_1)\dot{:}_{KB}$.

$CA_1\dot{:}_{KB}$	$CA_2\dot{:}_{KB}$	$CA_2|CA_1\dot{:}_{KB}$	$CA_1 \wedge CA_2\dot{:}_{KB}$	$CA_1 \vee CA_2\dot{:}_{KB}$	$(CA_1 \vee CA_2) \wedge (CA_2|CA_1)\dot{:}_{KB}$
p	q	1	p	q	q

Critical thinking and reflection are necessary components of developing $STTs$ consistent with professional practice. $STTs$ incorporate cognitive processes used to generate an appropriate intervention or enhance patient-centered health care. Moreover, the sort of medical attributes the physician is dealing with determines whether semantics in terms of physical probability P (object language) or mind state probability P_{KB} (metalanguage in terms of rational or subjective belief) will be considered. Noticeably, a physician practicing SMR inference can define the conditional $CA_2|CA_1$ on the basis of the geometrical representation of the corresponding case probabilities. For example, Table 3.7 shows that when the TVs of both CA_1 and CA_2 are 1 (or T), $CA_2|CA_1$ is assigned the truth-value 1 (T); whereas when CA_1 is 1 (T) and is 0 (F), $CA_2|CA_1$ is assigned the value 0 (F). In the other two cases in Table 3.7, the $CA_2|CA_1$ cannot be assigned a value in the stochastic sense, since the ratio in the corresponding geometrical representation is not defined (for details, see Christakos (2010)).

Summarizing our discussion on the subject so far, Table 3.10 presents a rough comparison between formal (or classical) and substantive (i.e., SMR) logic. Formal logic was developed in terms of an idealized language perfectly suitable for mathematics purposes. Certain *formal logic oddities* (FLO) do not concern mathematical thinking. Yet, in a previous section we saw that FLO are hard to tolerate when health care providers encounter empirical assertions (propositions, beliefs, judgments, claims). Substantive CD logic is FLO-free because it focuses on triadic case formulas with realistic assertions (a_A) about medical cases (metalanguage) rather than with cases themselves (object language). In addition, a substantive conditional is dialogically meaningful if the conclusion follows from the medical premises by means of biophysical and epidemiologic laws of nature and scientific understanding, a process that is often denoted by adding the subscript "KB" in medical reasoning operations. This is because in their practice, care

Table 3.10: Formal (classical) vs. substantive (SMR) logics.

Formal logic	Substantive logic
Mathematical reasoning	Everyday medical reasoning
$[\cdot]$ is CI	$(\cdot)\dot{:}_{KB}$ is CD^3
$TV[\cdot] = 0$ or 1	$STV[\cdot] \in [0,1]$
CTT (truth-functional)	STT (probability-functional)

[3]With $(\cdot)\dot{:}_{KB} \sim P_{KB} \equiv Pa_A$.

providers frequently accept or reject assertions with degrees of confidence less than certainty. Unlike formal logic, substantive reasoning incorporates a care provider's degrees of confidence about an *in situ* case and the associated uncertainties in terms of *KB*-based probability.

It is worth noticing that, starting from suitable *STT*s, several new probability relationships can be derived, beyond those listed in Table 3.8. In particular, with the help of *STT*s a valid logical equivalence, $\Phi_1 \equiv \Phi_1$, is established (Φ_1 and Φ_2 are logical functions; see Section 2.3.7), and subsequently the corresponding probabilities, $P_{KB}[\Phi_1] = P_{KB}[\Phi_2]$, are derived; i.e.,

$$\underbrace{[\Phi_1] \stackrel{STT}{\equiv} [\Phi_2]}_{Step\ 1} \Rightarrow \underbrace{P_{KB}[\Phi_1] = P_{KB}[\Phi_2]}_{Step\ 2}. \qquad (3.16)$$

Some examples of the above approach are given next.

Example 3.25: An illustration of the two steps of Eq. (3.16) in a particular case are shown below,

$$\left. \begin{array}{c} (CA_2 \vee \neg CA_2) \wedge CA_3 \stackrel{STT}{\equiv} CA_3 \\ \stackrel{Table\ 2.5}{=} (CA_2 \wedge CA_3) \vee (\neg CA_2 \wedge CA_3) \end{array} \right\}$$

$$\Rightarrow \begin{cases} P_{KB}[CA_3|CA_1] = P_{KB}[(CA_2 \wedge CA_3)|CA_1] \\ \qquad\qquad + P_{KB}[(\neg CA_2 \wedge CA_3)|CA_1], \end{cases}$$

where the first logical equivalence in *step* 1 is derived using *STT* and subsequently the second logical equality of *step* 1 is obtained using the first distribution relationship of Table 2.5. In *step* 2, the corresponding conditional probabilities of the logical formula in *step* 1 are shown to satisfy a useful relationship. Working along similar lines, many more case relationships, like,

$$P_{KB}[(CA_2 \wedge CA_3)|CA_1] = P_{KB}[CA_2|CA_1]P_{KB}[CA_3|(CA_1 \wedge CA_2)],$$

$$P_{KB}[CA_3|CA_1] = P_{KB}[CA_2|CA_1]P_{KB}[CA_3|(CA_1 \wedge CA_2)]$$
$$+ P_{KB}[\neg CA_2|CA_1]P_{KB}[CA_3|(CA_1 \wedge \neg CA_2)]$$

can be shown to be valid.

The preliminary probabilistic calculations of the above examples should prepare our readers for the stochastic medical inferences of Section 3.4.

3.3.4. *More on probability calculation: Is there a probameter?*

A probability statement makes sense only when there is an adequate understanding of the medical case it refers to. There exist several levels of comprehension, some of which are more incomplete than others. Accordingly, the value of a probability statement regarding a case is closely related to the comprehension level (e.g., an expert toxicologist is in a better position to make meaningful probability statements about poisoning problems than an expert podiatrist). Which is why a prime issue in decision-making is to link the theoretical probability notion with clinical reality.

But how can a rational physician measure probability? One can use a thermometer to physically measure body temperature, a high fidelity catheter-tip pressure transducer to measure pulmonary artery pressure, an echocardiogram to measure cardiac stroke volume, blood testing to measure cholesterol levels etc. But physicians do not have a probameter to physically measure probability, and as a result, it is not always obvious what the meaning of a numerical probability value is in medical practice. Indeed, health care providers are constantly reminded that the accurate calculation of the probability of a case assertion depends on their understanding of the particular health context and the ability to translate the assertion in rigorous quantitative terms. Generally, in order to calculate the probability of a clinical case the trained physician may consider two main approaches:

(1) ensuring adequate understanding of the underlying biomedical processes and clinical variables that can affect the case (theoretical approach, pathophysiological reasoning); or
(2) obtaining a satisfactory (quantitatively and qualitatively) experimental data base from which to draw phenomenological information about the case (empirical approach).

The first thing to notice is that the commonly used *frequency* interpretation of probability belongs to *approach* 2. Typically, a health statistician looks back through the medical records and literature to find cases relevant to the clinical case of interest and determine in what fraction of those cases the symptoms of the current case were observed and use this fraction as an initial estimate of the case probability. This practice is sometimes known in medicine as "representativeness heuristic" (Sox *et al.*, 2013). For example, Dr. A's estimate of the probability of cholecystitis in a patient depends on how closely the patient's symptoms fit the standard description of cholecystitis or resemble patients with cholecystitis from A's own clinical experience. Naturally, the patient's probability of cholecystitis is also influenced by the disease prevalence in the parent population considered. Although many physicians feel comfortable with this practice, nevertheless, it does not give them a reasonable estimate of the actual probability value for a number of reasons; for example, the patient's disease probability is influenced by the disease prevalence in the parent population considered, the cues that make up the standard disease description may be imperfect disease indicators, the standard disease predictors typically all occur together so that more than one predictor does not add information, misinterpreting random events as proving the success or failure of a therapeutic trial to test a diagnostic hypothesis, the fraction may fluctuate considerably as more cases are accumulated, and one usually wants to assign probabilities to particular medical cases that in the nature of things cannot be repeated in all their particularity as required by the frequency interpretation. Hence, it is often unclear how frequency probability should be applied to individual clinical cases. Other investigators estimate probabilities by using a comparison process, where the strength of evidence for a case hypothesis is compared with the strength of evidence for a set of alternatives. Support theory (Section 2.1), in particular, estimates the probability of a hypothesis H1 rather than H2 as the ratio of the evidential support for H1 over the sum of the supports for H1 and H2. Rather than focusing on case ratios, many other health care providers focus on logical arguments and take as their point of departure the viewpoint that some arguments are stronger, in varying degrees, than others.

In health care practice it is frequently more appropriate that physicians use a combination of *approaches* 1 and 2, since neither biological understanding nor clinical evidence are usually complete (one encounters many notions that are not defined with adequate precision, medical records may be mishandled due to the varied use physicians make of the same term and different doctors have assigned different levels of strength to the same evidence). Generally, the dynamics of a clinical situation play a key role in the calculation of the patient's probability to develop a specific disease.

Example 3.26: It is rather widely accepted that a smoker has a higher probability of getting *lung cancer* than a non-smoker. Yet, some smokers never get lung cancer, whereas some non-smokers do. The reason is that smoking is not the only factor that can cause lung cancer. Additional factors include age, pre-existing medical condition, genetics and exposure to other toxins (such as oxidants). The more factors a physician understands and incorporates into the medical-decision process, the more accurate the calculation of the case probability value will be.

Example 3.27: Consider the study of the *Black Death* epidemic (Christakos *et al.*, 2005). In order to calculate the probability value of the assertion

> "in Neuberg (Austria) the Black Death epidemic started on September 29, 1348",

the investigators studied all available historical sources, assessed the reliability of each one of them (e.g., whether other credible sources corroborate the reports of the specified source or not) and eliminated some of the sources for reasons of internal inconsistency, before reaching a final conclusion.

The reader may plausibly wonder:

> "In what sense is the numerical probability calculated by the above approaches an adequate assessment of physician's knowledge or belief?"

To answer this question, one can refer to formal vs. substantive *standards*. Obviously, the formal standard is that the probability satisfies certain mathematical conditions (Kolmogorov's axioms; Kolmogorov, 1933; Gnedenko and Khinchin, 2010). The substantive or *CD* standard, on the other hand, is provided by sound science and highly credible empirical evidence. The probability value η may be calculated directly from the relevant law of nature or scientific model M_{CA} obeyed by the case attribute CA_p, as follows

$$P_{KB}[CA_p : M_{CA}, M_0] = \eta, \quad (3.17)$$

where M_0 are case-specific conditions (boundary, initial) associated with the scientific law. That is, Eq. (3.17) calculates the case probability η so that the current objective knowledge is satisfied and the physician possesses it. The readers may appreciate that the substantive standard of the probability value η introduced in Eq. (3.17) differs from earlier standards, e.g., the standard based on *de Finetti's scoring rules* (de Finetti, 1974, 1975; Lindley, 1982). De Finetti's is a subjective approach (in the personal belief sense), in which a rather arbitrary "score function" is chosen (e.g., a quadratic function) that records one's personal assessment regarding case uncertainty, on the basis of which the η is subsequently calculated.

Example 3.28: Let us revisit Example 2.53, where exposure is measured by pollutant concentration PC_p (with possible values in the range 0 to 10 ppm), and DI_p denotes population disease incidence (with values in the range of 0 and 10%). The health investigator is interested about the numerical value η of the probability that the population incidence is 7%; i.e., $\eta = P_{KB}[DI_p = 7\%]$. Taking into account the simple "exposure–health" model of Eq. (2.23) that links PC_p and DI_p, Eq. (3.17) yields

$$P_{KB}[DI_p - 3.5PC_p = 0, M_0] = \eta, \qquad (3.18)$$

with case-specific condition $M_0 : P_{KB}[PC_p] = p_0 e^{-PC_p}$, $p_0 = e^{10}(e^{10} - 1)^{-1}$. Using the techniques of probability theory, Eq. (2.23) leads to

$$P_{KB}[DI_p] = 3.5^{-1} P_{KB}\left[PC_p = \frac{1}{3.5} DI_p\right]$$
$$= [3.5(e^{10} - 1)]^{-1} e^{10 - 3.5^{-1} DI_p}; \qquad (3.19)$$

and the numerical probability of interest is found to be $\eta = P_{KB}[DI_p = 7\%] = [3.5(e^{10} - 1)]^{-1} e^{10 - 3.5^{-1} \times 7} = 0.039$. In other words, the probability that the population incidence is $DI_p = 7\%$, given that the PC_p and DI_p jointly obey the "exposure–health" model of Eq. (2.23) with the case-specific condition M_0 is 3.9%.

When an investigator A practices SMR in a medical case, the KB of the case and the operationally formulated mental functions of the investigator should be integrated in A's reasoning and relevant triadic case formulas. This integration turns out to be of considerable significance in the theory of knowledge (teleology of reason, intentionality or adaptation). On occasion, a health care professional can use this approach in the rigorous calculation of the probability value η. If the professional's cognitive condition includes the intentionality function of maximizing the information state (*InfoState*) concerning a case CA_p, the η could be calculated as follows,

$$\begin{array}{ll} & M_{CA} \\ \text{Case premises} & CA_0 \\ & \max_{P_{KB}}[\textit{InfoState}|M_{CA}, CA_0] \\ \text{Care provider's} & \\ \text{conclusion} & \therefore P_{KB}[CA_p] = \eta \end{array} \qquad (3.20)$$

The *InfoState* is substantive information since its semantics include natural laws or scientific models (M_{CA}, CA_0) that are well-formed and meaningful.

InfoState can be expressed in various forms, such as information entropy, which means that the calculation of the probability value η in Eq. (3.20) involves a space–time synthesis of mental functions and scientific knowledge. Moreover, Eq. (3.20) may be seen as a rational formulation of the mental state of intentionality based on the investigator's cognitive situation (Christakos, 2010). In slightly different terms, given the availability of objective knowledge (natural law, *in situ* conditions), the triadic case formula involves a rational mind state, i.e., $r_A CA_p$, and the corresponding probability is $P[r_A CA_p] = \eta$.[16]

Example 3.29: If one applies approach (3.20) in the case of Example 3.28, one will obtain the same result, $\eta = 3.9\%$. Yet, approach (3.20) is more general than that of Example 3.28, because it can incorporate many more kinds of case-specific data and clinical evidence.

Furthermore, in several studies (e.g., mapping the space–time spread of an epidemic; Chapter 4) the same inference formulation as above can be used to derive the spatiotemporally varying probability densities of the disease maps, on the basis of which crucial predictions can be made and conclusions drawn concerning the handling of the health emergency, the measures that can provide adequate population treatment and the effective intervention that can prevent further spreading of the disease.

When a medical case involves assertions (beliefs, statements, views) with different probabilities, some of them can be calculated based on the physician's cognitive situation, whereas some others are beyond physician's control. This happens because, when considered jointly, the calculated probability values must satisfy certain mathematical rules that determine how the different probabilities must fit together. Indeed, the probability values assigned by an expert physician to different case diagnoses should be used together with the mathematical rules of probability to test the internal consistency of the expert's medical reasoning.

Example 3.30: A 22-year-old female presents herself to Dr. Finamore's office with pain in the right lower quadrant of her abdomen of 12 hours' duration. Her last normal menstrual period was four weeks ago.

---

[16] In the following section we will look at the subject of "mind states–probability assessment" from a different angle.

Dr. Finamore considers possible diagnoses:

$DD^{(1)}$: "the 22-year-old female has *gastroenteritis*"

with $P_{KB}[DD^{(1)}] = \eta_1$;

$DD^{(2)}$: "the 22-year-old female has *ectopic pregnancy*"

with $P_{KB}[DD^{(2)}] = \eta_2$ and

$(\neg DD^{(1)}) \wedge (\neg DD^{(2)})$: "none of the above is valid"

with $P_{KB}[(\neg DD^{(1)}) \wedge (\neg DD^{(2)})] = \eta_3$. In order to test the validity of Dr. Finamore's medical reasoning, he is further asked to assign a value to the probability of another possible diagnosis, say

$(\neg DD^{(1)}) \vee (\neg DD^{(2)})$: "either of the above is not valid."

If Dr. Finamore suggests a value $P_{KB}[(\neg DD^{(1)}) \vee (\neg DD^{(2)})] = \eta_4$ that does not satisfy the relation (addition rule of probability theory)

$$\eta_4 = 2 - (\eta_1 + \eta_2 + \eta_3),$$

then his reasoning is questionable. For numerical illustration, the physician could choose the probabilities $P_{KB}[DD^{(2)}|DD^{(1)}] = 0.6$, $P_{KB}[DD^{(1)}] = 0.5$ or $P_{KB}[DD^{(1)} \wedge DD^{(2)}] = 0.20$ separately, or even in pairs, but not simultaneously. That is, if Dr. Finamore chooses the first two probabilities, then $P_{KB}[DD^{(1)} \wedge DD^{(2)}] = 0.30$ (not 0.20) has to be the case, so that the physician's calculations obey the multiplication rule of probability theory (Table 3.1), $P_{KB}[DD^{(1)} \wedge DD^{(2)}] = P_{KB}[DD^{(1)}]P_{KB}[DD^{(2)}|DD^{(1)}]$.

Before leaving this section, we again stress that sound medical probabilities are the outcome of internally consistent thinking that fuses scientific knowledge, clinical experience and mathematical rigor.

3.4. Stochastic Medical Inferences

As noted earlier, the idea of integrating (*a*) medical observation with (*b*) logical rigor was deeply rooted in Aristotle's thinking, and it also plays a key role in *SMR* inference, where *part a* may include various knowledge bases (core and case-specific), and *part b* may contain distinct systems of logic (formal, substantive, conversational, dialogical). The introduction of the probability notion provides *SMR* with the necessary means to develop further many medical inferences in rigorous uncertainty terms. As mentioned earlier, many of the standard rules of mainstream logic can be reformulated to provide health care professionals with more effective tools of realistic medical and clinical inference in conditions of *in situ* uncertainty. A brief methodological note: many decision-makers argue that while pure epistemic logic is interested only in rational principles of believing, and could therefore be used with completely rational people for descriptive purposes, *stochastic medical inferences* under conditions of real world heterogeneity and clinical uncertainty have to go beyond mere description to explanation and prediction.

3.4.1. *From standard to stochastic syllogisms*

The above reference to stochastic medical inferences is exemplified in Table 3.11, where a number are displayed: CA_1 denotes the antecedent (premise) and CA_2 the consequent (conclusion). Among the most frequently used in clinical reasoning are *modus ponens, modus tollens*, affirming the consequent and denying the antecedent.

These inferences belong to the medical syllogism formulation of (2.16), which can be written in a stochastic reasoning setting as

$$\begin{array}{ll} \text{Case premises} & \begin{array}{l} a_A CA_1 \sim P_1 \\ a_A CA_2 \sim P_2 \\ \vdots \\ a_A CA_m \sim P_m \end{array} \\ \text{Care provider's} & \\ \text{conclusion} & \therefore \Phi[a_A CA_i; i = 1, \ldots, m] \sim P_\Phi \end{array} \quad (3.21)$$

where the premises involve sets of triadic case formulas, $a_A CA_i$ (CA_i, $i = 1, \ldots, m$, are vectors of case attributes, including disease symptoms, possible diagnoses, physical exams and test results), Φ denotes a function (logical, causal, probabilistic) connecting the premises in terms of biophysical laws, scientific theories, probability calculus, and logical connectives, and the Ps denote probability functions. The soundness of stochastic syllogisms of the form (3.21) generally depends on the validity of the underlying logical process but also on the validity of the underlying biomedical mechanisms and the quality of the probability estimates involved in the premises (see also, Section 4.5.1). In clinical practice, the estimation of the probabilities in the premises are influenced by personal experience (clinical findings from previous patients with the same clinical symptoms, signs, or test results), published experience

Table 3.11: Stochastic medical inferences (probabilistic syllogisms).

Stochastic modus ponens

Case premises	$a_A(CA_2\|CA_1)$	$(P_{KB}[CA_2\|CA_1] = p_1)$
	$a_A CA_1$	$(P_{KB}[CA_1] = p_2)$
Care provider's conclusion	$\therefore a_A CA_2$	$(P_{KB}[CA_2] \in [p_1 p_2, p_1 p_2 + 1 - p_2])$

Special cases

$a_A(CA_2\|CA_1)$	(p_1)	$a_A(CA_2\|CA_1)$	(p_1)	$k_A(CA_2\|CA_1)$	$(p_1 = 1)$
$k_A CA_1$	$(p_2 = 1)$	$k_A \neg CA_1$	$(p_2 = 0)$	$a_A \neg CA_1$	$(1 - p_2)$
$\therefore a_A CA_2$	(p_1)	$\therefore a_A CA_2$	$([0,1])$	$\therefore a_A CA_2$	$([p_2, 1])$

$k_A(CA_2\|CA_1)$	$(p_1 = 1)$	$k_A(CA_2\|CA_1)$	$(p_1 = 1)$	
$k_A \neg CA_1$	$(1 - p_2 = 1)$	$k_A CA_1$	$(p_2 = 1)$	$\equiv$ Modus ponens
$\therefore a_A CA_2$	$([0,1])$	$\therefore k_A CA_2$	(1)	

Case premises	$a_A(CA_1 \to CA_2)$	$(P_{KB}[CA_1 \to CA_2] = p_1)$
	$a_A CA_1$	$(P_{KB}[CA_1] = p_2)$
Care provider's conclusion	$\therefore a_A CA_2$	$(P_{KB}[CA_2] \in [\max(0, p_1 + p_2 - 1), p_1])$

Special cases:

$a_A(CA_1 \to CA_2)$	(p_1)	$k_A(CA_1 \to CA_2)$	$(p_1 = 1)$
$k_A CA_1$	$(p_2 = 1)$	$k_A \neg CA_1$	$(1 - p_2 = 1)$
$\therefore a_A CA_2$	(p_1)	$\therefore a_A CA_2$	$([0,1])$

$k_A(CA_1 \to CA_2)$	$(p_1 = 1)$	
$k_A CA_1$	$(p_2 = 1)$	$\equiv$ Modus ponens
$\therefore k_A CA_2$	(1)	

Stochastic modus tollens

Case premises	$a_A(CA_2\|CA_1)$	$(P_{KB}[CA_2\|CA_1] = p)$
	$a_A \neg CA_2$	$(P_{KB}[\neg CA_2] = q)$
Care provider's conclusion	$\therefore a_A \neg CA_1$	$(P_{KB}[\neg CA_1] \in [w, 1])$

$$w = \max\left\{\frac{1-p-q}{1-p}, \frac{p+q-1}{p}\right\}, \quad 0 < p, q < 1$$
$$= 1 - q, \quad p = 0,\ 0 < q \le 1$$
$$= q, \quad p = 1,\ 0 \le q < 1$$

(Continued)

Table 3.11: (*Continued*)

		Special cases:	
$k_A(CA_2\|CA_1)$	$(p_1 = 1)$	$k_A\neg(CA_2\|CA_1)$	$(1-p_1 = 1)$
$k_A CA_2$	$(1 - p_2 = 1)$	$k_A \neg CA_2$	$(p_2 = 1)$
$\therefore a_A \neg CA_1$	$([0, 1])$	$\therefore a_A \neg CA_1$	$([0, 1])$

Case premises	$a_A(CA_1 \to CA_2)$	$(P_{KB}[CA_1 \to CA_2] = p_1)$
	$a_A \neg CA_2$	$(P_{KB}[\neg CA_2] = p_2)$
Care provider's conclusion	$\therefore a_A \neg CA_1$	$(P_{KB}[\neg CA_1] \in [\max(0, p_1 + p_2 - 1), p_1])$

Special cases:

$a_A(CA_1 \to CA_2)$	(p_1)	$k_A(CA_1 \to CA_2)$	$(p_1 = 1)$
$k_A CA_2$	$(1 - p_2 = 1)$	$k_A CA_2$	$(1 - p_2 = 1)$
$\therefore a_A \neg CA_1$	$([0, p_1])$	$\therefore a_A \neg CA_1$	$([0, 1])$

$$\left.\begin{array}{ll} k_A(CA_1 \to CA_2) & (p_1 = 1) \\ k_A \neg CA_2 & (p_2 = 1) \\ \therefore k_A \neg CA_1 & (1) \end{array}\right\} \equiv \text{Modus tollens}$$

$$\left.\begin{array}{ll} k_A \neg(CA_1 \to CA_2) \equiv k_A(CA_1 \wedge \neg CA_2) & (1 - p_1 = 1) \\ k_A \neg CA_2 & (p_2 = 1) \\ \therefore k_A CA_1 & (1) \end{array}\right\} \equiv \text{Reduction}$$

Stochastic affirming the consequent

Case premises	$a_A(CA_2\|CA_1)$	$(P_{KB}[CA_2\|CA_1] = p_1)$
	$a_A CA_2$	$(P_{KB}[CA_2] = p_2)$
Care provider's conclusion	$\therefore a_A CA_1$	$\left(P_{KB}[CA_1] \in \left[0, \min\left\{\dfrac{p_2}{p_1}, \dfrac{1-p_2}{1-p_1}\right\}\right]\right)$

Special cases:

$k_A \neg(CA_2\|CA_1)$	$(1 - p_1 = 1)$	$k_A(CA_2\|CA_1)$	$(p_1 = 1)$
$k_A \neg CA_2$	$(1 - p_2 = 1)$	$k_A CA_2$	$(p_2 = 1)$
$\therefore a_A CA_1$	$([0, 1])$	$\therefore a_A CA_1$	$([0, 1])$

(*Continued*)

Case premises	$a_A(CA_1 \to CA_2)$	$(P_{KB}[CA_1 \to CA_2] = p_1)$
	$a_A CA_2$	$(P_{KB}[CA_2] = p_2)$
Care provider's conclusion	$\therefore a_A CA_1$	$(P_{KB}[CA_1] \in [1 - p_1, \min\{1, 1 - p_1 + p_2\}])$

Special cases:

$k_A(CA_1 \to CA_2)$		$k_A \neg(CA_1 \to CA_2)$	$(1 - p_1 = 1)$	
$k_A CA_2$	$(p_2 = 1)$	$\equiv k_A(CA_1 \land \neg CA_2)$		
$\therefore a_A CA_1$	$([0,1])$	$k_A \neg CA_2$	$(p_2 = 1)$	$\equiv$ Reduction
		$\therefore k_A CA_1$	(1)	

Denying the antecedent

Case premises	$a_A CA_2 \mid CA_1$	$(P_{KB}[CA_2 \mid CA_1] = p_1)$
	$a_A \neg CA_1$	$(P_{KB}[\neg CA_1] = p_2)$
Care provider's conclusion	$\therefore a_A \neg CA_2$	$(P_{KB}[\neg CA_2] \in [1 - p_2 - p_1(1 - p_2), 1 - p_1(1 - p_2)])$

Special cases:

$a_A CA_2 \mid CA_1$	(p_1)	$a_A CA_2 \mid CA_1$	(p_1)	$k_A CA_2 \mid CA_1$	$(p_1 = 1)$
$k_A CA_1$	$(1 - p_2 = 1)$	$k_A \neg CA_1$	$(p_2 = 1)$	$k_A \neg CA_1$	$(p_2 = 1)$
$\therefore a_A \neg CA_2$	$(1 - p_1)$	$\therefore a_A \neg CA_2$	$([0, 1])$	$\therefore a_A \neg CA_2$	$([0, 1])$

Case premises	$a_A(CA_1 \to CA_2)$	$(P_{KB}[CA_1 \to CA_2] = p_1)$
	$a_A \neg CA_1$	$(P_{KB}[\neg CA_1] = p_2)$
Care provider's conclusion	$\therefore a_A \neg CA_2$	$(P_{KB}[1 - p_1, \min\{1 - p_1 + p_2, 1\}])$

Special cases:

$a_A(CA_1 \to CA_2)$	(p_1)	$a_A(CA_1 \to CA_2)$	(p_1)	
$k_A CA_1$	$(1 - p_2 = 1)$	$k_A \neg CA_1$	$(p_2 = 1)$	
$\therefore a_A \neg CA_2$	$(1 - p_1)$	$\therefore a_A \neg CA_2$	$([1 - p_1, 1])$	
$k_A(CA_1 \to CA_2)$	$(p_1 = 1)$		$k_A(CA_1 \to CA_2)$	$(p_1 = 1)$
$k_A CA_1$	$(1 - p_2 = 1)$	$\equiv$ Modus ponens	$k_A \neg CA_1$	$(p_2 = 1)$
$\therefore k_A CA_2$	(1)		$\therefore a_A \neg CA_2$	$([0, 1])$

(Continued)

Table 3.11: (*Continued*)

Stochastic hypotheses		
Case premises	$a_A(CA_2|CA_1)$	$(P_{KB}[CA_2|CA_1] = p_1)$
	$a_A(CA_3|CA_1 \wedge CA_2)$	$(P_{KB}[CA_3|CA_1 \wedge CA_2] = p_2)$
Care provider's conclusion	$\therefore a_A(CA_3|CA_1)$	$(P_{KB}[CA_3|CA_1] \in [p_1 p_2, p_1 p_2 + 1 - p_1])$
Case premises	$a_A(CA_2|CA_1)$	$(P_{KB}[CA_2|CA_1] = p)$
	$k_A(CA_2 \to CA_3)$	$(\therefore CA_2 \to CA_3)$
Care provider's conclusion	$\therefore a_A(CA_3|CA_1)$	$(P_{KB}[CA_3|CA_1] \in [p, 1])$
Stochastic cautious monotonicity		
Case premises	$a_A(CA_2|CA_1)$	$(P_{KB}[CA_2|CA_1] = p_1)$
	$a_A(CA_3|CA_1)$	$(P_{KB}[CA_3|CA_1] = p_2)$
Care provider's conclusion	$\therefore a_A(CA_3|CA_1 \wedge CA_2)$	$\left(P_{KB}[CA_3|CA_1 \wedge CA_2] \in \left[\max\left\{0, \dfrac{p_1+p_2-1}{p_1}\right\}, \min\left\{\dfrac{p_2}{p_1}, 1\right\}\right]\right)$
Stochastic sequential		
Case premises	$a_A(CA_1 \to CA_2)$	$(P_{KB}[CA_1 \to CA_2] = p_1)$
	$a_A(CA_2 \to CA_3)$	$(P_{KB}[CA_2 \to CA_3] = p_2)$
Care provider's conclusion	$\therefore a_A(CA_1 \to CA_3)$	$(P_{KB}[CA_1 \to CA_3] \in [p_1 + p_2 - 1, 1]),$
		$p_1 + p_2 \geq 1$

Special cases:

$$\left. \begin{array}{ll} k_A(CA_1 \to CA_2) & (p_1 = 1) \\ k_A(CA_2 \to CA_3) & (p_2 = 1) \\ \therefore k_A(CA_1 \to CA_3) & (1) \end{array} \right\} \equiv \text{Chain rule}$$

Case premises	$a_A(CA_1 \to CA_2)$	$(P_{KB}[CA_1 \to CA_2] = p)$
Care provider's conclusion	$\therefore a_A(CA_2 \to CA_1)$	$(P_{KB}[CA_2 \to CA_1] \in [1-p, 1])$

Special cases:

$k_A(CA_1 \to CA_2)$	$(p = 1)$	
$\therefore a_A(CA_2 \to CA_1)$	$([0, 1])$	

(*Continued*)

Table 3.11: (Continued)

Stochastic disjunction

Case premises	$a_A\,CA_1$	$(P_{KB}[CA_1] = p_1)$
	$a_A\,CA_2$	$(P_{KB}[CA_2] = p_2)$
	$a_A(CA_1 \wedge CA_2)$	$(P_{KB}[CA_1 \wedge CA_2] = p_3)$
Care provider's conclusion	$\therefore a_A(CA_1 \vee CA_2)$	$(P_{KB}[CA_1 \vee CA_2] = p_1 + p_2 - p_3)$

Special cases:

$$\left.\begin{array}{ll} k_A(CA_1) & (p_1 = 1) \\ k_A(CA_2) & (p_2 = 1) \\ k_A(CA_1 \wedge CA_2) & (p_3 = 1) \\ \therefore k_A(CA_1 \vee CA_2) & (1) \end{array}\right\} \equiv \text{Sequence syllogism}$$

Case premises	$a_A\,CA_1$	$(P_{KB}[CA_1] = p_1)$
	$\ldots$	
	$a_A\,CA_n$	$(P_{KB}[CA_n] = p_n)$
Care provider's conclusion	$\therefore a_A \vee_{i=1}^{n} CA_i$	$\left(P_{KB}[\vee_{i=1}^{n} CA_i] \in \left[\max\{p_i\}, \min\left\{1, \sum_{i=1}^{n} p_i\right\}\right]\right)$

Special cases:

$$\left.\begin{array}{ll} k_A\,CA_1 & (p_1 = 1) \\ \ldots & \\ k_A\,CA_n & (p_n = 1) \\ \therefore k_A \vee_{i=1}^{n} CA_i & (P_{KB}[\vee_{i=1}^{n} CA_i] = 1) \end{array}\right\} \equiv \text{Sequence syllogism}$$

Stochastic disjunction-material conditional

Case premises	$a_A(CA_1 \vee CA_2)$	$(P_{KB}[CA_1 \vee CA_2] = p_1)$
	$a_A(CA_1 \rightarrow CA_2)$	$(P_{KB}[CA_1 \rightarrow CA_2] = p_2)$
Care provider's conclusion	$\therefore a_A(CA_2)$	$(P_{KB}[CA_2] = p_1 + p_2 - 1)$

Special cases:

$$\left.\begin{array}{ll} k_A(CA_1 \vee CA_2) & (p_1 = 1) \\ k_A(CA_1 \rightarrow CA_2) & (p_2 = 1) \\ \therefore k_A\,CA_2 & (1) \end{array}\right\} \equiv \text{Reduction syllogism}$$

(reports reflecting the concentrated experience of many experts during long periods of time, published studies on disease prevalence, statistical determined disease predictors, regression), clinical care setting, and special or unusual patient attributes. The introduction of the notion of probability has empowered CD logic with one more essential tool. More specifically, to the triadic elements, $Quid$ (What) — $Quis$ (Who) — $Quomodo$ (How), one adds the fourth element,

Qua (Which) are the chances of a_A: Probability $P[a_A CA]$.

For example, $qua\ coniectura$ (which are the chances of), or $qua\ verisimilitudine$ (which likelihood to assign). Several examples of syllogism (3.21) in medical research and clinical practice will be discussed in the following pages.

Since in practice the premises involve different case assertions, so does the conclusion $\Phi[a_A CA_i]$. Depending on the a_A-forms (k_A, r_A, s_A, b_A or g_A), different probabilities of occurrence $P[\Phi]$ are associated with the conclusion. In Table 3.11, the approach of Eq. (3.21) is implemented to account for uncertainties in the premises, the conclusion Φ and the inference-relationship between them. In the SMR setting, the stochastic inference process involves two main stages:

(1) probability values are assigned to the premises, and
(2) the probability of the conclusion is derived in an internally consistent manner using rules of substantive logic and probability calculus.

All of the conditional inferences, in particular, have in common three elements: the conditional premise "if–then", the categorical premise "given", and the conclusion "infer". We are now ready to accustom ourselves to some of the most important stochastic medical inferences.

The readers are reminded that in its standard or classical logic form, $modus\ ponens$ and $modus\ tollens$, respectively, are called the inference rules:

Case premises	$CA_2 \backslash CA_1$	$CA_2 \backslash CA_1$
	CA_1	$\neg CA_2$
	and	
Care provider's conclusion	$\therefore CA_2$	$\therefore \neg CA_1$

where "$\cdot \backslash \cdot$" $\equiv$ "$\cdot | \cdot$," "$\cdot \to \cdot$," or "$\cdot \leftrightarrow \cdot$". An interesting observation concerning $stochastic\ modus\ ponens$ and $stochastic\ modus\ tollens$ in Table 3.11 is that when the premises are given for sure ($p_1 = p_2 = 1$), the conclusion of the statistical conditional holds with probability 1, which is not true for the material conditional. When calculating the probability bounds in Table 3.11 and in similar cases, one may find useful relationships such as

$$P_{KB}[CA_2|CA_1] = p \therefore P_{KB}[\neg CA_1|\neg CA_2] \in [0,1],$$

and

$$P_{KB}[\neg CA_1|\neg CA_2] = p \therefore P_{KB}[CA_2|CA_1] \in [0,1].$$

Remarkably, even when the standard inference rules are not logically valid, the corresponding stochastic inferences still make sense, which is yet another reason for clinical practitioners and medical researchers to appreciate the greater theoretical power and practical flexibility of stochastic inference over classical inference in real world conditions. This is the case, e.g., of affirming the consequent and denying the antecedent inferences, respectively,

$$\begin{array}{c|cc} \text{Case premises} & CA_1 \rightarrow CA_2 & CA_1 \rightarrow CA_2 \\ & CA_2 & \neg CA_1 \\ \hline \text{Care provider's conclusion} & \therefore CA_1 & \therefore \neg CA_2 \end{array}.$$

As we see in Table 3.11, although the classical inferences above do not yield any valid conclusions, the corresponding stochastic inferences in terms of probability can be derived in a coherent manner. The *stochastic affirming the consequent* inference involves the "inverse" conditional probability $P_{KB}[CA_2|CA_1] = p_1$, bringing to mind the *Bayesian* approach

$$P_{KB}[CA_1] = P_{KB}[CA_1|CA_2]\frac{p_2}{p_1}, \qquad (3.22a)$$

with $p_2 = P_{KB}[CA_1]$, which, however, differs from the conclusion shown in Table 3.11,

$$P_{KB}[CA_1] \in \left[0, \min\left\{\frac{p_2}{p_1}, \frac{1-p_2}{1-p_1}\right\}\right]. \qquad (3.22b)$$

(The readers may find it interesting to consider, say, the possibility $\frac{p_2}{p_1} > 1$.) In practice, the care provider could decide to which one of the syllogisms of Table 3.11 the case belongs, and then implement the appropriate inference. If the care provider's case does not belong to any of the syllogisms listed in Table 3.11, one could derive new syllogisms using the same basic approach that led to the derivation of the inferences of Table 3.11. In fact, beyond the list of medical inferences in Table 3.11, several others will be derived accordingly in the examples that follow. These new inferences can be added to the list of Table 3.11.

Example 3.31: To put matters in a health care perspective, here we follow the thinking process of physician A who faces a medical case, in which DS is the disease symptom and DD is the disease diagnosis. Appropriately obtained physician probability estimates (based on previous experience, published results, and empirical evidence) include the conditional probability of the symptom given the disease, $P_{KB}[DS|DD] = p_1$, and the probability of the symptom being present, $P_{KB}[DS] = p_2$. The physician needs to calculate the probability of the disease diagnosis being true, $P_{KB}[DD]$. Given the premises of the particular case, the physician considers that affirming the consequent is an appropriate inference for calculating $P_{KB}[DD]$. In light of Table 3.11, the affirming the consequent probability for the statistical conditional is

$$P_{KB}[DD] \in \left[0, \min\left\{\frac{p_2}{p_1}, \frac{1-p_2}{1-p_1}\right\}\right]. \quad (3.23a)$$

Some interesting observations can be made. When the symptom probability p_2 is greater than the diagnosis probability p_1, and p_2 approaches 1, the probability interval of DD approaches 0; i.e., $\frac{1-p_2}{1-p_1} \to 0$ and, hence, $P_{KB}[DD] \approx 0$. The maximum value of the upper bound of $P_{KB}[DD]$ is obtained when $p_2 = p_1$, since then $P_{KB}[DD] \in [0, 1]$. Under the same conditions, the affirming the consequent probability for the material conditional (i.e., given $P_{KB}[DD \to DS] = p_1$) is $P_{KB}[DD] \in [1 - p_1, \min\{1, 1 - p_1 + 1\}]$ or

$$P_{KB}[DD] \in [1 - p_1, 1]. \quad (3.23b)$$

Once more, different disease probabilities are obtained for the same kind of inference when the physician's reasoning involves statistical vs. material conditionals.

As noted earlier, the list of Table 3.11 includes inferences beyond the common Bayesian approach. This is very valuable, since the restrictive assumptions of the Bayesian approach can make it rather unrealistic for several real medical problems.[17]

---

[17] Certain limitations of the Bayesian formulation were mentioned in Section 3.3.2 earlier. Moreover, it is often assumed that the symptoms associated with a disease are conditionally independent (see, e.g., Eq. (3.50)).

Table 3.12: A numerical comparison of stochastic medical inferences.

Inference	Conditional type	
	Statistical ($\mid$)	Material ($\rightarrow$)
Modus ponens $\left.\begin{array}{l}P_{KB}[CA_2\backslash CA_1] = p_1 = 0.8 \\ P_{KB}[CA_1] = p_2 = 0.4\end{array}\right\}$	$P_{KB}[CA_2] \in [0.32, 0.92]$	$P_{KB}[CA_2] \in [0.2, 0.8]$
Denying the antecedent $\left.\begin{array}{l}P_{KB}[CA_2\backslash CA_1] = p_1 = 0.8 \\ P_{KB}[\neg CA_1] = p_2 = 0.4\end{array}\right\}$	$P_{KB}[\neg CA_2] \in [0.12, 0.52]$	$P_{KB}[\neg CA_2] \in [0.2, 0.6]$
Affirming the consequent $\left.\begin{array}{l}P_{KB}[CA_2\backslash CA_1] = p_1 = 0.8 \\ P_{KB}[CA_2] = p_2 = 0.4\end{array}\right\}$	$P_{KB}[CA_1] \in [0, 0.5]$	$P_{KB}[CA_1] \in [0.2, 0.6]$
Modus tollens $\left.\begin{array}{l}P_{KB}[CA_2\backslash CA_1] = p_1 = 0.8 \\ P_{KB}[\neg CA_2] = p_2 = 0.4\end{array}\right\}$	$P_{KB}[\neg CA_1] \in [0.25, 1]$	$P_{KB}[\neg CA_1] \in [0.2, 0.8]$

Example 3.32: Table 3.12 presents numerical comparisons between some of the stochastic medical inferences of Table 3.11. Furthermore, it is worth noticing that for stochastic *modus ponens* inferences with $P_{KB}[CA_1] = 1$, both the statistical and material conditionals give $P_{KB}[CA_2] = 0.8$. When the probabilities of both premises are equal, (=0.5), the statistical conditional gives $P_{KB}[CA_2] = [0.25, 0.75]$, and the material conditional gives $P_{KB}[CA_2] \in [0, 0.5]$. Here, it is the probability range (upper bound–lower bound) that is the same for both conditionals (0.5).

Furthermore, the readers can derive *SMR* inferences for several combinations of the premises displayed in Table 3.11. For illustration, let us look at a few characteristic health care argumentations.

Example 3.33: A physician evaluates a patient's case following a procedure that includes (i) principles of medical practice of general applicability (PMP, Section 3.6.2); (ii) interviewing and taking patient's history; (iii) performing physical examination; (iv) consulting published results; (v) selecting diagnostic tests and interpreting the results; (vi) making the final diagnosis and suggesting a treatment. Steps (i)–(iv) lead to the case premises,

Premise 1: $P_{KB}[CA_1 \vee CA_2] = p_1$;
Premise 2: $P_{KB}[CA_1 \rightarrow CA_2] = p_2$.

The physician seeks the disease probability $P_{KB}[CA_2]$. The argumentation process leading to its calculation is tabulated below:

Step	Inference	Explanation
1	$P_{KB}[CA_1 \to CA_2]$ $= p_2 = P_{KB}[CA_2|CA_1]P_{KB}[CA_1] + 1 - P_{KB}[CA_1]$	From *premise* 2 and probability theory
2	$P_{KB}[CA_1 \vee CA_2]$ $= p_1 = P_{KB}[CA_1] + P_{KB}[CA_2]$ $- P_{KB}[CA_2|CA_1]P_{KB}[CA_1]$	From *premise* 1 and probability theory
3	$P_{KB}[CA_2] = p_1 + p_2 - 1$	Adding *steps* 1–2

The second health care argumentation process includes the premises

Premise 1: $P_{KB}[CA_2] = p_1$;
Premise 2: $P_{KB}[CA_1|CA_2] = p_2$;
Premise 3: $P_{KB}[CA_1|\neg CA_2] = p_3$.

The physician seeks the probability $P_{KB}[CA_2|CA_1]$. The argumentation process leading to its calculation is tabulated below:

Step	Inference	Explanation
1	$P_{KB}[CA_2|CA_1] = P_{KB}[CA_1|CA_2]\dfrac{P_{KB}[CA_2]}{P_{KB}[CA_1]}$	From *premises* 1–2 and probability theory
2	$P_{KB}[CA_1] = P_{KB}[CA_1|CA_2]P_{KB}[CA_2]$ $+ P_{KB}[CA_1|\neg CA_2]P_{KB}[\neg CA_2]$ $= p_1 p_2 + p_3(1 - p_1)$	From *step* 1, *premise* 3 and probability theory
3	$P_{KB}[CA_2|CA_1] = \dfrac{p_1 p_2}{P_{KB}[CA_1]} = \dfrac{p_1 p_2}{p_1 p_2 + p_3(1 - p_1)}$	Adding *steps* 1–2
4	$P_{KB}[CA_1 \to CA_2] = 1 - (1 - p_1)p_3$	From *step* 3 and probability theory
5	$P_{KB}[CA_1 \leftrightarrow CA_2] = 1 - (1 - p_1)p_3 - (1 - p_2)p_1$	From *step* 3 and probability theory

At this point it would be interesting to compare the typical conclusions reached by using two different kinds of inference.

Example 3.34: Example 3.22 investigated the case of an American hospital confronted with Ebola disease (ED). It was mentioned that in the US the base-rate chance of someone having ED is $P_{KB}[ED] = 0.0001\%$. An accurate Ebola test (ET) was available in the hospital with probabilities of disease detection $P_{KB}[ET|ED] = 99.99\%$ and $P_{KB}[ET|\neg ED] = 0.5\%$. The inference used by the hospital physicians in Example 3.22 led to the calculation of the conditional Ebola probability as $P_{KB}[ED|ET] = 0.02\%$. For validation purposes, here we decide to use the stochastic inference described in the second part of Example 3.33 above (which can be added to the list of Table 3.11 above),

$$\begin{array}{ll} \textit{Case premises} & \begin{array}{l} P_{KB}[ED] = p_1 \\ P_{KB}[ET|ED] = p_2 \\ P_{KB}[ET|\neg ED] = p_3 \end{array} \\ \\ \textit{Physician's} & \therefore P_{KB}[ED|ET] = \dfrac{p_1 p_2}{p_1 p_2 + (1-p_1)p_3} \\ \textit{conclusion} & \end{array} \qquad (3.24)$$

with $p_1 = 0.0001\%$, $p_2 = 99.99\%$, and $p_3 = 0.5\%$. The present inference calculates the conditional Ebola probability,

$$P_{KB}[ED|ET] = \frac{(0.1 \times 10^{-5}) \times (99.99 \times 10^{-2})}{(0.1 \times 10^{-5}) \times (99.99 \times 10^{-2}) + (1 - 0.1 \times 10^{-5}) \times 0.5 \times 10^{-2}},$$
$$= 0.02\%,$$

which is the same result as in Example 3.22.

Let us conclude this section by revisiting the *ethical dilemma* that confronted the Mohists in ancient China concerning a doctor's supervision of the execution of assassins. In Sections 1.1.5 and 2.6.2 we discussed briefly some early developments in Chinese logic. The discussion is continued in the example below, in which, in order to reconcile the execution of assassins with love for all men a Mohist physician may view probability as a quantitative measure of a person's humanity level and use a practical inference as appropriate.

Example 3.35: As in Example 2.49, let iA, iB and iM denote, respectively, "is an assassin", "is a Buddhist monk" and "is a man". In order to reconcile the execution of assassins with love for all men the physician can express quantitatively the four-stage process of Mohist reasoning

(Example 2.49) using the *stochastic modus ponens inference* of Table 3.11 as follows:

Case premises
$$P_{KB}[iM|iA] = p_1$$
$$P_{KB}[iA] = p_2$$

Mohist's conclusion
$$\therefore P_{KB}[iM] \in [p_1 p_2, p_1 p_2 + 1 - p_2]$$

The Mohist physician makes two essential assumptions: (*a*) because of the low humanity level of iA, the probability of iM^{18} given that iA is practically zero, is $p_1 \approx 0$; and (*b*) the probability p_2 of the particular person being iA is assumed to be a certainty (e.g., there is overwhelming evidence about it) is $p_2 \approx 1$. As a result, in the case of the assassin the stochastic *modus ponens* inference gives the probability that the assassin is actually a man as $P_{KB}[iM] \approx 0$. Accordingly, the Mohist physician's response to the ethical dilemma is that when supervising the execution of an assassin the likelihood that one is supervising the execution of a decent human being is zero. As should be expected, the probability of the assassin's actually being a man depends critically on the probability p_2 of the person being an assassin, i.e., $P_{KB}[iM] \leq 1 - p_2$, which would raise significant ethical issues concerning the amount of evidence required to establish a person's guilt etc. (see *assumption b* above). For comparison, consider next the case of a Buddhist monk. The inference remains valid by replacing iA with iB. In particular, the probability p_1 of being a man (iM) given that he is a Buddhist monk (iB) is practically certainty, $p_1 \approx 1$; and the probability p_2 of his being a monk (iB) is 1. The *modus ponens* inference now yields $P_{KB}[iM] \approx 1$. That is, the probability that by supervising the execution of a monk the physician is supervising the execution of a man is certainty. In sum, according to the Mohist line of thought expressed in probability terms it is almost certainly ethically acceptable to supervise the execution of an assassin but not that of a Buddhist monk.

3.4.2. *Premise strengthening, internally consistent and uninformative inferences*

The term *premise strengthening* is used to denote the situation in which the current case probabilities can be updated given an additional information source.

---

[18] "Man" here is understood as a decent human being.

Example 3.36: Consider a medical case with the associated disease symptoms, $DS^{(1)}$ and $DS^{(2)}$, and the disease diagnosis, DD. The probabilities $P_{KB}[DD|DS^{(1)}] = p_1$ (*premise* 1) and $P_{KB}[DS^{(2)}|DS^{(1)}] = p_2$ (*premise* 2) are known. The physician needs to calculate what is the probability of diagnosis when the two symptoms are considered together. The answer is provided by the following premise strengthening inference (which can be added to the list of Table 3.11):

$$
\begin{array}{ll}
\text{Case premises} & \begin{array}{l} P_{KB}[DD|DS^{(1)}] = p_1 \\ P_{KB}[DS^{(2)}|DS^{(1)}] = p_2 \end{array} \\
\text{Physician's} & \therefore\ P_{KB}[DD|DS^{(1)} \wedge DS^{(2)}] \\
\text{conclusion} & \in \left[\max\left\{0, \frac{p_1 + p_2 - 1}{p_2}\right\}, \min\left\{\frac{p_1}{p_2}, 1\right\} \right]
\end{array}
$$

(3.25)

Obviously, the medical inference (3.25) is generally valid for any case attributes.

Stochastic CD inferences, such as those in Table 3.11, can be derived in terms of *interval* probability values, as well. A list of such inferences is displayed in Table 3.13. In many cases, the health care provider may reach a conclusion (diagnosis, prognosis, treatment) using different kinds of inferences, although the latter may require different premises. In any case, for *validation* purposes it is imperative that for the same medical case the conclusions drawn by different inferences be *internally consistent*. Which is why many decision-makers follow the above validation approach to improve their confidence about their case conclusions. In the example that follows we reconsider Examples 3.22 and 3.34 assuming a different set of premises, and we reach a conclusion that is consistent with that of the earlier example.

Example 3.37: Let us have yet another look at the Ebola case, where now the following premises are considered, instead:

Premise 1: $P[ET|ED] \in [99.89 \times 10^{-2}, 99.99 \times 10^{-2}]$
Premise 2: $P[ET] \in [0.45 \times 10^{-2}, 0.5 \times 10^{-2}]$.

These premises are consistent with the ones considered in the previous examples. The stochastic affirm the consequent syllogism in terms of interval probabilities applies (Table 3.13), where $p_1 = 99.89\%$, $p_2 = 99.99\%$,

Table 3.13: Stochastic medical inferences involving probability intervals.

Stochastic modus ponens

Case premises	$a_A(CA_2\|CA_1)$	$(P_{KB}[CA_2\|CA_1] \in [p_1, p_2])$
	$a_A CA_1$	$(P_{KB}[CA_1] \in [q_1, q_2])$
Care provider's conclusion	$\therefore a_A CA_2$	$(P_{KB}[CA_2] \in [p_1 q_1, 1 - q_1 + p_2 q_2])$

Stochastic modus tollens

Case premises	$a_A(CA_2\|CA_1)$	$(P_{KB}[CA_2\|CA_1] \in [p_1, p_2])$
	$a_A \neg CA_2$	$(P_{KB}[\neg CA_2] \in [q_1, q_2])$
Care provider's conclusion	$\therefore a_A \neg CA_1$	$(P_{KB}[\neg CA_1] \in [\max\{w_1, w_2\}, 1])$

$$w_1 = \frac{1 - p_1 - q_2}{1 - p_1} \quad (p_1 + q_2 \le 1); = \frac{q_2 - p_1}{p_1}$$
$$(p_1 + q_2 > 1)$$

$$w_2 = \frac{1 - p_2 - q_1}{1 - p_2} \quad (p_2 + q_1 \le 1); = \frac{p_2 + q_1 - 1}{p_2}$$
$$(p_2 + q_1 > 1)$$

Stochastic affirming the consequent

Case premises	$a_A(CA_2\|CA_1)$	$(P_{KB}[CA_2\|CA_1] \in [p_1, p_2])$
	$a_A CA_2$	$(P_{KB}[CA_2] \in [q_1, q_2])$
Care provider's conclusion	$\therefore a_A CA_1$	$(P_{KB}[CA_1] \in [0, \min\{w_1, w_2\}])$

$$w_1 = \frac{1 - q_2}{1 - p_1} \quad (p_1 \le q_2); = \frac{q_2}{p_1} \quad (p_1 > q_2)$$

$$w_2 = \frac{1 - q_1}{1 - p_2} \quad (p_2 \le q_1); = \frac{q_1}{p_2} \quad (p_2 > q_1)$$

Stochastic denying the antecedent

Case premises	$a_A(CA_2\|CA_1)$	$(P_{KB}[CA_2\|CA_1] \in [p_1, p_2])$
	$a_A CA_2$	$(P_{KB}[\neg CA_1] \in [q_1, q_2])$
Care provider's conclusion	$\therefore a_A CA_1$	$(P_{KB}[\neg CA_2] \in [(1 - p_2)(1 - q_2), 1 - p_1(1 - q_2)])$

Stochastic monotonicity

Case premises	$P_{KB}[CA_2\|CA_1] \in [p_1, p_2]$
	$P_{KB}[CA_3\|CA_1] \in [q_1, q_2]$
Care provider's conclusion	$\therefore P_{KB}[CA_2 \wedge CA_3\|CA_1] \in [\max\{0, p_1 + q_1 - 1\}, \min\{p_2, q_2\}]$

$q_1 = 0.45\%$, and $q_2 = 0.5\%$. Hence,

$$P_{KB}[ED] \in [0, 0.5\%], \qquad (3.26)$$

which is consistent with the conclusion of Examples 3.22 and 3.34. To continue our syllogism, given $P[ET] \in [0.45\%, 0.5\%]$, one finds $1 - P[ET] = P[\neg ET] \in [99.5\%, 99.55\%]$. And a valid syllogism is the stochastic *modus tollens* (Table 3.13),

$$\begin{array}{ll} \text{Ebola Case} & P_{KB}[ET|ED] \in [99.89\%, 99.99\%] \\ \text{premises} & \underline{P_{KB}[\neg ET] \in [99.5\%, 99.55\%]} \\ \text{Physician's} & \\ \text{conclusion} & \therefore \ P_{KB}[\neg ED] \in [99.5\%, 1] \end{array} \qquad (3.27)$$

Therefore, $1 - P[\neg ED] = P[ED] \in [0, 0.5\%]$, i.e., the same conclusion as that of (3.26) — a useful validation process. As the readers can see, the stochastic *modus tollens* syllogism of Table 3.13 reduces to that of Table 3.11 if one selects $p_1 = p_2 = p$ and $q_1 = q_2 = q$, which is appropriate from a validation viewpoint.

Let us take stock. Depending on the known premise probabilities of the particular case, the health care providers can implement the appropriate inference to derive their conclusions regarding the case at hand. More than one type of inference can be used by a care provider as a useful validation process. All this means that it is not necessary to always be limited to the same type of inference. This flexibility can be of significant value, since in many medical cases the realistic implementation of the mainstream approach is confronted with serious difficulties (the prior probabilities required for the implementation of the prior-posterior rule are not available, technical issues, singularities, biophysical impossibility of Bayesian reversibility).

A health care provider using quantitative decision-making techniques should keep in mind that there are inferences that are *uninformative*, in the sense that the probability values of the case conclusions merely belong to the interval $[0, 1]$, and they do not offer any new information to the care provider. We already encountered some consequential uninformative cases in the previous pages. Noticeably, in Bayesian decision analysis (Section 3.3.3) uninformative priors have been considered in medical investigations that are not always adequately justified. In the HIV study by Ades *et al.* (2008), for example, there was sufficient data to inform all the parameters of interest, so the only information introduced in the priors was deliberately "vague" and it was specified only that each basic probability parameter can lie in the range 0 to 1. Here are some more examples.

Example 3.38: Consider, once more, the earlier *Ebola* example, where information concerning the high accuracy of the Ebola test is at Dr. Xin's disposal, i.e.,

Premise: $P_{KB}[ET|ED] = 99.99\%$.

Given this information, the case probability $P_{KB}[\neg ED|\neg ET]$ is sought. The inference process leading to its calculation is tabulated below:

Step	Inference	Explanation
1	$P_{KB}[\neg ET|ED] = 1 - P_{KB}[ET|ED] = 0.01\%$	From *premise* and probability theory (normality relationship)
2	$P_{KB}[\neg ED|\neg ET] = 1 - P_{KB}[ED|\neg ET]$	From probability theory (normality relationship)
3	$P_{KB}[\neg ED|\neg ET] = 1 - P_{KB}[\neg ET|ED]\dfrac{P_{KB}[ED]}{P_{KB}[\neg ET]}$ $= 1 - 0.0001 \times \dfrac{P_{KB}[ED]}{P_{KB}[\neg ET]}$	From *steps* 1–2, and probability theory (Bayes theorem)
4	$P_{KB}[\neg ET] = P_{KB}[\neg ET|ED]P_{KB}[ED]$ $+ P_{KB}[\neg ET|\neg ED]P_{KB}[\neg ED]$[19]	From probability theory (total probability rule)
5	$\min P_{KB}[\neg ED|\neg ET] = 1 - \dfrac{0.0001 \times 1}{0.0001} = 0$	From *steps* 3–4
6	$\max P_{KB}[\neg ED|\neg ET] = 1 - \dfrac{0.0001 \times 0}{P_{KB}[\neg ET]} = 1$	From *steps* 3–4
7	$P_{KB}[\neg ED|\neg ET] \in [0,1]$	From *steps* 3 and 5–6

In other words, despite the fact that Dr. Xin knows that the probability of the available test detecting Ebola, given that the patient has the disease, is

---

[19]For $P_{KB}[ED] = 1$, $P_{KB}[\neg ET] = P_{KB}[\neg ET|ED] = 0.0001$; and for $P_{KB}[ED] = 0$, $P_{KB}[\neg ET] = P_{KB}[\neg ET|\neg ED]$.

very high, she knows nothing about the probability that the patient does not have the disease given that the test does not detect it ($P_{KB}[\neg ED|\neg ET] \in [0,1]$), which may seem, to some extent, of counterintuitive.

The uninformative inference of the Ebola disease above is actually valid for any case attributes CA_1 and CA_2, i.e.,

$$\begin{array}{ll} \textit{Case premises} & P_{KB}[CA_2|CA_1] = p \\ \textit{Physician's conclusion} & \therefore P_{KB}[\neg CA_1|\neg CA_2] \in [0,1] \end{array} \qquad (3.28)$$

for any $p \in [0,1]$; as well as the symmetric inference,

$$\begin{array}{ll} \textit{Case premises} & P_{KB}[\neg CA_1|\neg CA_2] = p \\ \textit{Physician's conclusion} & \therefore P_{KB}[CA_2|CA_1] \in [0,1] \end{array} \qquad (3.29)$$

Moreover, other uninformative inferences that can emerge in medical decision-making include the following *hypothetical* syllogisms,

$$\begin{array}{ll} \textit{Case premises} & \begin{array}{l} P_{KB}[CA_3|CA_2] = p_1 \\ P_{KB}[CA_2|CA_1] = p_2 \end{array} \\ \textit{Physician's conclusion} & \therefore P_{KB}[CA_3|CA_1] \in [0,1] \end{array} \qquad (3.30)$$

and

$$\begin{array}{ll} \textit{Case premises} & \begin{array}{l} P_{KB}[CA_2|CA_1] \in [p_1,p_2] \\ P_{KB}[CA_3|CA_2] \in [p_1,p_2] \end{array} \\ \textit{Physician's conclusion} & \therefore P_{KB}[CA_3|CA_1] \in [0,1] \end{array} \qquad (3.31)$$

for any $p_1, p_2 \in [0,1]$.

A word of warning: uninformative medical inferences may be the result of the popular premise strengthening process discussed earlier. In clinical practice and population exposure assessment, a decision-maker routinely attempts to improve the current probabilities by updating them in light of new evidence. If, however, the inferences employed by the care professional are uninformative, the attempt to improve decision-making is futile.

Example 3.39: Assume that in view of the existing core knowledge G the current probability a physician assigns to the diagnosis DD is $P_{KB}[DD|G] = p_1$, $p_1 \in [0,1]$. Subsequently, the evidence denoted as S becomes available. Given this new information, the inference below is valid,

$$\begin{array}{ll} \text{Case premises} & \dfrac{P_{KB}[DD|G] = p_1}{}, \\ \text{Physician's} & \therefore\ P_{KB}[DD|G \wedge S] \in [0,1] \\ \text{conclusion} & \end{array} \qquad (3.32)$$

which is obviously uninformative. In addition, inference (3.32) remains uninformative if $P_{KB}[DD|G] = p_1$ is replaced by $P_{KB}[DD|G] \in [p_1, p_2]$. As a result, when in a case only the probability p_1 (or, the probability interval $[p_1, p_2]$) is available, no informative inference is generally possible by incorporating new evidence S. Clearly, additional premises are needed that can turn the uninformative inference (3.32) into a useful one.

In sum, the common point concerning uninformative inferences is that they often generate results that are of limited value since they do not lead to new knowledge.

3.5. Probability, Uncertainty and Information of Diagnoses or Prognoses Sets

We start this section with some additional terminology concerning the triadic case formula, $a_A CA$, of a typical medical case (as introduced in Section 2.3.7). As usual, a_A denotes an assertion of the care provider A regarding a case CA. By definition, a_A involves a disease *diagnoses set*, DD-set of the form

$$\Pi[a_A(V_j DD^{(j)})], \text{ or briefly, } \Pi_{DD}[a_A],$$

where $V_j DD^{(j)} = DD^{(1)} \vee \ldots \vee DD^{(n)}$; and an observed *symptoms set*, DS-set

$$\Pi[a_A(\Lambda_i DS^{(i)})], \text{ or briefly, } \Pi_{DS}[a_A],$$

where $\Lambda_i DS^{(i)} = DS^{(1)} \wedge \ldots \wedge DS^{(m)}$. The associated probability, uncertainty and information of the DD-sets that $V_j DD^{(j)}$ will turn out to be the case as asserted by A are, respectively,

$$P[a_A(V_j DD^{(j)})], \quad U[a_A(V_j DD^{(j)})] \text{ and } I[a_A(V_j DD^{(j)})],$$

or simply $P_{DD}[a_A]$, $U_{DD}[a_A]$ and $I_{DD}[a_A]$ (which measures the information provided before the event, so-called a_A-informativeness).

For the purposes of this book, the notion of information is twofold:

(1) *Technical* information at the syntactic level, i.e., it deals with uninterpreted mathematical rules and symbols (which are seen as carriers of information). Accordingly, technical information has been applied particularly successfully in terms of computers that are syntactic devices.
(2) *Substantive* information understood as semantic content, i.e., it is well-formed (data properly put together according to rules that govern the health system of interest) and meaningful (data compliant with the meaning of the system).

The readers should notice that, at a first glance, a key difference between these notions of information is that "technical" is information without medical meaning. However, several scientists argue that technical information is not meaningless; rather, it is just not yet meaningful. When meaning is assigned to the rules and symbols of technical information, the latter could turn into substantive medical information.

Example 3.40: No matter whether the corresponding question is

"What is the patient's weight?"

or

"How heavy is this desk?"

an equiprobable answer

"60 kg"

contains the same amount of technical information in both cases. On the other hand, substantive information is concerned with questions such as:

(1) When is a diagnosis informative?
(2) How is clinical information related to knowledge?

(As an exercise, the readers may formulate and appreciate similar examples of technical vs. substantive information in their medical research or clinical practice.)

Frequently, the health care providers applying SMR inference in medical practice implicitly make use of the following three basic postulates regarding $P_{DD}[a_A]$, $I_{DD}[a_A]$, $\Pi_{DD}[a_A]$ and $U_{DD}[a_A]$:

Postulate 1: The fewer diagnoses (possibilities) a DD-set involves, technically, the smaller the probability that it will turn out to be the case, *viz.*,

$$\begin{array}{ll} \text{Case premises} & \Pi_{DD}[a_A] \downarrow \\ \text{Care provider's conclusion} & \therefore\ P_{DD}[a_A] \downarrow \end{array} \qquad (3.33)$$

where the symbol "$\downarrow$" means "decreases".[20] Eq. (3.33) is also supported by probability theory: if $j = 1,\ldots,n$ and $j' = 1,\ldots,n'$ with $n < n'$, then

$$P[a_A(V_j DD^{(j)})] < P[a_A(V_{j'} DD^{(j')})].$$

Postulate 2: The fewer diagnoses a DD-set allows, technically the more informative it is, technically *viz.*,

$$\begin{array}{ll} \text{Case premises} & \Pi_{DD}[a_A] \downarrow \\ \text{Care provider's conclusion} & \therefore\ I_{DD}[a_A] \uparrow \end{array} \qquad (3.34)$$

where the symbol "$\uparrow$" denotes "increase".

Postulate 3: The more diagnoses (possibilities) an assertion a_A allows, technically, the smaller is the uncertainty about the outcome, *viz.*,

$$\begin{array}{ll} \text{Case premises} & \Pi_{DD}[a_A] \uparrow \\ \text{Care provider's conclusion} & \therefore\ U_{DD}[a_A] \downarrow \end{array} \qquad (3.35)$$

i.e., the more possibilities are allowed, the smaller the uncertainty about the prognosis being correct, which is a postulate with a Popperian flavor.

---

[20] So to speak, the more tickets one buys, the higher one's probability that one will win.

From *Postulates* 1–3, certain useful conclusions are drawn:

Case premises	$P_{DD}[a_A] \downarrow$	
Care provider's conclusion	$\therefore I_{DD}[a_A] \uparrow$	(3.36)
Case premises	$I_{DD}[a_A] \uparrow$	
Care provider's conclusion	$\therefore U_{DD}[a_A] \uparrow$	(3.37)
Case premises	$U_{DD}[a_A] \uparrow$	
Care provider's conclusion	$\therefore P_{DD}[a_A] \downarrow$	(3.38)

In other words, uncertainty is proportional to information and inversely proportional to probability, whereas information is inversely proportional to probability.

Let us now turn our attention from the general a_A-form to the particular forms a case assertion takes in clinical practice and medical research, i.e., g_A (mere guess), b_A (personal belief), s_A (substantiated belief), r_A (rational reasoning) and k_A (objective knowledge). As noted earlier, in addition to the physician's competency and related epistemic strength, there are other factors that affect the form of the assertion that is suitable for the particular case. Indeed, how much is *at stake* in a medical case may have a direct impact (a) on the *informativeness* of the physician's assertions (i.e., how much information an assertion is assumed to contain) and (b) on the associated *quality of evidence* available to the physician in that case.

The informativeness of an assertion to the physician, before the event, is proportional to the associated uncertainty of the assertion itself turning out to be correct, after the event. In light, then, of Eq. (2.10) and (3.37), the technical informativeness of an assertion form may increase from g_A to k_A, *viz.*, in general

$$I[g_A] < I[b_A] < I[s_A] < I[r_A] < I[k_A]. \tag{3.39}$$

Clearly, physicians are most confident about the case when they can allow themselves the form k_A (objective knowledge), and least confident when they limit themselves to the use of the form g_A (mere guessing). Concerning the required level of the physician's confidence, how confident a physician ought to be about a particular case (including whether a physician ought to adopt an objective knowledge of full belief) depends on the difficulty level of the case, the available resources, how much is at stake etc.

Moreover, earlier analysis implied that in S/TRF terms, the larger the number of possible worlds (realizations) in which a case attribute is considered valid, the higher the epistemic strength of the corresponding assertion; symbolically, $n[g_A] > n[b_A] > n[s_A] > n[r_A] > n[k_A]$. Dividing each term of the last relationship by the number of all possible realizations yields

$$P[g_A] > P[b_A] > P[s_A] > P[r_A] > P[k_A], \qquad (3.40)$$

i.e., the larger the number of possible worlds in which a case attribute is considered valid, the higher the probability that the corresponding assertion will turn out to be case (which is in accordance with *Postulate* 1 earlier). Notice the inverse order of Eqs. (2.10) and (3.40). That is to say, when the physician's uncertainty about a disease symptom decreases, the probability of the disease's presence increases.

Example 3.41: In a medical decision-making study (Redelmeier *et al.*, 1995), physicians were asked to review the following case:

> "A 22-year-old Hollywood actress presents herself to the emergency department with pain in the right lower quadrant of her abdomen of 12 hours' duration. Her last normal menstrual period was four weeks ago".

Two groups with the same number of physicians, selected at random, were involved in this study. *Group* 1 was asked to estimate probabilities for three possible disease diagnoses:

$DD^{(1)}$: *gastroenteritis*,
$DD^{(2)}$: *ectopic pregnancy*,
$DD^{(0)}$: *none of the above*.

At the same time, *group* 2 was asked to do the same for six diagnoses:

$DD^{(1)}$: *gastroenteritis*,
$DD^{(2)}$: *ectopic pregnancy*,
$DD^{(3)}$: *appendicitis*,
$DD^{(4)}$: *pyelonephritis*,
$DD^{(5)}$: *pelvic inflammatory disease*,
$DD^{(0)}$: *none of the above*.

The two tasks differed only in that the residual diagnosis ("none of the above") in *group 1* was partially unpacked in the list of *group 2*. All the physicians were told that the patient had only one condition (the judged probabilities should add up to 1). Logically, the probability of the residual diagnosis in *group 1* should equal the sum of the probabilities of the corresponding diagnoses in *group 2*. However, it was found that the average probability assigned by the physicians of *group 1* to the residual diagnosis was

$$P_{KB}[DD^{(0)}] = 0.50;$$

which was smaller than the residual diagnosis probability in *group 2*,

$$P_{KB}[DD^{(3)}, DD^{(4)}, DD^{(5)}, DD^{(0)}] = 0.69.$$

Evidently, unpacking the residual diagnosis reminded physicians of diseases they might have overlooked, or increased the salience of diagnoses that they had considered. Also, the readers may notice that the average probability assigned to "gastroenteritis" was substantially higher in *group 1* than in *group 2*,

$$P_{KB}[DD^{(1)}] = 0.31 \ (group \ 1),$$
$$P_{KB}[DD^{(1)}] = 0.16 \ (group \ 2).$$

An interesting implication of the last finding is that

$$P_{KB}[\neg DD^{(1)}] = 1 - P_{KB}[DD^{(1)}] = P_{KB}[DD^{(2)}, DD^{(0)}]$$
$$= 0.69 \ (group \ 1)$$
$$P_{KB}[\neg DD^{(1)}] = P_{KB}[DD^{(2)}, DD^{(3)}, DD^{(4)}, DD^{(5)}, DD^{(0)}]$$
$$= 0.82 \ (group \ 2).$$

In other words, the following inequality is valid:

$$P_{KB}[DD^{(2)}, DD^{(0)}] < P_{KB}[DD^{(2)}, DD^{(3)}, DD^{(4)}, DD^{(5)}, DD^{(0)}]; \tag{3.41}$$

or in terms of the corresponding uncertainties,

$$U_{KB}[DD^{(2)}, DD^{(3)}, DD^{(4)}, DD^{(5)}, DD^{(0)}] < U_{KB}[DD^{(2)}, DD^{(0)}]. \tag{3.42}$$

The analysis above provides additional evidence that the application of probability in medical sciences requires a good understanding of the mathematical theory and considerable concrete experience with real situations.

The inverse relationship between uncertainty and probability is linked to some useful mathematical definitions of medical uncertainty in terms of probability and *vice versa*. Among them, the following two formulations are of significant interest:

$$U[a_A] = P[\neg a_A] = 1 - P[a_A] \tag{3.43}$$

and

$$U[a_A] = \lambda \log_\beta P^{-1}[a_A], \tag{3.44}$$

where λ is a numerical constant that depends on the logarithmic base β. According to Eq. (3.43), a health care provider's uncertainty about a medical assertion may be formally seen as one's certainty about the negation of the same assertion. Eq. (3.44), on the other hand, is linked to the well-known Shannon information theory (for more technical details the readers are referred to Christakos (2010)). The logical relationships between uncertainty, probability and information discussed earlier are actually represented in the above definitions. Two limited cases can be immediately noticed. First, clearly $U[a_A] = 0$ when the assertion is considered to be a certainty. Second, $U[a_A] = 1$ when it is certain that the assertion will turn out not to be the case.

3.6. Diagnosis Ranking and Symptom Confirmation Strength

Let us take stock. In Section 2.1 it was pointed out that among the main components of medical reasoning are high-quality disease symptoms (DS) and disease diagnoses (DD) that enable a rigorous case prognosis (CP), i.e., prediction of the course of a disease. These three factors are closely connected. A DS, for example, may occur to more than one DD, whereas DD can be the guiding light for CP. In this part of the book we will present a series of parameters for quantitatively assessing the state of a medical case (estimating a disease's acceptance level, measuring a symptom's confirmation or disconfirmation strength).

3.6.1. Quantitative case parameters

The actual process of diagnosis-making is basically a cognitive process that generates and evaluates assertions (beliefs, claims, statements, perspectives) concerning a possible disease. Naturally, the *DS-set* Π_{DS} introduced in Section 3.5 does not include every single detail of the patient's situation. Instead, Π_{DS} focuses on the symptoms and signs pertaining to the case. It may include physical, mental and emotional symptoms; symptoms that are commonly linked to particular diseases; symptoms linked to patient's complaints; strange and apparently peculiar symptoms or vague and undefined symptoms. Subsequently, Π_{DS} is associated with a disease diagnosis, *DD-set* Π_{DD}, on the basis of which a case prognosis, *CP-set*, is suggested by the health care provider.

A particular diagnosis $DD^{(q)}$ explains a set of symptoms to a certain degree that should satisfy commonly accepted principles of medical practice. The so-called $DS^{(w)}$-*profile* is a set of disease diagnoses associated with a specific symptom $DS^{(w)}$ denoted as $\pi_{DD}[DS^{(w)}]$. A small π is used to declare that only those $DD^{(j)}$ of the *DD-set* Π_{DD} are considered that co-occur with $DS^{(w)}$ (clearly, $\pi_{DD}[DS^{(w)}] \subseteq \Pi_{DD}$). Correspondingly, the $DD^{(q)}$-*profile* is a set of disease symptoms associated with the specific diagnosis $DD^{(q)}$ denoted as $\pi_{DS}[DD^{(q)}]$. As before, a small π is used to declare that only those $DS^{(i)}$ of the *DS-set* Π_{DS} are considered that co-occur with $DD^{(q)}$. Intuitively, this analysis can encode the clinicians' knowledge of "diagnosis–symptom relationships" into quantitative parameters and relationships with an adequate degree of accuracy. These are introduced next.

Health care providers need quantitative tools that illuminate precepts and case notions and offer emergent themes that are generalizable to other cases. *SMR* provides quantitative tools for improving one's ability to understand and manage essential issues of symptom extracting and diagnosis making. The *frequency of co-occurrence* of a certain symptom $DS^{(w)} \in \Pi_{DS}$ with a certain diagnose $DD^{(q)}$ is analytically defined as

$$\theta[DS^{(w)} \wedge DD^{(q)}] \in [0,1]. \qquad (3.45)$$

Equation (3.45) establishes the mapping $\theta : \Pi_{DS} \times \Pi_{DD} \to [0,1]$, where $\Pi_{DS} \times \Pi_{DD}$ belongs to the *G-KB* of the case.

Being able to rank symptoms properly is the key to an adequate case analysis often leading to the inner sanctum of a case. *Ranking degree* (*RD*) is the degree of acceptance of a certain diagnosis $DD^{(q)} \in \Pi_{DD}$ on the basis of a given symptom DS_w compared to the other diagnoses of the set Π_{DD}, i.e.,

$$RD(DD^{(q)}|DS^{(w)}) = \frac{P_{KB}[DD^{(q)} \wedge DS^{(w)}]}{\Sigma_j P_{KB}[DD^{(j)} \wedge DS^{(w)}]}. \quad (3.46)$$

Alternatively, the ranking degree can be used to measure the confirmation power or strength of a symptom. Equation (3.46) can be potentially extended to express the degree of acceptance of a diagnosis $DD^{(q)}$ based on a given DS-set, Π_{DS}, compared to other diagnoses of the set Π_{DD}, by summing over all $DS^{(i)} \in \Pi_{DS}$, i.e.,

$$RD(DD^{(q)}|\Pi_{DS}) = \Sigma_i RD(DD^{(q)}|DS^{(i)}).$$

When physicians study a case they routinely take into consideration the confirmation (and disconfirmation) strength of the patient's symptoms with respect to the possible diagnoses. Undoubtedly, physicians seek symptoms that have a quality to them and are the facts of the case under consideration. The *confirmation strength* (*CS*) of any case symptom $DS^{(w)} \in \Pi_{DS}$ for the particular diagnosis $DD^{(q)}$ vs. for diagnosis $DD^{(r)}$ is given by

$$CS(DD^{(q)}, DD^{(r)}|DS^{(w)}) = \frac{P_{KB}[DD^{(q)}|DS^{(w)}]}{P_{KB}[DD^{(r)}|DS^{(w)}]}. \quad (3.47)$$

Another way to look at Eq. (3.47) is as the *degree of acceptance* of $DD^{(q)}$ (conditional on $DS^{(w)}$) in comparison with another $DD^{(r)}$. Equation (3.47) can be generalized to include any set of symptoms. So, the *CS* of the symptoms set Π_{DS} for diagnosis $DD^{(q)}$ vs. for diagnosis $DD^{(r)}$ is:

$$CS(DD^{(q)}, DD^{(r)}|\Pi_{DS}) = \frac{P_{KB}[DD^{(q)}|\Lambda_i DS^{(i)}]}{P_{KB}[DD^{(r)}|\Lambda_i DS^{(i)}]}$$

$$= \frac{P_{KB}[\Lambda_i DS^{(i)}|DD^{(q)}]}{P_{KB}[\Lambda_i DS^{(i)}|DD^{(r)}]} \frac{P[DD^{(q)}]}{P[DD^{(r)}]}. \quad (3.48)$$

Again, Eq. (3.48) is a measure of the *CS* of the patient's symptoms with respect to the distinct diagnoses $DD^{(q)}$ and $DD^{(r)}$, i.e., the degree of the acceptance of a certain diagnosis $DD^{(q)}$ conditional on the set of patient's symptoms Π_{DS} in comparison with another diagnosis $DD^{(r)}$.

Furthermore, Eq. (3.48) is known as the *Bayesian Odds* (*BO*) function of disease $DD^{(q)}$ vs. disease $DD^{(r)}$, i.e.

$$BO(DD^{(q)}, DD^{(r)}|\Pi_{DS}) = CS(DD^{(q)}, DD^{(r)}|\Pi_{DS}). \qquad (3.49)$$

According to this perspective, the *BO* is the *posterior odds* of the medical case, $\frac{P_{KB}[\Lambda_i DS^{(i)}|DD^{(q)}]}{P_{KB}[\Lambda_i DS^{(i)}|DD^{(r)}]}$ is the likelihood ratio and $\frac{P_{KB}[DD^{(q)}]}{P_{KB}[DD^{(r)}]}$ is the prior odds of the case. Here, $P_{KB}[DD^{(q)}]$ and $P_{KB}[DD^{(r)}]$ are the chances of the respective disease diagnosis being correct before observing any symptoms $DS^{(i)}$ ($i = 1, \ldots, m$). As above, Eq. (3.49) measures the confirmation strength of the patient's symptoms with respect to distinct diagnoses against another one. One may also talk about the *degree of covering* of a given symptom set Π_{DS} by a diagnosis $DD^{(q)}$. If the disease symptoms $DS^{(i)}$ ($i = 1, \ldots, m$) are *conditionally independent*, Eq. (3.48) reduces to

$$CS(DD^{(q)}, DD^{(r)}|\Pi_{DS}) = \frac{P_{KB}[DD^{(q)}]}{P_{KB}[DD^{(r)}]} \Lambda_i \frac{P_{KB}[DS^{(i)}|DD^{(q)}]}{P_{KB}[DS^{(i)}|DD^{(r)}]}. \qquad (3.50)$$

i.e., the combined contribution of all symptoms on making a particular diagnosis is equal to the product of the contributions of the individual symptoms.

On the other hand, the *disconfirmation degree* of a case diagnosis can be measured on the basis of the anticipated symptoms that are not currently exhibited by the patient. Thus, the *exclusion strength* (*ES*) is the degree of exclusion of a diagnosis $DD^{(q)}$ on the basis of a given symptom $DS^{(w)}$ that was expected to be exhibited by the patient, but is not exhibited, i.e.,

$$ES[DD^{(q)}|DS^{(w)}] = 1 - \frac{P_{KB}[\neg(DS^{(w)} \land DD^{(q)})]}{\sum_i P_{KB}[\neg(DS^{(i)} \land DD^{(q)})]}. \qquad (3.51)$$

Equation (3.51) can be also written as

$$ES[DD^{(q)}|DS^{(w)}] = 1 - \frac{1 - P_{KB}[DS^{(w)} \land DD^{(q)}]}{\sum_i (1 - P_{KB}[DS^{(i)} \land DD^{(q)}])}. \qquad (3.52)$$

Just as in the case of *RD*, Eq. (3.51) can be extended to express the exclusion strength of a disease $DD^{(q)}$ on the basis of a given set Π_{DS} that were expected to be exhibited by the patient, but they are not exhibited by summing over all $DS^{(i)}$, i.e.,

$$ES(DD^{(q)}|\Pi_{DS}) = \Sigma_i ES(DD^{(q)}|DS^{(i)}).$$

The case parameters defined in Eqs. (3.45)–(3.52) have significant value for developing robust and general conclusions about a disease or condition. They can be employed by the health care professionals to structure the clinical, exposure and epidemiological evidence and gain insight to assist decision-makers in making better decisions.

In sum, the CS concept is motivated by the problem of how much a specified diagnosis is acceptable in comparison to another diagnosis (conditional on a set of symptoms); the RD notion is motivated by the problem of how much a specified diagnosis is acceptable given a symptom set, and the ES concept is motivated by the problem of how to interpret the expected symptoms that are not exhibited by the patient. Hence, in quantitative medical reasoning the task of diagnosis may be, in many cases, reduced to calculating the above parameters, and then collecting them into an acceptability measure for each diagnosis based upon a given set of symptoms. Yet this is not an easy task, whereas time pressure is usually an additional burden to the intellectual difficulties.

3.6.2. *Principles of medical practice and their quantitative expressions*

When physicians encounter a case in medical research and clinical practice they usually follow certain rules and principles of general applicability, also known as *principles of medical (PMP)* (Niknam and Niknam, 2008). The quantitative reasoning notions and tools above are designed to properly satisfy these rules. In Table 3.14 we list some examples of these rules, and describe their meaning in the quantitative reasoning context. Note that the validity of these principles, in their quantitative formulation, can be rigorously shown in terms of the case parameters introduced earlier.

The readers may notice that DMP is an empirical principle that points out the effects of changing the size of a DD-set. UP emphasizes the absolute confirmation strength of a unique symptom–disease association. CCP discusses the increased confirmation strength of the co-existence of more than one symptom. FIP establishes a symptom inclusion rule associated with the symptom's frequency. SMP refers to the effect of a changing DS-set size. SNP points out the case in which a disease occurs only if a specific symptom is present (the *SNP* formulation follows the *modus tollens* rule). *CDP* discusses the increased disconfirmation strength of the non-existence of more than one symptom. *FEP* establishes a symptom exclusion rule associated with the symptom's frequency of non-existence.

In conclusion, several empirical principles used by health care professionals in their medical and clinical practice can be justified in terms of quantitative case parameters such as the above. The common point of this quantification effort is this: while one is free to pursue mathematical investigations, the quantitative results obtained will be marginal and hence ignored if they do not correspond to the real world experience of the care professionals.

Table 3.14: Principles of medical practice (*PMP*) and their quantitative formulation.

PMP	Description	Formulation
Diagnosis Multiplicity Principle (DMP)	The more (fewer) diagnoses one associates with a given $DS^{(w)}$, the lower (higher) is the confirmation strength of $DS^{(w)}$ for a certain $DD^{(q)}$.	$\pi_{DD}[DS^{(w)}]_{\uparrow(\downarrow)}$ $DD^{(q)}, DD^{(r)} \in \Pi_{DD}, \quad q \neq r$ $\therefore CS[DD^{(q)}, DD^{(r)}\|DS^{(w)}]_{\downarrow(\uparrow)}$
Uniqueness Principle (UP)	If a given $DS^{(w)}$ is observed in only one $DD^{(q)}$, the existence of $DS^{(w)}$ absolutely confirms $DD^{(q)}$.	$\pi_{DS}[DD^{(q)}] \equiv DS^{(w)}$ $DD_r \in P_{DD}, \quad r \neq q$ $\therefore CS[DD^{(q)}, DD^{(r)}\|DS^{(w)}]_{\max}$
Conjunction Confirmation Principle (CCP)	If $DS^{(w)}$, $DS^{(w')}$ independently confirm the existence of $DD^{(q)}$ to certain degree, their co-existence confirms $DD^{(q)}$ more than each symptom separately.	$CS(DD^{(q)}, DD^{(r)}\|DS^{(w)} \wedge DS^{(w')})$ $> CS(DD^{(q)}, DD^{(r)}\|DS^{(w)})$ $> CS(DD^{(q)}, DD^{(r)}\|DS^{(w')})$
Frequency Inclusion Principle (FIP)	The more (less) often a certain $DS^{(w)}$ occurs in a given $DD^{(q)}$, the larger (smaller) is the confirmation strength of $DS^{(w)}$ for $DD^{(q)}$.	$\theta(DS^{(w)} \wedge DD^{(q)})_{\uparrow(\downarrow)}$ $\therefore CS(DD^{(q)}, DD^{(r)}\|DS^{(w)})_{\uparrow(\downarrow)}$

(*Continued*)

Table 3.14: (*Continued*)

PMP	Description	Formulation
Symptoms Multiplicity Principle (SMP)	The more (fewer) symptoms occur with a certain $DD^{(q)}$, the lower (higher) is the exclusion strength of a not-exhibited $DS^{(o)}$ for $DD^{(q)}$.	$\pi_{DS}[DD^{(q)}]_{\uparrow(\downarrow)}$ $\therefore ES(DD^{(q)}\|DS^{(o)} \notin \Pi_{DS})_{\downarrow(\uparrow)}$
Symptom Necessity Principle (SNP)	If a certain $DS^{(w)}$ is always occurring with a given $DD^{(q)}$ and $DS^{(w)}$ is not found in the particular patient, the $DD^{(q)}$ should be excluded.	$DD^{(q)} \in \Pi_{DD} \to DS^{(w)} \in \Pi_{DS}$ $\neg(DS^{(w)} \in \Pi_{DS})$ $\therefore \neg(DD^{(q)} \in \Pi_{DD})$
Conjunction Disconfirmation Principle (CDP)	If the independent non-existence of $DS^{(o)}$, $DS^{(o')}$ disconfirm $DD^{(q)}$ to certain degrees, non-existence of both symptoms disconfirms $DD^{(q)}$ more than each one independently.	$ES(DD^{(q)}\|DS^{(o)} \wedge DS^{(o')})$ $> ES(DD^{(q)}\|DS^{(o)})$ $> ES(DD^{(q)}\|DS^{(o')})$
Frequency Exclusion Principle (FEP)	The more (less) often a certain $DS^{(w)}$ occurs in a given $DD^{(q)}$, the higher (lower) is the exclusion strength for the non-existence of $DS^{(o)}$.	$\theta(DS^{(w)} \wedge DD^{(q)})_{\uparrow(\downarrow)}$ $\therefore ES(DD^{(q)}\|DS^{(o)})_{\uparrow(\downarrow)}$

3.7. The Trouble with Medical Probability

Undoubtedly, probability assessment and interpretation is vital in medical judgment and decision-making. The matter, however, is not a walk in the park, so to speak. We already discussed in Section 3.2 several cases of clinical practice and medical research where the interpretation of probability is not straightforward, and is often misunderstood. Probability assessment and interpretation in a clinician's everyday routine can be a slippery affair. For one thing, it is not unusual that a clinician fails to properly translate an assertion in probabilistic terms (which can lead to poor interpretation of disease symptoms, wrong diagnoses and inappropriate treatments), or that a care provider processes false information (e.g., it is not surprising that some experts prefer their own theories and models to competing ones that come closer to the truth). The readers may appreciate some examples of probability inferences that clearly depend on the physician's cognitive adequacy about the case at hand.

Example 3.42: Consider the rare yet deadly *hemorrhagic* disease (*HD*). Assume that the physician's diagnosis is that

"Mr. Drosinis suffers from *HD*",

and the case-specific information is that

"diagnostic hemorrhagic test (*HT*) is used".

The test is highly accurate: if one has *HD*, the *HT* will be positive with probability 99%, and if one does not have *HD*, the *HT* will be negative with probability 99%. Given the high accuracy of *HT*, it is sometimes inferred that if Mr. Drosinis tests positive then his chances of having *HD* are 99%, i.e.,

$$P_{KB}[HD|HT] = 99\%.$$

However, this is false. Rather, 99% is the probability that a test is positive if Mr. Drosinis has *HD*, i.e.,[21]

$$P_{KB}[HT|HD] = 99\%.$$

In fact, the rarer the disease, the lower the probability that a positive test result implies that one actually has the disease, despite the test's high accuracy. To see this in numerical terms, first assume that 1 out of every 1000 people has *HD*, i.e., $P_{KB}[HD] = 0.1\%$. In this case, it is calculated that a patient is 9% likely to have the disease if tested positive, or

$$P_{KB}[HD|HT] = 9\%.^{22}$$

---

[21] Analogous was the case of the Ebola disease test in Example 3.22.

[22]
$$P_{KB}[HD|HT] = P_{KB}[HT|HD]P_{KB}[HD]P_{KB}^{-1}[HT] - 0.99 \times 0.001 \times 0.011^{-1} = 0.09,$$

with

$$P_{KB}[HT] = P_{KB}[HT|HD]P_{KB}[HD] + P_{KB}[HT|\neg HD]P_{KB}[\neg HD]$$
$$= 0.99 \times 0.001 + 0.01 \times 0.999 = 0.011.$$

Next, assume that 2 out of every 1000 people have the disease, $P_{KB}[HD] = 0.2\%$, which yields

$$P_{KB}[HD|HT] = 18\%,$$

meaning that in light of a positive test, Mr. Drosinis is 18% likely to have the disease. The moral here is that physicians must have a good understanding of the relevant probabilities and their context when they present test results and discuss their diagnosis with patients.

It should be noticed that several studies on hypothesis generation have revealed that physicians generate fewer hypotheses than are possible for the particular case; the number of hypotheses that are actually considered is constrained by cognitive limitations and task characteristics, and the number of hypotheses is further reduced when the physicians are under time pressure, and the generated hypotheses have the highest *a priori* probability. Accordingly, the number and *a priori* probabilities of the hypotheses considered by the physician will affect the perceived probability of the hypotheses under consideration.

Example 3.43: Studies have shown that physicians whose cognitive condition includes a set of reliable diagnostic tests and a sound understanding of the notion of probability can readily calculate the correct likelihood that a patient has *Parkinson's disease*. On the other hand, physicians whose condition does not include a good understanding of probability often fail to calculate the correct likelihood because they commit fallacies and make mistakes. These include the *base rate fallacy* in which physicians focus on irrelevant data and neglect to account for the prior probability of the case, and the tendency to look for confirming rather than disconfirming evidence. Another common mistake is that, supposedly, the conjunction of two statements has a greater probability than one of the statements. In particular, consider the diagnosis

"Ms Vergara suffers from Parkinson's disease (PD)",

and the case-specific evidence

"Ms Vergara is older than 65 years (S)".

In light of the above, it often seems natural to some physicians to assume that $P[PD \wedge S] > P[PD]$, which is actually false. This mistake is due to the confusion of the notion of "conjunction" ($\wedge$) with that of "conditional" $(\cdot|\cdot)$, for which it is indeed valid that $P[PD|S] > P[PD]$. More precisely,

"the probability to have PD given that one is S" is greater than "the probability to have PD", which in turn is greater than "the probability to have PD and to be S".

Physicians on the verge of making a decision about a case should be aware that some of the standard argumentation rules in Table 2.10 apply in terms of probabilities, whereas some others do not.

Example 3.44: Assume that DS and DD denote a disease symptom and a diagnosis, respectively. A valid probabilistic formulation of the standard logic scheme *modus ponens* is as follows:[23]

$$\begin{array}{ll} \text{Case premises} & \dfrac{P_{KB}[DD|DS] \gg}{DS} \\ \text{Physician's} \\ \text{conclusion} & \therefore P_{KB}[DD] \gg \end{array} \quad (3.53)$$

In other words, given a high ($\gg$) conditional probability of a diagnosis *assuming* the presence of a symptom, and subsequently the symptom being *indeed* the case, the probability of the diagnosis is also high. On the contrary, the following probabilistic formulation of *modus tollens*,

$$\begin{array}{ll} \text{Case premises} & \dfrac{P_{KB}[DD|DS] \gg}{\neg DD}, \\ \text{Physician's} \\ \text{conclusion} & \therefore P_{KB}[DS] \ll \end{array} \quad (3.54)$$

does not apply; i.e., if the conditional probability of a diagnosis assuming the presence of a symptom is high, and subsequently the diagnosis turns out not to be the case, then it is not necessarily valid that the probability of the symptom is small. As should be expected, the validity of Eqs. (3.53)–(3.54) can also be shown using Table 3.11.

The probability of an assertion or a mind state, Eq. (3.3), includes the possibility that different probabilities are assigned to alternative

---

[23] As usual, the symbol $\gg$ means "considerably high" and the $\ll$ means "considerably low".

representations of the same clinical case in terms of a_A. The readers may be reminded that along similar lines is the concern of the so-called support theory (Section 2.1), in which probability is assigned not to events (object language in the present book) but rather to descriptions of events (metalanguage) that in support theory are called "hypotheses". As was mentioned in Section 2.1, the support theory refers to the interplay between object language and metalanguage in a probabilistic milieu, and assumes not that each hypothesis refers to a unique event, but that a given event can be described by more than one hypothesis. Viewed in this methodological framework, the support model is a special case of Eq. (3.3).

The above and similar examples demonstrate that although a physician's uncritically subjective assessment may be a fact in clinical practice, it usually turns out to be unreliable. There are several reasons for this unreliability: a physician's focusing on a single possibility, discounting unspecified possibilities or underestimating the impact of examination findings on probability estimation. Furthermore, it may be worth noticing that within the general framework of Eq. (3.3), *fuzzy* assertions (involving diagnoses, symptoms, beliefs) can be interpreted in terms of the associated *KB*s possessed by the medical investigator. The following example attempts to illustrate this point.

Example 3.45: Consider the case in which little Alex is brought to the physician's office and needs to be assigned a prior probability to the initial diagnosis

"Alex is malnourished (MN)".

As it stands, some physicians consider the term "malnourished" a fuzzy one, because it may refer to a person living in a

"poor neighborhood, in which case the probability $P_{KB_1}[MN]$ would be high";

"in a middle-class neighborhood, in which case the probability $P_{KB_2}[MN]$ is considerably smaller" or

"in a rich neighborhood, meaning that the probability $P_{KB_3}[MN]$ is expected to be even smaller"

(perhaps for different reasons). The distinct *KB*s of the individual physicians in charge of the medical case are closely relevant to the initial diagnosis they are likely to make, in which case the formulation introduced by

Eq. (3.3) provides the means for an effective and rigorous interpretation of fuzzy propositions and data in terms of varying KBs.

Continuing research on the implementation of Eq. (3.3) focuses on developing efficient methods for encompassing core theory and relevant data, creating links between medical reasoning and disease assessment, not ignoring contrary evidence and dealing with the potential existence of unanticipated knowledge (Section 5.3).

3.8. Translating Medical Assertions into Probabilistic Terms

At this point, most health care providers would agree that in their daily routine the assertions, assessments and decisions contain an amount of uncertainty that should be translated into rigorous probabilistic terms. In many medical studies, a physician A finds it convenient to probalify reasoning by assuming that asserting that CA_p holds at location–time p implies that its probability is high enough, and *vice versa*. In symbolic terms,

$$a_A CA_p \leftrightarrow P_{KB}[CA_p] \geq \zeta_a, \qquad (3.55)$$

where $P_{KB}[CA_p] = P[a_A CA_p]$, and ζ_a, termed the justification degree of the case, is a number properly chosen by the health care provider, depending on the clinical or exposure environment. In many cases, a reasonable range of values is $0.5 < \zeta_a \leq 1$.[24] The degree ζ_a may be linked, e.g., to the level of certainty at which a clinician is willing to start disease treatment (the so-called treatment-threshold probability, which is calculated by properly assessing a treatment's harms and benefits, accounting for the patient's views etc.). Under these conditions, A's assertion about CA_p is valid iff its probability is at least as high as ζ_a, and, accordingly, the latter may be seen as a measure of the required degree of case justifications. In large-scale health systems, the justification degree may be also linked to the so-called critical threshold of the disease, i.e., a disease shows quantitatively different behavior above and below a threshold (Volz and Meyers, 2009). Below the critical threshold the occurrence of, say, an epidemic outbreak is considered unlikely, whereas above the threshold the outbreak is possible to a varying degree.

---

[24] Note that Eq. (3.55) directly implies that $a_A(\neg CA_p) \leftrightarrow P_{KB}[CA_p] \leq 1 - \zeta_a$.

Table 3.15: Possible $\zeta_a - a_A$ correspondence.

Case justification degree	a_A-form
$\geq \zeta_k$	k_A
$[\zeta_r, \zeta_k)$	r_A
$[\zeta_s, \zeta_r)$	s_A
$[\zeta_b, \zeta_s)$	b_A
$[\zeta_g, \zeta_b)$	g_A

Table 3.16: Mr. Yu's $\zeta_a - a_A$ (BP) correspondence.

BP case justification degree	a_A-form
≥ 0.9	k_A
$[0.8, 0.9)$	r_A
$[0.7, 0.8)$	s_A
$[0.6, 0.7)$	b_A
$[0.5, 0.6)$	g_A

Evidently, the subscript "a" means that the value of ζ_a is linked to the form of the assertion a_A, particularly, the larger the ζ_a value the epistemically stronger the a_A form. This observation leads to a practical way to choose among a_A-forms: assuming a possible ranking of ζ_a-values

$$\zeta_g < \zeta_b < \zeta_s < \zeta_r < \zeta_k,$$

the correspondence between ζ_a and a_A in Table 3.15 is plausible.

Example 3.46: For numerical illustration, physician A is concerned about a patient's vital statistics, in particular with

"Mr. Yu's blood pressure (BP)".

Physician A assumes the case-specific ζ_a-a_A association shown in Table 3.16. If for the particular case the calculated degree is $\zeta_a = 0.94$, in light of Eq. (3.55), the physician is likely to be justified in choosing $a_A \equiv k_A$, i.e.,

"A knows that 'Mr. Yu has high BP'" ($k_A BP$).

If, on the other hand, the assumed justification degree is $\zeta_a = 0.65$, the physician may rather prefer that $a_A \equiv b_A$, i.e.,

"A believes that 'Mr. Yu has high BP'" ($b_A BP$).

We mentioned earlier that when confronted with an intuitive or empirical argument that seems plausible, a health care provider could test its validity by translating it in SMR inference terms, and seeing whether the resulting formulation is valid or not. In light of Eq. (3.55), a useful method of testing medical arguments encountered in a case study (which may include theoretical assertions, empirical evidence and intuitive assessments) is described next. Say, the medical argument to be tested is expressed in terms of the relevant case attribute vector $\boldsymbol{CA}$, which includes one or more case attributes $\{CA_{i,p}; i = 1, 2, \ldots, n\}$, and is written as $Arg(\boldsymbol{CA})$, where Arg means "argument expressed in terms of the case attributes inside the parenthesis".

(1) The $Arg(\boldsymbol{CA})$ is translated in terms of the appropriate a_A-forms. Whether a_A will be selected to denote specifically k_A, r_A, s_A, b_A or g_A may depend, *inter alia*, on the biomedical knowledge, expertise, justification degree ζ_a and the critical threshold associated with the case.
(2) In light of Eq. (3.55), the assertion probabilities $P_{KB}[\boldsymbol{CA}] = P[a_A\boldsymbol{CA}]$ of $Arg(\cdot)$ in *step 1* are formulated following the probability analysis of Section 3.4 (see also Tables 3.11 and 3.13).
(3) If all the corresponding probability values are $\geq \zeta_a$, the original argument has successfully passed the test (it is ζ_a-*valid*). If the probability values are $< \zeta_a$, the argument has failed the test (ζ_a-*invalid*), and must be revised accordingly.

Symbolically, the above method may be represented as

$$Arg[\boldsymbol{CA}] \xrightarrow{step\ 1} a_A[\boldsymbol{CA}], \zeta_a$$
$$\xrightarrow{step2} P_{KB}[\boldsymbol{CA}] \xrightarrow{step\ 3} \begin{cases} \geq \zeta_a, \text{ Valid} \\ < \zeta_a, \text{ Invalid} \end{cases}. \quad (3.56)$$

When applying this method, the physician should understand explicitly the reasons for following the suggested *steps* 1–3. On occasion, it is appropriate that the medical thinking be recursive, i.e. to take itself into consideration (e.g., the physician would ask oneself, "What thought am I thinking now?"). Recursive thinking is related to metacognition — a method of introspection in which a care provider contemplates or reflects on one's own thinking. Representing the ability to realize which thinking mode one is in, metacognition is a vital component of the diagnostic process. As noted in Kassirer (2010), certain clinical reasoning studies cite the view that metacognition may be an effective strategy for avoiding cognitive errors (Graber, 2003). To gain some insight, let us study a few examples.

Example 3.47: Earlier we considered the intuitive argument that (see Section 2.3.5)

"physician A asserts that disease diagnosis DD is not the case"

is equivalent to

"it is not the case that A asserts that DD".

The two assertions of the argument can be written as $a_A \neg DD$ and $\neg a_A DD$. Accordingly, the argument to be tested is symbolically written as

$$Arg[DD, \neg DD]: a_A \neg DD \equiv \neg a_A DD \ (\textit{step 1 of method (3.56)}).$$

The corresponding probabilistic relationship is

$$P[a_A \neg DD] \equiv P[\neg a_A DD] \ (\textit{step 2}).$$

Using the probability formulation to test the validity of this argument, one starts by letting $a_A \neg DD$ be the case, which means that $a_A \neg DD \leftrightarrow P[a_A \neg DD] = P_{KB}[\neg DD] \geq \zeta_a$; and since from basic probability theory $P_{KB}[\neg DD] = 1 - P_{KB}[DD]$, the physician can easily deduce that

$$P[a_A(DD)] = P_{KB}[DD] = 1 - P_{KB}[\neg DD] \leq \zeta_a \ (\textit{step 3}).$$

That is, $a_A DD$ is not valid, which, in turn, implies that

$$\neg a_A DD \textit{ is the case.}$$

Therefore, with the help of the method of Eq. (3.56), what was previously merely a plausible argument has now been proven to be rigorously valid.

Example 3.48: Let us revisit the environmental health argument made in Example 2.56. According to the method of Eq. (3.56), above, each term of Eq. (2.26) should be considered valid across space–time iff the associated probability is high enough. The probabilities of the first two terms,

$$P_{KB}[CA_{2,\boldsymbol{p}}|CA_{1,\boldsymbol{p}}] \quad \text{and} \quad P_{KB}[\neg(\neg CA_{3,\boldsymbol{p}}|CA_{1,\boldsymbol{p}})],$$

could indeed be high. However, the probability of the third term is

$$P_{KB}[CA_{2,\boldsymbol{p}}|CA_{1,\boldsymbol{p}} \wedge CA_{3,\boldsymbol{p}}] = 0,$$

since given $CA_{3,\boldsymbol{p}}$ (i.e., Mr. Carsteanu leaves the city during a high pollution period), the state $\neg CA_{2,\boldsymbol{p}}$ is the case (i.e., it is not valid that he gets sick by experiencing high pollutant levels), which invalidates the inference of

Eq. (2.26) in the present environmental health situation. These findings emphasize the constant need for care providers to test the validity of *ad hoc* inferences before accepting them as valid in the case of interest.

Example 3.49: Another situation of an apparently straightforward inference is as follows:

$r_A \neg (IHIV \wedge DHIV)$: "it is a rational belief that it is not valid that Mr. Chernov is 'immune to HIV' and at the same time he 'died of HIV',"

in which case, if it is claimed that

$r_A DHIV$: "it is a rational belief that Mr. Chernov 'died of HIV',"

then

$r_A \neg IHIV$: "it is a rational belief that it is not the case that Mr. Chernov is 'immune to HIV'."

The above inference seems dialogically meaningful. Yet, the implementation of method (3.56) warns the investigator that the inference is not necessarily valid, unless the $r_A DHIV$ becomes certain knowledge.[25]

To see why this happens, let us consider the general expression of the above syllogism for any two case attributes CA_1 and CA_2, as follows:[26]

$$\begin{array}{ll} \text{Case premises} & \begin{array}{c} r_A \neg (CA_1 \wedge CA_2) \\ \underline{r_A CA_2} \end{array} \\ \begin{array}{l} \text{Physician's} \\ \text{conclusion} \end{array} & \therefore \ r_A \neg CA_1 \end{array} \qquad (3.57)$$

The method of Eq. (3.56) shows that this result is not generally valid in substantive logic terms. The readers may notice that the premises of inference (3.57) are linked to probabilities, as follows:

$$r_A \neg (CA_1 \wedge CA_2) \leftrightarrow P_{KB}[\neg(CA_1 \wedge CA_2)] \geq \zeta_r (\in [0.8, 0.9)),$$

say, and

$$r_A CA_2 \leftrightarrow P_{KB}[CA_2] \geq \zeta_r.$$

---

[25] The readers may recall that when $k_A CA \equiv CA$ one talks about complete knowledge of the case (all rigorous conditions of an objective assessment of the situation are fulfilled).
[26] Recall de Morgan's rule, $\neg(CA_1 \wedge CA_2) = (\neg CA_1) \vee (\neg CA_2)$.

From basic probability theory,

$$P_{KB}[\neg(CA_1 \wedge CA_2)] = 1 - P_{KB}[CA_1 \wedge CA_2] \geq \zeta_r.$$

The same theory shows that

$$P_{KB}[CA_1 \wedge CA_2] \leq P_{KB}[CA_1] = 1 - P_{KB}[\neg CA_1],$$

or that

$$P_{KB}[\neg CA_1] \leq 1 - P_{KB}[CA_1 \wedge CA_2] \geq \zeta_r$$

holds true, which does not exclude the possibility that

$$P_{KB}[\neg CA_1] \leq \zeta_r.^{27}$$

As a result, $r_A \neg CA_1$ can be invalid, thus contradicting the conclusion of inference (3.57). If, however, $r_A CA_2$ becomes a certainty, the inference (3.57) becomes

$$\begin{array}{cc} \text{Case premises} & \begin{array}{c} r_A \neg (CA_1 \wedge CA_2) \\ CA_2 \\ \hline \end{array} \\ \text{Physician's conclusion} & \therefore \; r_A \neg CA_1 \end{array}, \qquad (3.58)$$

i.e., a valid inference. This is demonstrated with reference to method (3.56). Since in inference (3.58) the second premise is certainly valid, the first premise gives

$$r_A \neg (CA_1 \wedge CA_2) \leftrightarrow P_{KB}[\neg(CA_1 \wedge CA_2)] = P_{KB}[\neg CA_1] \geq \zeta_r,$$

i.e., $r_A \neg CA_1$ is now valid.

The take home message is that a case inference may be valid in terms of an a_A-form but not in terms of another form. A list of intuitive arguments whose validity can be proven using the method of Eq. (3.56) is given in Table 3.17. It is worth noticing that translating states of mind (assertion form) into probabilities has many other applications. As we will see in

---

[27]For numerical illustration, let us focus on the limiting case $P_{KB}[\neg(CA_1 \wedge CA_2)] = \zeta_r = 0.82$, i.e., the probability a care provider rationally assigns to $\neg(CA_1 \wedge CA_2)$ is equal to the justification degree of the current case. Then, $P_{KB}[CA_1 \wedge CA_2] = 1 - 0.82 \leq P_{KB}[CA_1]$, or, $P_{KB}[\neg CA_1] \leq 0.82$, which means that $r_A \neg CA_1$ is not valid at the $\zeta_r = 0.82$ degree, and, hence, the reasoning of Eq. (3.57) does not apply here.

Table 3.17: Intuitive arguments involving statistical conditionals.

Case premises	$r_A(CA_1|(CA_2 \wedge CA_3))$	Case premises	$a_A \neg (CA_1 \wedge CA_2)$
Care provider's conclusion	$\therefore r_A(CA_1 \vee (\neg CA_2| CA_3))$		$k_A CA_2$ or CA_2
		Care provider's conclusion	$\therefore a_A \neg CA_1$
Case premises	$r_A(CA_1|(CA_2 \vee CA_3))$	Case premises	$r_A((CA_1 \vee CA_2)|CA_3)$
			$r_A(\neg CA_1|CA_3)$
Care provider's conclusion	$\therefore \neg CA_2$ $\therefore r_A(CA_1|CA_3)$	Care provider's conclusion	$\therefore r_A(CA_2|CA_3)$
Care provider's conclusion	$\therefore \neg [r_A(CA_1|CA_2) \wedge r_A(\neg CA_1|CA_2)]$	Case premises	$a_A \neg CA_1$
		Care provider's conclusion	$\therefore \neg a_A CA_1$
Case premises	$r_A((CA_1 \wedge CA_2)|CA_3)$	Case premises	$r_A(CA_1|(CA_2 \vee CA_3))$
			$r_A(CA_1|(\neg CA_2 \vee CA_3))$
Care provider's conclusion	$\therefore r_A(CA_1|CA_2 \wedge CA_3)$	Care provider's conclusion	$\therefore r_A(CA_1|CA_3)$
Case premises	$r_A((CA_1 \wedge CA_2)|CA_3)$	Case premises	$r_A(CA_1|CA_2)$
			$r_A(CA_1|CA_3)$
Care provider's conclusion	$\therefore r_A(CA_1|CA_3)$ $\wedge r_A(CA_2|CA_3)$	Care provider's conclusion	$\therefore r_A(CA_1|(CA_2 \vee CA_3))$

Section 4.4, e.g., the method provides a useful spatiotemporal criterion of disease causation (etiology) under conditions of *in situ* uncertainty.

The inferences discussed above demonstrate that the combination of qualitative and quantitative analyses in medical reasoning is more powerful than their being used alone. Among other things, qualitative methods are used to develop and refine health care professionals' hypotheses, whereas subjective evaluation can help professionals discover more about their subject area. As the study in the example below illustrates, one may discover relationships between variables that health investigators currently neglect.

Example 3.50: Using qualitative methods, investigators made the key discovery that only the children of wealthy families developed childhood liver cancer in the Philippines during the 1950s. As was discussed in Example 1.1, motivated by such qualitative findings a systematic research was conducted concerning the relationship between diet and diseases of affluence (cancer, diabetes, heart disease). During that time period peanut butter in the Philippines was contaminated with a mold called aflatoxin, a highly carcinogenic substance that was responsible for the development of cancer tumors in the population.

In sum, quantitative analysis allows rigorous description of the medical case of interest, whereas qualitative analysis provides significant insight. Actually, the case's motto is: calculation without comprehension is fruitless, and comprehension without calculation is ambiguous.

3.9. Space–Time Reasoning Dynamics

It has been said, and with good reason, that there is no information change without a changing world. As our discussion so far has emphasized, the aim of medical inference is to be *dynamic* rather than static, i.e., a case assertion $a_A C A_p$ may vary with time and place ($p = (s, t)$), depending on changes in the health care provider's epistemic condition and the case's context. What the reasoning dynamics actually change is not an ontic state but only a mind state (a case assertion, a theory, a model). A physician's mind state is being revised, but this does not necessarily have any effect on the ontic facts (i.e., expressing factual situation).

3.9.1. *Changes in assertions and substantive conditionals*

Let us take stock. Stochastic reasoning in medical research and clinical practice considers different kinds of mind state changes (containing different types of a_A-forms). These changes, which must assure internal consistency among the available *KB*s and care provider's belief systems, are generally expressed in terms of the *substantive conditional* in the broad sense, denoted as "$\cdot \backslash \cdot$" (Section 2.6.1). The "$\cdot \backslash \cdot$" is useful in the interpretation of key reasoning terms, including "implication," "causation" and "synthesis". The precise formulation of "$\cdot \backslash \cdot$" in *SMR* inference depends on case's contentual and contextual environment. For instance, the presence of a biophysical law signifies a causal connection between exposure and health effect, in which case the "$\cdot \backslash \cdot$" must account for that law. In the space–time synthesis milieu, causality assumes various forms (Section 4.4). Under certain considerations, a suitable choice of "$\cdot \backslash \cdot$" is the material conditional ($\cdot \rightarrow \cdot$), with one important distinction: while in formal logic the "$\cdot \rightarrow \cdot$" is content-free, *SMR* inference assigns to it context and content, and then proceeds with the relevant probabilistic calculations.

Table 3.18: Substantive conditional inferences.

Case premises	$CA_1 \therefore CA_2$		$CA_2 \backslash CA_1$
Care provider's conclusion	$\therefore CA_2 \backslash CA_1 \dot{:} KB$		$\therefore (CA_2 \vee CA_3) \backslash CA_1 \dot{:} KB$
Case premises	$CA_2 \backslash CA_1$ $CA_3 \backslash CA_1$		$(CA_1 \vee \neg CA_2) \backslash CA_1$
Care provider's conclusion	$\therefore (CA_2 \wedge CA_3) \backslash CA_1 \dot{:} KB$		$\therefore (\neg CA_1) \backslash CA_3 \dot{:} KB$
Case premises	$CA_{2,p} \backslash CA_{1,p}$ $\neg(\neg CA_{3,p} \backslash CA_{1,p})$		$CA_{2,p} \backslash CA_{1,p}$ $CA_{2,p} \backslash CA_{3,p}$
Care provider's conclusion	$\therefore CA_{2,p} \backslash (CA_{1,p} \wedge CA_{3,p}) \dot{:} KB$		$\therefore CA_{2,p} \backslash (CA_{1,p} \vee CA_{3,p}) \dot{:} KB$
Case premises	$\left. \begin{array}{l} CA_2 \backslash CA_1 \\ CA_1 \backslash CA_2 \\ \therefore CA_3 \backslash CA_1 \end{array} \right\} \leftrightarrow CA_3 \backslash CA_2 \dot{:} KB$		CA_1 $CA_2 \backslash CA_1$ $\leftrightarrow CA_1 \wedge CA_2 \dot{:} KB$
Care provider's conclusion			

Example 3.51: In order to establish $a_A CA_{1,p} \to a_A CA_{2,p}$ for the attributes $CA_{1,p}$ and $CA_{2,p}$, a physician has to conceive $a_A CA_{1,p}$ and $a_A CA_{2,p}$, and to show that from the conditions for $a_A CA_{1,p}$ one obtains the conditions specified by $a_A CA_{2,p}$, according to reasoning rules that preserve biophysical continuity and consistency from $CA_{1,p}$ to $CA_{2,p}$. As we saw earlier (Section 2.6), other examples of "$\cdot \backslash \cdot$" include the equivalence conditional ($\cdot \leftrightarrow \cdot$), counterfactual conditional ($\cdot \Rightarrow \cdot$) and statistical conditional ($\cdot | \cdot$). For rigor, many experts argue that in several medical studies the substantive conditional is assumed to satisfy the eight inferences listed in Table 3.18.

If the case-specific KB (S) includes new elements that are consistent with the core KB (G), the latter is expanded with the addition of the former. If, on the other hand, the S-KB includes clinical and other elements that contradict the G-KB, either these elements must be abandoned or the G must be revised to account for the new elements. In any case, integrated changes in G and S must result in a rational and coherent study of the medical case.

Example 3.52: If a medical doctor's view about the cause of lung cancer changes given new findings about smoking, this is a rational change. If, on the other hand, the doctor's view about cancer changes because of new findings concerning bird migration in Africa, it is difficult to consider it as a rational change.

As we continued to lay the land in medical thinking, the readers realized that reasoning dynamics in conditions of uncertainty can provide an essential set of relationships between case attributes that vary across space–time and/or *in situ* assertions about these case attributes, such as those listed in Table 2.5. So, starting from a valid clinical argument other arguments can be generated that are the same or more useful than the previous one. The matter is illustrated with the help of an example.

Example 3.53: Given that $DD^{(1)}$ and $DD^{(2)}$ are two different disease diagnoses, assume that a health care professional proposes that

$DD^{(1)} \vee (\neg DD^{(2)})$: "either diagnosis $DD^{(1)}$ is valid or diagnosis $DD^{(2)}$ is not valid".

In view of de Morgan's rule of Table 2.5, this argument yields

$\neg(\neg DD^{(1)} \wedge DD^{(2)})$: "it is not the case that $DD^{(1)}$ is not valid and $DD^{(2)}$ is valid".

Although the new argument is logically equivalent to the previous one, it offers additional insight by looking at the medical case from another angle. Also, given $DD^{(1)}$ and based on the addition relation of the same table, it holds that

$DD^{(1)} \vee DD^{(2)}$: "either diagnosis $DD^{(1)}$ or $DD^{(2)}$ is valid".

Lastly, by means of double negation,

$DD^{(1)} \therefore \neg(\neg DD^{(1)})$: "diagnosis $DD^{(1)}$ implies that it is not true that $DD^{(1)}$ is not valid".

Another thing worth recalling is that, given the space–time dependence and associated uncertainty characterizing a disease's evolution, its attribute CA_p is represented mathematically as a S/TRF (Section 2.2.2), which means that each assertion in Tables 2.9 and 2.10 is assigned a probability value. In particular, the probability dynamics of medical statements about

CA_p often assume the symbolic representation

$$P_{KB'}[CA_p] = P_{KB}[CA_p \backslash S] = \eta, \tag{3.59}$$

where the physician's epistemic state concerning CA_p has changed due to new data S. The probability function, $P_{KB}[\cdot \backslash \cdot]$, is a one-place function and the conditional event, $CA_p \backslash S$, is its argument. Since S must be physically and logically consistent with the core knowledge G, Eq. (3.59) may express *updating, revision* or *expansion* of the case. In medical diagnosis, e.g., belief revision is often caused by some piece of crucial data that contradicts the existing belief system about the disease. This state change may affect not only the physician's cognitive condition (knowledge currently possessed by the physician about the case), but may have consequences for other belief systems as well.

Example 3.54: Before seeing the patient, and in light of the existing G-KB, the physician in Example 2.9 identified the two most likely diagnoses. Assuming that the physician has a good understanding of probability, initial probabilities are assigned to each diagnosis based on G. During the subsequent medical interview, physical examination and supplemental testing (all these belong to S-KB), the initial probabilities undergo certain (upwards or downwards) adjustments depending on what the patient says and on the examination and testing results. Moreover, the patient's course of symptoms over time may provide a new element of S-KB that can cause to revision of the diagnostic probabilities. That is, medical diagnosis is essentially a feedback process.

The following example is from environmental health, and deals with some interesting updating effects of experimentation on the probability of population exposure.

Example 3.55: Regional public health management assumed that the population exposure distribution (pollutant concentration, PC_p) occurred in the geographical region during a specified time period (jointly denoted by p, as usual). In the same environment,

"an experiment e_1 reports that $PC_p \in \Delta$" (KB_1),

where Δ denotes an interval of possible concentration values. In the real world, the KB_1-based probability of the care provider's belief $b_A(PC_p \in \Delta)$

can be generally different from the actual probability, i.e.,

$$P[b_A(PC_p \in \Delta)] = P_{KB_1}[PC_p \in \Delta] \neq P[PC_p \in \Delta].$$

Standard health statistics implicitly assume that $P_{KB_1}[KB_1] = P[PC_p \in \Delta] = 1$. This is an interesting equation linking a metalanguage probability $P_{KB_1}[\cdot]$, with an object language probability $P[\cdot]$. In other words, the first probability is relativized to KB_1 whereas the second probability is non-relativized to any epistemic condition but is an unknown characteristic of nature. An exposure probability $P_{KB}[PC_p \leq \gamma_p]$, considered in an initial knowledge environment KB (γ_p is an exposure threshold at p), can change in light of a new piece of knowledge like KB_1. Accounting for KB_1, the updated exposure probability will be $P_{KB}[PC_p \leq \gamma_p|KB_1]$, which is generally different from $P_{KB}[PC_p \leq \gamma_p|PC_p \in \Delta]$. Readers may recall that KB_1 denotes that "experiment e_1 reports that $PC_p \in \Delta$", whereas "$PC_p \in \Delta$" by itself means that this is actually the case in nature. This observation is worth the readers' attention. Another interesting result is obtained if, in addition to KB_1, the care provider considers a second piece of information, an

"experiment e_2 reports that $PC_p \in \Delta$" (KB_2).

The probability equation of population exposure

$$P_{KB}[PC_p \leq \gamma_p|KB_1 \wedge KB_2] = P_{KB}[PC_p \leq \gamma_p|KB_1]$$

is valid if one assumes that $P[PC_p \in \Delta] = P[KB_2] = 1$ (which, though, is not generally valid in environmental health practice). Remarkably, the above probability equation is also obtained from "premise strengthening" analysis (Section 3.4.2). In particular, in Example 3.36 let us replace DD with $PC_p \leq \gamma_p$, $DS^{(1)}$ with KB_1, and $DS^{(2)}$ with KB_2, in which case the first premise of Eq. (3.25) becomes $P_{KB}[PC_p \leq \gamma_p|KB_1] = p_1$. We also notice that $P[KB_2] = 1$ implies $P_{KB}[DS^{(2)}|DS^{(1)}] = p_2 = P[KB_2|KB_1] = 1$, and the conclusion of Eq. (3.25) becomes

$$P_{KB}[PC_p \leq \gamma_p|KB_1 \wedge KB_2] \in [\max\{0, p_1\},$$
$$\min\{p_1, 1\}] \equiv p_1 = P_{KB}[PC_p \leq \gamma_p|KB_1].$$

This is, clearly, the same as the probability equation above.

It is worth revisiting the S/TRF concept in light of the probability dynamics formulation (3.59). Assume that $CD^{(i)}$ denotes one case diagnosis from the set W of all possibilities allowed by the physician's current knowledge environment,

$$W : \{CD^{(i)}; i = 1, 2, \ldots, n\},$$

i.e., W is directly linked to KB. Let $P_{KB}[CD^{(i)}] = P_W[CD^{(i)}]$ be the corresponding KB (or W)-based probability. For any other case diagnosis $CD^{(j)}$, $j \neq i$, consider the associated relative probabilities

$$\pi^i = P[CD^{(i)}|(CD^{(i)}, CD^{(j)})]^{28} \quad \text{and} \quad \pi^j = P[CD^{(j)}|(CD^{(i)}, CD^{(j)})],$$

such that $\pi^{(i)} + \pi^{(j)} = 1$ (probability summation property). Then, it can be shown that,

$$\frac{P_{KB}[CD^{(i)}]}{P_{KB}[CD^{(j)}]} = \frac{\pi^{(i)}}{1 - \pi^{(i)}}. \quad (3.60)$$

for any j. Equation (3.60) is a useful formula, and slightly surprising result, stating that the ratio of the probability of a specified case diagnosis $CD^{(i)}$ over that of any other possibility $CD^{(j)}$ is independent of the physician's current KB, and depends only on the relative probability $\pi^{(i)}$.

Example 3.56: Mr. Dreyfus' physician considers five possible diagnoses concerning his health situation

$$W : \{CD^{(i)}; i = 1, 2, \ldots, 5\}.$$

The physician's experience seems to favor the case diagnosis

"Mr. Dreyfus has flu" $(CD^{(1)})$,

with $P_{KB}[CD^{(1)}] = P_{\{CD^{(i)};\ i=1,2,\ldots,5\}}[CD^{(1)}] = P[CD^{(1)}|\{CD^{(1)},\ldots, CD^{(5)}\}]$. Let $\pi^{(1)}$ be the probability of $CD^{(1)}$ given that at the current study phase two possibilities are considered, $\{CD^{(1)}, CD^{(2)}\}$,[29] i.e.,

$$\pi^{(1)} = P_{\{CD^{(1)}, CD^{(2)}\}}[CD^{(1)}] = P[CD^{(1)}|(CD^{(1)}, CD^{(2)})].$$

Then, $1 - \pi^{(1)} = P[CD^{(2)}|(CD^{(1)}, CD^{(2)})] = \pi^{(2)}$ and Eq. (3.60) reduces to

$$\frac{P_{KB}[CD^{(1)}]}{P_{KB}[CD^{(2)}]} = \frac{\pi^{(1)}}{1 - \pi^{(1)}}$$

---

[28] Noticeably, one could also write $P[CD^{(i)}|\{CD^{(i)}, CD^{(j)}\}] = P_{\{CD^{(i)}, CD^{(j)}\}}[CD^{(i)}]$.
[29] Say, $CD^{(2)} \equiv \neg CD^{(1)}$, meaning that $CD^{(2)}$ represents a non-flu diagnosis.

for any KB (this includes KBs associated with the cognitive condition of experienced and novice care providers alike).

Before concluding this section, we remind our readers that in the discussion of Section 2.6 the statistical conditional seems to have two theoretical advantages over classical material conditional, in certain cases: it accounts for *in situ* uncertainty and it is free of some counterintuitive features of material conditional, including monotonicity. Unlike its classical counterpart, the substantive conditional considered in SMR inference (including the stochastic material one) is nonmonotonic[30] expressing uncertain relations between the "if" and the "then" part of a conditional medical assertion. Interestingly, the substantive conditional $CA_{2,p} \backslash CA_{1,p}$ may be interpreted as a "high" conditional probability assertion if the probability $P_{KB}[CA_{2,p} \backslash CA_{1,p}]$, is "high". This comment introduces us to the next section.

3.9.2. *Probability dynamics and hypothesis confirmation*

A physician practicing SMR may confirm a medical diagnosis or clinical hypothesis of *in situ* health dynamics by checking the validity of the associated uncertainty or probability dynamics. In other words, as part of the medical decision process in conditions of uncertainty a physician may create a sort of correspondence,

$$\text{In Situ Health Dynamics} \leftrightarrow \text{Uncertainty (Probability) Dynamics},$$

(3.61)

which could confirm or disconfirm a qualitative assertion in terms of its associated quantitative uncertainty or probability functions.

Example 3.57: Assume that $PC_p = 0.11\,ppm$ denotes daily exposure to ozone (O_3) for an individual living at

p: "Hangzhou city during 2012".

Let e_1 and e_2 denote two experiments that both support the O_3 exposure situation. Following the analysis of previous sections, suppose that an

---

[30]See also Section 3.9.3.

investigator A makes what seems to be a commonsensical belief syllogism,

$$\begin{array}{l}\text{Ozone exposure}\\ \text{case premise}\end{array} \quad \dfrac{b_A(PC_p = 0.11|e_1 \vee e_2)}{\therefore \ b_A(PC_p = 0.11|e_1) \wedge b_A(PC_p = 0.11|e_2)}, \qquad (3.62)$$
$$\begin{array}{l}\text{Investigator's}\\ \text{conclusion}\end{array}$$

i.e.,

"A's belief that the specified O_3 exposure occurs given that either experiment e_1 suggests that this is the case or experiment e_2 suggests that this is the case, implies the conjunction of the beliefs that the O_3 exposure occurs given e_1 and that it occurs given e_2".

The corresponding probability formulation of the health dynamics (3.62) is

$$P_{KB}[PC_p = 0.11|(e_1 \vee e_2)] = P_{KB}[PC_p = 0.11|e_1]P_{KB}[PC_p = 0.11|e_2], \qquad (3.63)$$

which, however, is not generally valid. Accordingly, A's apparently commonsensical belief about O_3 exposure may be not true, in general.

Several other possibilities exist. Within the context of *in situ* health dynamics, a physician can implement *SMR* to potentially replace the rigidity of formal logic with the flexibility of *uncertainty reasoning*. To be more specific, it has been shown that

$$\text{if } \Phi_1 \therefore \Phi_2, \text{ then } U_{KB}[\Phi_2] \leq U_{KB}[\Phi_1], \qquad (3.64)$$

where Φ_1 and Φ_2 are logical functions (Section 2.3.7) of different case assertions (exposure, disease symptoms and diagnoses, medical tests and treatments); and $U_{KB} = 1 - P_{KB}$ denotes the uncertainty of a physician's assertion.[31] As it turns out, Eq. (3.64) has significant consequences for inferences involving clinical premises and medical conclusions.

Uncertainty functions associated with some of the syllogisms of Table 3.11 are displayed in Table 3.19. To throw additional light on Eq. (3.64) and its application, let us consider a few rather simple but instructive cases.

Example 3.58: In Eq. (3.64), assume that $DS^{(1)}$ and $DS^{(2)}$ denote two different disease symptoms, so that $\Phi_1 \equiv DS^{(1)}$ and $\Phi_1 \equiv DS^{(1)} \vee DS^{(2)}$.

---

[31] As usual, $U_{KB} \equiv U_{a_A}$.

Table 3.19: Uncertainty inequalities associated with some rules of Table 3.13.

Modus tollens:	*Modus ponens*:
$U_{KB}[\neg CA_1] \leq U_{KB}[CA_1 \to CA_2]$ $+ U_{KB}[\neg CA_2]$	$U_{KB}[CA_2] \leq U_{KB}[CA_1 \to CA_2] + U_{KB}[CA_1]$
Excluded middle:	*Absorption*:
$U_{KB}[CA_2] \leq U_{KB}[CA_1 \to CA_2]$ $+ U_{KB}[\neg CA_1 \to CA_2]$	$U_{KB}[CA_1 \to (CA_1 \land CA_2)]$ $\leq U_{KB}[CA_1 \to CA_2]$
Simplification:	*Conjunction*:
$U_{KB}[CA_1] \leq U_{KB}[CA_1 \land CA_2]$	$U_{KB}[CA_1 \land CA_2] \leq U_{KB}[CA_1] + U_{KB}[CA_2]$
Contradiction:	*Disjunctive syllogism*:
$U_{KB}[\neg CA_1] \leq U_{KB}[CA_1 \to CA_2]$ $+ U_{KB}[CA_1 \to \neg CA_2]$	$U_{KB}[CA_2] \leq U_{KB}[CA_1 \lor CA_2] + U_{KB}[\neg CA_1]$

It is immediately seen that the entailment in the left part of Eq. (3.64) is valid

$$DS^{(1)} \therefore (DS^{(1)} \lor DS^{(2)}),$$

since from probability theory it is known that the inequality in the right part, $U_{KB}[DS^{(1)} \lor DS^{(2)}] \leq U_{KB}[DS^{(1)}]$, holds true.[32] Furthermore, in order to demonstrate the invalidity of the inference

$$DS^{(1)} \therefore (DS^{(1)} \land DS^{(2)}),$$

instead of showing that $DS^{(1)}$ can be valid but $DS^{(1)} \land DS^{(2)}$ invalid (which may involve searching a large number of possible scenarios), the health care provider merely needs to demonstrate that the corresponding inequality in the right part of Eq. (3.64) is violated, since probability theory shows that

$$U_{KB}[DS^{(1)} \land DS^{(2)}] \geqq U_{KB}[DS^{(1)}].$$

The reasoning (3.64) can be extended in different ways. For example, based on an earlier result (Christakos, 2010: 310), the uncertainty of the conclusion of a substantive reasoning process cannot exceed the sum of the

---

[32] This is easy to prove with the help of probability theory, using the fact that $U_{KB} = 1 - P_{KB}$.

uncertainties of its premises, i.e.,

$$\text{if } \boldsymbol{CA} \therefore CA^{(k)} \text{ then } U_{KB}[CA^{(k)}] \geq U_{KB}[\boldsymbol{CA}] \equiv \Sigma_{i=1}^{m} U_{KB}[CA^{(i)}], \tag{3.65}$$

where, as usual, $\boldsymbol{CA} \equiv \{CA_i^{(i)}, i = 1, 2, \ldots, m\}$. The readers may find it worth noting that the same result is also derived as a consequence of our discussion in Section 3.2.2 — the probability of an assertion is the sum of the probabilities of the possible realizations (worlds) in which the assertion happens to be valid.

3.9.3. The case of non-monotonic medical reasoning

Non-monotonic reasoning investigates decision-making that allows retracting diagnostic conclusions in the light of new evidence. This kind of thinking is employed by *SMR* inference, as well as by logics that study the formalization of commonsensical argumentation. An intriguing situation emerges when a new piece of valid information can mislead care providers to reject accurate knowledge. Let us illustrate the notion of non-monotonicity with the help of an example.

Example 3.59: A team of health care providers plans to visit the Polish town Kostrzyn. Poland belongs to the European Union (EU), and, of course, so does Kostrzyn, denoted as $EU(Ko)$. Since the

"flu vaccine FV_1 is available in EU member countries" $EU(FV_1)$,

admittedly at varying prices, the team prepared a health management plan based on the belief that

"Kostrzyn citizens used FV_1" $UFV_1(Ko)$,

where the FV_1 price in Poland is €45. In view of the above, the following inference seems to be valid

Polish study premises 1–2	$P_{KB}[UFV_1(Ko)|EU(Ko)] = 0.95$ $EU(Ko)$
Care provider's conclusion 1	$\therefore P_{KB}[UFV_1(Ko)] = 0.95$

But during their last team meeting before departure, the care providers were given the new piece of information that "cheaper vaccine FV_2 is available in parts of Poland, including Kostrzyn", $AFV_2(Ko)$

(the FV_2 price is €35, although it is not as effective as the first vaccine). Given $AFV_2(Ko)$, the experts revised the previous considerations to infer that

Polish study
Premises 3–5
$$\frac{AFV_2(Ko) \\ P_{KB}[UFV_1(Ko)|AFV_2(Ko)] = 0.01 \\ P_{KB}[EU(Ko)|AFV_2(Ko)] = 0.99}{}$$

Care provider's conclusion 2
$\therefore P_{KB}[UFV_1(Ko)|EU(Ko) \wedge AFV_2(Ko)] \leq 0.01$

But the care professionals team was unaware of the geographical fact that

"Kostrzyn is on the German border",

$GB(Ko)$ so most people still get the better FV_1, at the cheap price of €35. This fact corrects the last inference as follows

Polish study premise 6
$$\frac{GB(Ko)}{}$$

Care provider's conclusion 3
$\therefore P_{KB}[UFV_1(Ko)|EU(Ko) \wedge GB(Ko)] = 0.95$

This is the same probability value as in earlier *conclusion* 1 (0.95) and drastically different than the probability value in *conclusion* 2 (0.01). In view of the otherwise valid evidence $AFV_2(Ko)$, the team falsely revised its initial highly probable information concerning $UFV_1(Ko)$. This rather paradoxical situation is frequently described by the motto, "one knows more by knowing less".

Reasoning dynamics in clinical practice and medical research include an explicit way of structuring and evaluating one's thinking and acting. This can effectively help a physician realize which case aspects are linked to uncertain knowledge, how this uncertainty propagates across space and time and, most frequently, what the above imply for the case's conclusions. Accordingly, the physician must be fully aware of what decision-making attempts to achieve in the specified medical case, what is its relevance for resolving key issues and how it can help improve public and private medical practice. Later, we will examine the case of self-reference, i.e., the ability of a subject (patient) to speak of oneself.

3.10. Medical Syllogisms Involving Likelihood Ratios

Sometimes physicians like to express certain of their decision support calculations in terms of the case *likelihood ratios*. Such ratios are defined as

$$L_R^+ = \frac{P_{KB}[DT|DD]}{P_{KB}[DT|\neg DD]}, \quad L_R^- = \frac{P_{KB}[\neg DT|DD]}{P_{KB}[\neg DT|\neg DD]}, \quad (3.66\text{a–b})$$

where, as earlier, DT denotes diagnostic test and DD disease diagnosis. The likelihood ratios offer a measure of how much a physician's uncertainty may change after new test data become available; L_R^+ and L_R^- represent, respectively, a positive and a negative test result in a diseased person. For example, the corresponding likelihood ratios of the case sensitivity analysis of Table 3.6 vary between 20 and 1000 — a large variation indeed. Useful ratio values have been tabulated and are available in the medical literature that characterize clinical findings for various diseases. For illustration, Table 3.20 was developed by Diamond and Forester (1979) and is concerned with the varying amounts of ST segment depression during an exercise stress test and the corresponding likelihood ratio values of coronary artery disease.

Some convenient clinical formulas and medical syllogisms are obtained in terms of the tabulated likelihood ratios, such as

Case premises $\quad P_{KB}[DD] = p \qquad\qquad\qquad P_{KB}[DD] = p$
$\qquad\qquad\qquad L_R^+ \qquad\qquad\qquad\qquad\qquad L_R^-$

Physician's conclusion $\quad \therefore P_{KB}[DD|DT] = \frac{pL_R^+}{1+p(L_R^+-1)} \quad$ and $\quad \therefore P_{KB}[DD|\neg DT] = \frac{pL_R^-}{1+p(L_R^- -1)}$

(3.67a–b)

where the post-test probabilities were calculated using Bayes rule. When in a medical case the pre-test probability increases, so do both post-test

Table 3.20: Exercise stress electrocardiogram test results and associated coronary artery disease likelihood ratios.

Medical test result (mm ST segment depression)	<1.00	[1.00 – 1.50)	[1.50 – 2.00)	[2.00 – 2.50)	≥ 2.50
Likelihood ratio for coronary artery disease	0.40	2.09	4.50	10.80	38.00

probabilities above. The likelihood ratio difference of a case is

$$\Delta L = L_R^+ - L_R^- = \frac{P_{KB}[DT|DD]}{P_{KB}[DT|\neg DD]} - \frac{1 - P_{KB}[DT|DD]}{1 - P_{KB}[\neg DT|\neg DD]}$$

Notice that if $L_R^+ > 1$ then $\Delta L > 0$. The reverse holds when the likelihood ratio is less than 1. In addition to the statistical post-test case probability of Eq. (3.12), the material and equivalent post-test case probabilities of Eqs. (3.13)–(3.14) can be expressed in terms of the likelihood ratios, e.g.,

$$P_{KB}[DT \to DD] = 1 + P_{KB}[DT|\neg DD](p-1)$$

$$= 1 + \frac{P_{KB}[DT|DD]}{L_R^+}(p-1)$$

$$P_{KB}[DT \leftrightarrow CD] = pP_{KB}[DT|DD] + (P_{KB}[DT|\neg DD] - 1)(p-1)$$

$$= 1 - p + P_{KB}[DT|\neg DD](pL_R^+ + p - 1)$$

(3.68a–d)

Similar formulas are valid in terms of L_R^-. Next we revist and extend some numerical results, discussed in Sox et al. (2013).

Example 3.60: Mr. Oldershaw is a middle-aged man with a history of atypical angina pectoris. His physician Dr. A is examining him for the possibility that he suffers from *coronary artery* disease (CD). After taking Mr. Oldershaw's history into consideration A assigns a prior (pre-test) probability $P_{KB}[CD] = p = 0.70$, which is a representative value for a man of Mr. Oldershaw's age and with his medical history (Weiner et al., 1979). The physician ordered a stress exercise stress electrocardiogram test (ST) for Mr. Oldershaw, and the results indicated 1.45 mm ST segment depression. According to Table 3.20, $L_R^+ = 2.09$. Then, using syllogism (3.67a) above, physician A concludes that the post-test probability of CD is,

$$P_{KB}[CD|ST] = \frac{0.70 \times 2.09}{1 + 0.70 \times (2.09 - 1)} = 0.83,$$

which is considerably larger that the pre-test probability $P_{KB}[CD] = 0.70$ due to the positive ST result.

Example 3.61: Mr. O'Rorke's physician, Dr. A, observes a lung mass on his chest radiograph. According to published results, Mr. O'Rorke's pre-test probability of lung cancer (LC) is $P_{KB}[LC] = p = 0.40$. Also, the probability that a lung mass (LM) is present in persons with LC is $P_{KB}[LM|LC] = 0.6$ (chest radiograph sensitivity), whereas the probability that a lung mass is not present in persons who do not have LC is $P_{KB}[\neg LM|\neg LC] = 0.96$ (1-chest x-ray specificity). On the basis of this information, the likelihood ratio is easily obtained from Eq. (3.66a) as $L_R^+ = 15$. Subsequently the post-test LC probability is found from Eq. (3.67a) to be

$$P_{KB}[LC|LM] = \frac{0.40 \times 15}{1 + 0.40 \times (15 - 1)} = 0.909.$$

The corresponding material and equivalent conditional probabilities of CD can be also calculated as, from Eq. (3.68b) and (3.68d) respectively

$$P_{KB}[LM \rightarrow LC] = 1 + \frac{P_{KB}[LM|LC]}{L_R^+}(p-1) = 1 + \frac{0.6}{15}(0.40 - 1) = 0.976$$

$$P_{KB}[LM \leftrightarrow LC] = 1 - p + P_{KB}[LM|\neg LC](pL_R^+ + p - 1) = 0.816.$$

As was mentioned earlier (Section 3.3.2), which one of the above post-test probabilities to use in a specific clinical case is a matter for a capable physician to decide. Using stochastic theory it can be shown that it holds that $P_{KB}[LM \rightarrow LC] \geq P_{KB}[LC|LM], P_{KB}[LM \leftrightarrow LC]$, where $P_{KB}[LC|LM]$ may be larger or smaller than $P_{KB}[LM \leftrightarrow LC]$, depending on the particular case conditions. Interestingly, when the probability of a failed diagnostic test is smaller than the sum of the differences "pre-test disease probability minus the joint disease-test probability" and "post-test disease probability minus the joint disease-test probability," then $P_{KB}[LM \leftrightarrow LC] \leq P_{KB}[LC|LM]$.[33] These conditions are satisfied in the above LC case, since $P_{KB}[\neg LM] = 0.736 < P_{KB}[LC] - P_{KB}[LC \wedge LM] + P_{KB}[LC|LM] - P_{KB}[LC \wedge LM] = 0.4 + 0.909 - 2 \times 0.24 = 0.829$. Accordingly, as was expected it is indeed true that $P_{KB}[LM \leftrightarrow LC] < P_{KB}[LC|LM]$, i.e., $0.816 < 0.909$.

3.11. Summing Up: Checking the Validity of Medical Arguments

In the previous sections we discussed certain quantitative approaches for checking the validity of medical arguments. A summary of these approaches

---

[33]The converse inequality is valid that when the probability of a failed diagnostic test is larger than the sum of the differences pre-test and post-test disease probabilities minus the joint disease-test probabilities.

is as follows:

$$Arg[CA_p] ---> P_{KB}[CA_p] \begin{cases} \xrightarrow{P_{KB}\text{-}approach} Table\ 3.11 \begin{cases} Valid\ Arg \\ Invalid\ Arg \end{cases} \\ \xrightarrow{\zeta_a\text{-}approach} Eq.\ (3.55) \begin{cases} Valid\ Arg \\ Invalid\ Arg \end{cases} \\ \xrightarrow{U_{KB}\text{-}approach} Eq.\ (3.64) \begin{cases} Valid\ Arg \\ Invalid\ Arg \end{cases} \end{cases}$$

Basically, the medical argument in question, $Arg[CA_p]$, is first expressed in terms of the corresponding probabilities, $P_{KB}[CA_p]$. Subsequently, one of the following approaches could be used to check the validity of the probabilistic expression:

(1) the P_{KB}-approach, which seeks to test if the expression directly satisfies the rules of the probability theory (e.g., Table 3.1) along the methodology lines of Table 3.11; or
(2) the ζ_a-approach, which is based on the correspondence introduced in Eq. (3.55), see also Eq. (3.56); or
(3) the U_{KB}-approach, which is based on the logic-uncertainty criterion introduced in Eq. (3.64).

The above approaches seek to improve the rationality and effectiveness of the procedures and heuristics that physicians use to (i) detect/assess case symptoms, (ii) generate/confirm diagnoses, (iii) estimate the risks/benefits of medical tests and case treatments, and (iv) judge the prognostic significance of items (i)–(iii). For purposes of illustration the P_{KB}-approach was used in Examples 3.31–3.39, the ζ_a-approach was used in Examples 3.47–3.49 and the U_{KB}-approach was selected in Example 3.58.[34]

One approach may be chosen over another for different reasons, including the following:

(a) One approach is more appropriate or easier to apply than the others, given the form of the medical argument of interest.
(b) The goals of the medical case may favor one approach over the other. For example, if the case needs an assessment of the epistemic strength

---

[34] Several other examples of all three approaches are discussed throughout the book, and additional examples of the third approach are discussed in Christakos (2010).

or the justification degree of an argument, the ζ_a-approach may be preferred.
(c) Beyond the final outcome, which is testing the validity of an argument, each approach provides different kinds of additional information about the case.
(d) More than one approach can be used for validation purposes, as well as for comparing the relative merits of the different approaches.

We will revisit the topic in the section on medical causality of the following chapter.

3.12. Self-Referential Medical Assertions and Cognitive Favorability

In public and private medical practice one encounters what may be called *self-referential* (*SR*) assertions, i.e., assertions made by patients about themselves. Let us consider an example.

Example 3.62: Rev. Vardas asserts that

"he has a *stomach pain* (*SP*)", denoted as $a_V SP$.

In such cases, and assuming that the patient is an honest person, Rev. Vardas' physician may admit the validity of patient's assertions of the form

$$a_V SP \therefore SP$$
$$SP \therefore a_V SP, \tag{3.69a–b}$$

i.e., if Rev. Vardas claims to have a stomach pain, then he does indeed has a stomach pain, Eq. (3.69a). In other words, the above reasoning implies that it is in principle impossible for a patient to hold inconsistent beliefs about his stomach pain. The converse Eq. (3.69b) may also be valid, i.e., if Rev. Vardas has a stomach pain, then he believes indeed that he has it.

It is worth noticing that some clinical statements that are valid in the context of *non-self-referential* (*NSR*) assertions[35] may be not so for *SR* medical assertions. We hope that an example will illustrate matters.

---

[35] These are primarily the assertions discussed in previous sections, like judgments made by a physician about a patient's health state.

Example 3.63: A patient possibly suffers from *Huntington's disease* (HD). Consider the NSR case in which

"physician A withholds judgment (w_A) concerning the possibility that the patient suffers from HD ($w_A HD$)".

This assertion has an equivalent formulation as follows:

$$w_A HD \text{ iff } (\neg a_A HD) \wedge (\neg a_A \neg HD). \tag{3.70}$$

Due to uncertainty, the physician believes HD with probability, say, 0.55, and $\neg HD$ with probability 0.45. On the other hand, a rational agent may find it difficult to admit that the state w_A above is valid in the SR case considered earlier. For illustration, consider again Example 3.62, in which "Rev. Vardas believes that he has a stomach pain" (SP). Indeed, by combining the kind of inference of Eqs. (3.69a–b) in terms of SP with *modus ponens*, we get a contradiction as follows

$$
\begin{array}{ll}
& a_A \neg SP \therefore \neg SP \\
\text{Case premises} & (\neg a_A \neg SP) \therefore SP \\
& (\neg a_A SP) \wedge (\neg a_A \neg SP) \\
\hline
\text{Care provider's} & \\
\text{conclusion} & \therefore \neg SP \wedge SP
\end{array}
$$

That is, since the conclusion, $\neg SP \wedge SP$, is logically false, the $w_A SP$ in Eq. (3.70) is false, and the $\neg w_A SP$ should be valid instead. Similar is the case of the assertion $a_A SP \wedge a_A \neg SP$, i.e., a sensible patient like Rev. Vardas cannot claim that he has a stomach pain and at the same time claim that he does not have one. As a matter of fact, the above arguments can be rigorously proven in the SR setting for any case attribute CA.

NSR vs. SR situations like the above, or even more complicated ones, are not unusual between doctors and their patients. In these and similar clinical cases, the health care professionals may discover that Eqs. (3.69a–b) would actually lead to some noticeable results and assessments. *Inter alia*, it can be shown that for SR assertions satisfying Eqs. (3.69a–b), in principle it is assumed valid that $a_A CA \therefore k_A CA$, which was not necessarily valid for most NSR assertions considered in previous sections. Our readers may have noticed that a SR case assertion is actually a special case of a meta-assertion in which the content of the case assertion in the metalanguage and the content of the assertion in the object language are actually the same.

Such a case assertion is in effect referring to itself. Not surprisingly, meta-assertions of this type could lead to certain paradoxes in decision analysis, and, hence, should be carefully considered by the physicians when they arise in clinical practice.

When studying a case, a health care provider may need to choose between two or more assertions. In logic terms, "assertion concerning CA_1, $a_A CA_1$, is favored by physician A over the assertion about CA_2, $a_A CA_2$", i.e.,

$$a_A CA_1 F a_A CA_2 \tag{3.71}$$

where F denotes "is favored over". In SMR terms this means that the uncertainty of $a_A CA_1$ is lower than that of $a_A CA_2$, i.e.,

$$a_A CA_1 F a_A CA_2 \leftrightarrow U[a_A CA_1] < U[a_A CA_2].^{36} \tag{3.72}$$

The readers may compare Eq. (3.72) with Eq. (3.64). Expression (3.72) may be generalized to any two logical functions Φ_1 and Φ_2,

$$\Phi_1 F \Phi_2 \leftrightarrow U[\Phi_1] < U[\Phi_2].^{37} \tag{3.73}$$

On the basis of the above, certain useful results are derived that are of considerable use in medical research and clinical practice, such as

$$a_A CA_1 F a_A CA_2 \leftrightarrow (a_A \neg CA_2) F (a_A \neg CA_1), \tag{3.74}$$

i.e., "$a_A CA_1$ is favored by A over $a_A CA_2$" iff "$a_A \neg CA_2$ is favored by A over $a_A \neg CA_1$". An interesting transitivity property of favorability is

$$[\neg(a_A CA_1 F a_A CA_2) \wedge \neg(a_A CA_2 F a_A CA_3)] \rightarrow \neg(a_A CA_1 F a_A CA_3), \tag{3.75}$$

i.e., if it is not valid that "$a_A CA_1$ is favored by A over $a_A CA_2$" and also it is not valid that "$a_A CA_2$ is favored by A over $a_A CA_3$" then it is not valid that "$a_A CA_1$ is favored by A over $a_A CA_3$". For example, if "it is not true that A favors the belief of an infection diagnosis over the belief of a neoplasm diagnosis" and also "it is not true that A favors the belief of a neoplasm diagnosis over the belief of a collagen vascular diagnosis" then "it

---

[36] Equivalently, the probability of $a_A CA_1$ is higher than that of $a_A CA_2$, $P[a_A CA_1] > P[a_A CA_2]$.

[37] For illustration, expression (3.72) is obtained from (3.73) by letting $\Phi \equiv a_A CA_1$, $\Phi_2 \equiv a_A CA_2$.

is not true that A favors the belief of an infection diagnosis over the belief of collagen vascular diagnosis".

3.13. Not Just a Set of Guidelines

The discussion in the preceding sections demonstrated most emphatically that rigorous medical reasoning in conditions of realistic uncertainty is not just a set of guidelines aimed at rigorous decision-making based on the evaluation of relevant arguments, but includes much more that could profoundly alter the way many health professionals view quality of care and patient safety. Accordingly, stochastic medical reasoning (SMR) seeks to integrate in a systematic manner

(1) diverse *knowledge bases* (core and case-specific) and
(2) *logics* (formal and interpretive) with
(3) *mind-body* associations and
(4) a variety of *professional skills*,

and all the above in a realistic *in situ* context. This kind of reasoning is of the utmost importance to scientific and rational thinkers, and, accordingly, its key features cannot be overlooked by physicians examining disease and injury.

Another point that the book would like to bring to the readers' attention is that decision-making should not be viewed as a purely empirical process. Instead, it also contains non-empirical, analytical or *a priori* (sometimes referred to as transcendental) elements in medical knowledge, which do not come from experience, but are nevertheless legitimately applied to the data or contents of knowledge furnished by the health care provider's experience. Sound *in situ* medical judgment depends essentially on the health care professional's conception of the relations between

(a) assertions (mental states),
(b) logical and physical connectives,
(c) language and metalanguage and
(d) probability theory.

Whether or not decision-making turns out to be sterile for all practical purposes of clinical practice and medical research, and to generate results that are logically incompatible, depends fundamentally on the above conception. Therefore, this book constantly and consistently argues that a rigorous and efficient way to obtain a deeper understanding of the underlying logical structure of a physician's everyday language is to learn how to put it into the *SMR* terms discussed in Chapters 2 and 3.

Before leaving this chapter, we would like to bring to our readers' attention the fact that many health care professionals find it constructive to take advantage of the dualistic opposition of the objectivist and subjectivist perspectives of science. Specifically, objectivism accepts knowledge claims as potentially true and warranted on objective evidence, whereas subjectivism grounds medical knowledge in perception, phenomenology and social construction. Although these two perspectives differ in their ontologies (the reality of constructs and relations) and methodologies (how these relations can be observed), both views accept that reliable medical knowledge for rigorous decision-making is, indeed, possible.

Chapter 4

Space–Time Medical Mapping and Causation Modeling

4.1. Techniques With a "Health Warning"

The quantitative techniques used in medical sciences should be generally considered as approximations of reality, because disease varies with sex, age, anatomy, pathophysiology, season, climate, weather and several other domestic, social and hygienic circumstances by which the health status of individuals or communities is influenced, and which are in continuous state of mutation. If experience is any guide, several of these techniques should come with a "health warning", so to speak. Such are, among others, techniques implemented by means of "black-box" computational schemes, technical "guidelines", "off-the-shelf" software and the like. For better or for worse, the content-independent notion (CI; see Section 2.3.4) captures, indeed, the essence of many health statistics and biostatistics techniques (Gutiérrez et al., 2005; Helwade and Subramanyam, 2010; Chien and Bangdiwala, 2012).[1] In which case, at the very least, health care professionals should be fully aware of the precise domain of applicability of a technique, and also be conscious of the knowledge and skills that are required to practice the technique. Let us put matters in perspective with the help of an example.

Example 4.1: Regression generates optimal (in the $MMSE$[2] sense) sampling designs for an unsampled case attribute. The estimation error variance is data-independent, and all that is needed is an adequate covariance

---

[1] Nevertheless, the readers should be reminded that the CI notion would not be favored, say, by Immanuel Kant for whom thoughts without content are empty and intuitions without concepts are blind.

[2] Minimum mean squared error.

model. In this setting, regression is a *CI* technique. Similar is the case of correlation-based disease causation (or etiology), which is basically a *CI* implication. We will have more to say about regression and etiology in the following sections.

The analysis in the previous chapters has shown that a more sound approach to medical decision-making may require the reshaping of a health care professional's thinking. Unlike standard logic and *CI* statistics, stochastic medical reasoning (*SMR*) is not just an instrument that can be used indifferently, but a theoretical science in its own right that is substantively linked to other scientific fields (physical, environmental, social; Tamerius *et al.*, 2006 and Yu *et al.*, 2011a). This involves, among other things, a richly structured high-quality epistemology that describes case entities, medical concepts and semantic relationships. Yet, the readers may recall that pure epistemic logic is interested only in rational principles of believing, and could therefore be used with completely rational people for descriptive purposes. Medical investigations in conditions of *in situ* and clinical uncertainty, on the other hand, often need to go beyond mere description to valid inferences, prediction and mapping. Inference matters were discussed in Chapters 2 and 3, whereas prediction and mapping will be studied next.

4.2. Space–Time Disease Mapping

New disease causing agents (such as viruses and bacteria) are the result of an increasing number of populations and cultures whose environmental conditions are favorable to their creation and diffusion across space–time. Larger cities mean high density of humans and the commingled animal species that sustain them (Tiefelsdorf, 2007; Dhondt *et al.*, 2011 and Luo and Opaluch, 2011). These are conditions ripe for the evolution of disease agents whose effect on humankind, and sometimes animals, could be disastrous. Under the circumstances, of significant interest is *disease* or *medical mapping*, e.g., mapping of population density distributions sufficient to support the growth of agents, travel vectors (human and animal migration) enabling them to move within populations and environmental factors in those population areas that create a hospitable environment for disease evolution (Griffith and Christakos, 2007; Choi *et al.*, 2008; Reiczigel *et al.*, 2010 and Liao *et al.*, 2011a–b). As a matter of fact, with the help of technology (*TGIS*, visualization tools etc.), this kind of mapping has important applications in medical science in the broad sense (including clinical medicine, environmental health and geomedicine). Using mapping to deliver space–time intelligence to medical professionals could advance the way they view health care and improve quality of life.

4.2.1. Objectives of medical mapping

In many *in situ* cases, medical mapping seeks to generate informative representations of the space–time distribution of disease case attributes based on the available *KB*s, quantitative tools and technology. Figure 4.1 provides an illustrative visualization of the medical mapping problem. In this figure, CA_1 is the case attribute of interest and CA_2 is a secondary attribute related to CA_1 via the empirical function $F[\cdot]$. The objective is to derive estimates at any space–time point of interest; e.g., a CA_1 estimate, denoted as $C\hat{A}_1$, is sought at the coordinates–time $(s_{1,k},\ s_{2,k},\ t_k)$ in Fig. 4.1. The available *KB*s include core knowledge (G, see pile of books at the top of Fig. 4.1) in the form of biophysical laws, scientific models and theoretical associations, and site-specific knowledge (S, see documents, graphs, numerical and interval values, and functions in Fig. 4.1) in the form of empirical evidence, hard and soft (or uncertain) data at a number

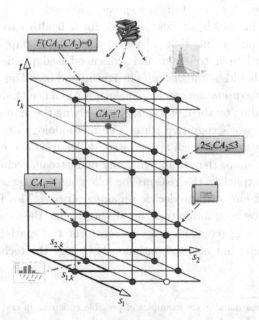

Figure 4.1: A visual representation of the medical space–time mapping problem.

of observation points (Mikler *et al.*, 2006; Thayer *et al.*, 2006; Lee and Rogerson, 2007; Ugarte *et al.*, 2012).

Medical or disease mapping is part of the crucial process of understanding the evolution and spread mechanisms of a range of chronic and epidemic diseases, especially during an era of globalization characterized by vastly increasing international trade and widespread emigration and immigration. What used to be rather local diseases (e.g., rooted in specific local communities) are rapidly expanded to the global scale. Diseases are appearing where they have never been before and/or are reappearing in regions of the world where they have been eliminated, sometimes leading to epidemic or even pandemic situations. Mapping is very useful in cases like epidemic outbreaks (i.e. there are more cases of a disease than expected in a given area, or among a specific group of people, over a particular period of time) or in endemics (i.e., a population has a high level of the disease all the time).[3] Medical mapping notions and techniques offer an efficient way to observe changes in the health state of populations that take place in composite space–time domains. People appreciate images and other visual representations, in which case *SMR*-based mapping could give them the means to think about them critically.

Although breadth is usually an advantage, a mapping technique does not need to cover all cases of disease spread and epidemic distribution one may encounter in health studies. Instead, for a health care professional, quality is more important, that is, one would desire a mapping technique that is grounded in a coherent and evidence-based predictive scientific theory. A map should be *theoretically* insightful and based on rich *empirical* findings. Map *interpretation* can help make sense of a health situation, yet it may vary depending on the methodology underlying the map's construction and the investigator's cognitive state. The methodology determines what aspects of reality are captured, appreciated, understood and communicated by means of the map. Hence, applying any mainstream technique without understanding explicitly the reasons for following its suggested steps can cause significant harm to the decision-making process. This is where *SMR* could prove to be a valuable tool, for it shows that the issue is not only to select a mapping technique, say, to represent the spatial pattern of a specified epidemic, but also to continuously change the technique's steps,

---

[3] Giardiasis and even malaria are examples of possible endemics in certain parts of the world.

structure and form to match the changing needs of the study as the disease spreads across space–time.

4.2.2. The fundamental mapping equations

Multidisciplinary public health studies may range from situations characterized by complete collaboration and ideal communication between the participants to situations characterized by limited (or inexistent) communication and conflicting goals of the participants. In the public health setting, stochastic medical reasoning and inference is concerned with how social environments determine the ways in which disease spreads, and how epidemics can be prevented or controlled. Stochastics theory gives medical reasoning a vocabulary on which to base its own methodology in which the aspatial methods of clinical practice are transformed into the idea of the three unites — of *case*, *place* and *time*, and their substantive interrelations.

In a similar setting, aspatial health statistics are not good at predicting fundamental breaks from the past. For example, when earlier patterns between dependent and explanatory variables break down, statistical predictions tend to fare poorly. This is one of the reasons why a set of stochastic mapping techniques have been developed (such as *BME*, *MCME*, *ECME*[4] etc.) (Christakos et al., 2002, and references therein) that are basically different solutions of the fundamental set of equations

$$\int d\mathbf{CA}\,(\mathbf{g}-\bar{\mathbf{g}})f_G(\mathbf{CA},\boldsymbol{\mu},\mathbf{g}) = 0$$
$$\int d\mathbf{CA}\,\boldsymbol{\xi}_S\, f_G(\mathbf{CA},\,\boldsymbol{\mu},\mathbf{g}) - A[f_K(\mathbf{CA},\mathbf{p})] = 0,$$

(4.1a–b)

where $\mathbf{g}$ is a mathematical vector with elements representing the *G-KB* available; $\bar{\mathbf{g}}$ is the stochastic expectation of $\mathbf{g}$; $\boldsymbol{\mu}$ is a vector with elements that depend on the location–time coordinates and assign proper weights to the elements of $\mathbf{g}$ (i.e., they assess the relative significance of each $\mathbf{g}$ element in the composite disease map sought); f_G is the probability density function (*PDF*) of disease distribution at each space–time point (the subscript means that the disease *PDF* is built on the basis of core knowledge), and its shape depends on the expected information form (f_G is a function of $\mathbf{g}$ and $\boldsymbol{\mu}$); the vector $\boldsymbol{\xi}_S$ represents specificatory data; $A[\cdot]$ is a mapping operator and

---

[4]The abbreviations mean Bayesian Maximum Entropy (*BME*), Material Conditional ME (*MCME*) and Equivalence Conditional ME (*ECME*).

f_K is the updated case *PDF* at each space–time point of the domain of interest (the subscript K means that the construction of f_K is based on the total *KB*, which is the result of internally consistent blending of G and S).

Just as the informativeness of an image increases as the image becomes more detailed, so the salience and perceived likelihood of a disease map increases as Eqs. (4.1a–b) incorporate more core scientific knowledge and case-specific data. Furthermore, these equations encourage health investigators to learn how to structure their thinking in several different ways, which helps gain new insights about health. A practitioner encounters a diversified range of views, opinions and treatment methods, which are properly blended with matured judgment and experience. Epidemiologists will work out the relationships between the available sources of data from routinely collected government and commercial statistics, regular and *ad hoc* surveys. The relationships will be embedded in coherent models incorporating the totality of evidence. New evidence will be routinely appended to the evidence base, and the models will be used for the design of new research, changes in statistics collected by government, and as a basis for policy models. Equations (4.1a–b) are the result of a fusion of normative aspects (relevant to the process of choosing among competing criteria of space–time mapping) and empirical aspects (arising because disease mapping involves specific data and generates particular patterns across space–time). The truth or falsity of normative statements in the above sense usually cannot be tested by observation, but a physician must use appeals to biomedical theories, logic and values to argue for a health statement's acceptance. On the other hand, empirical statements do not appeal to values (they do not contain a preference for one outcome over another) and are tested through observation. Some regression mapping techniques appeal to a normative criterion of minimizing the statistical error variance across space–time, and then implement it using empirical evidence (Demyanov *et al.*, 2001; Lin *et al.*, 2011; Harris *et al.*, 2011). These techniques depend on two factors: a preference for a particular criterion and the choice of a particular empirical dataset. The potential limitations of these techniques have been discussed in the relevant literature.

The other group of mapping techniques (based on Eqs. (4.1a–b)) are associated with different forms of f_G and $A[\cdot]$. In the case of *BME*-based medical mapping,

$$f_G = e^{\mu^T g} \quad \text{and} \quad A[\cdot] = A \tag{4.2}$$

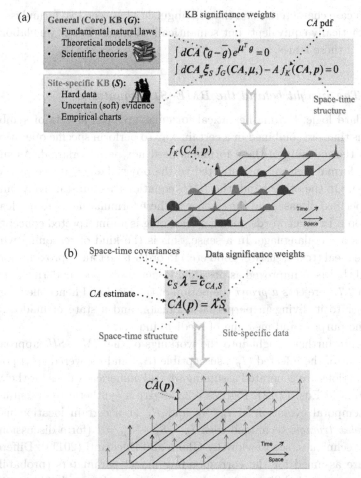

Figure 4.2: Visual representation of the (a) space–time *BME* and (b) *STSK* (ST Simple Kriging) techniques.

(A is here a constant), and Eqs. (4.1a–b) reduce to the set of *BME* equations shown in Fig. 4.2a. The *BME* formulation can account for the generalized *SIR* model (2.5.1), or other disease models, via the vector g of the *G-KB*, in which case one talks about the *BME–SIR* approach to disease spread modeling and prediction. The space–time synthesis in the *BME* equations is non-monotonic, i.e., new *KB*s can potentially change the justification degrees of previous assertions (beliefs, diagnoses, predictions etc. based on older *KB*s). This is because, *inter alia*, *SMR* goes beyond the rigidity of deduction that characterizes purely formal constructions, whereas *BME* creates a case characterization that distinguishes between two formulas that are syntactically equivalent, but semantically distinct. Below we elaborate further on these issues.

4.2.3. The insight behind the BME–SIR equations

In standard logic, the mathematical formulas are seen as sets of symbols and signs that are combined in a certain way to perform specific operations, whereas the meaning of these formulas need not be understood. As such, this is a formal conception restricted to the object language (see previous chapters). On the other hand, the *BME* equations are substantively linked to the specified disease, and the essence of these formulas needs to be clearly understood. In other words, medical reasoning is an interpreted conception that uses a metalanguage. In a sense, this is the kind of reasoning which seeks the ideal (truth, justice, good etc.) in order to get an approximation of the ideal that is as improved as possible. Remarkably, postmodern thinking (Section 2.7.4) rejects *a priori* the notion of the ideal and hence, never gets any closer to it, living in perpetual confusion and a state of inadequacy about the purpose and meaning of medical inquiry.

To gain further insight into the workings of the *BME–SIR* approach, simulations of the infected (I_p), susceptible (S_p) and recovered (R_p) population fractions are generated assuming several numerical values for the *SIR* parameters in Eqs. (2.12). Figures 4.3a–d give a synthetic representation of the temporal evolution of I_p, S_p and R_p at a certain location using the *simplex triangle* technique with $I_p + S_p + R_p = 1$ (for a discussion of simplex techniques, see Pawlowsky-Glahn and Buccianti (2011)). Different values are assumed for the corresponding model paramaters (probability of recovery a, probability of infection transmission b, population fraction q residing at the domain of interest and kernel bandwidth β). In Fig. 4.3a the distance between dots represents a time interval starting from the lower right corner of the triangles. Some intuitive results are quantitatively displayed in Fig. 4.3a. For higher b values: (*i*) the maximum infected population fraction is also higher (and reached at an earlier stage); accordingly, (*ii*) the reduction of the susceptible fraction is faster with time; and (*iii*) the faster the increase of the recovered population fraction. Moreover, as b increases, the limit over time of the susceptible fraction (that is, the population fraction that finally remains unaffected by the disease) tends to be closer to zero. In Fig. 4.3b, the a variation leads to the "inverse" *SIR* behavior from the b variation in Fig. 4.3a. For smaller

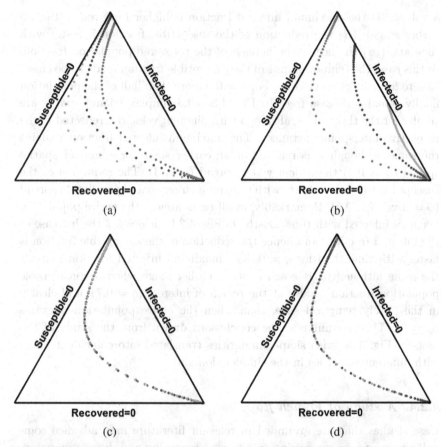

Figure 4.3: Simplex triangle representation of the temporal evolution of infected, susceptible and recovered population fractions at a location:

(a) For different transmission probabilities, $b = 0.1$ (right dotted line), $= 0.2$ (middle dotted line), $= 0.3$ (left dotted line); recovery probability, $a = 0.1$, population fraction residing at domain of interest, $q = 0.7$, and kernel bandwidth, $\beta = 0.5$.

(b) For different recovery probabilities, $a = 0.3$ (left dotted line), $= 0.4$ (middle dotted line), $= 0.6$ (right dotted line); $b = 0.4$, $q = 0.7$, and $\beta = 0.5$.

(c) The plots (dotted lines) for the purely temporal model and the spatiotemporal models (kernel bandwidths $\beta = 0.5$, and $= 3$) coincide; $a = 0.1$, $b = 0.4$, and $q = 0.7$.

(d) The plots (dotted lines) for the purely temporal model and the spatiotemporal models for different population fractions residing at domain of interest $q = 0$, $= 0.7$; $\beta = 0.5$, $a = 0.1$, and $b = 0.4$.

a values: (i) the maximum infected fraction is higher (and reached at an earlier stage); (ii) the reduction of the susceptible fraction is faster with time and (iii) the faster the increase of the recovered population fraction. In this case, the temporal limit of the susceptible fraction tends to be closer to zero for smaller values of a. For $a = 0.6$ more than half of the population finally remains disease free. In Fig. 4.3c, the simplex triangle paths are similar for the three cases, although with different velocities reflected in the respective inter-point distances. The maximum infected fraction remains the same, although it is reached at an earlier stage for a smaller spatial spread ($\beta = 0.5$) than for a wider spread ($\beta = 3$). The reduction of the susceptible fraction is slower with time for a larger spatial population spread ($\beta = 3$ vs. $\beta = 0.5$). Remarkably, in all cases almost the entire population becomes infected with time. Lastly, in Fig. 4.3d, for $q = 0.7$ the increase of the infected fraction, and hence the reduction of the susceptible fraction is faster with time than for $q = 0$. The maximum infected fraction remains the same although it is reached at an earlier stage when a considerable population fraction resides at the region of interest ($q = 0.7$), and clearly in the purely temporal case, than when the entire population migrates ($q = 0$). This is similar to the conclusions drawn from the study of the plots of Fig. 4.3d with shape similarities translated into coincident paths with different velocities in the simplex domain.

4.2.4. A study of French flu

Case studies that are grounded in relevant literature may advance compelling arguments concerning *in situ* case reasoning and disease mapping. This is the purpose of the health studies published by Choi et al. (2008); Yu et al. (2009; 2010; 2011a–b; 2012; 2013); Reiczigel et al. (2010); Liao et al. (2011a, 2011b); Chen et al. (2012a, 2012b); Kolovos et al. (2012) and Yu and Wang (2010, 2013). The example below discusses one of these studies, pertaining to flu spread in France.

Example 4.2: Kolovos et al. (2012) studied a French population during flu season using the *BME–SIR* technique. A central modeling thesis in the study of space–time disease distribution and epidemic spread is that individuals or communities that generally have the potential to affect one another upon interaction have the potential of being characterized and studied as organized collectives of interacting entities. Specifically, the focus of the French flu study is the population ratio of new infecteds (RNI), $\rho_{s,t}$, in a region with geographical coordinates s and at time t. The *RNI* is the ratio of the number of new infecteds ($I_{s,t}$) over the region's population. The space–time $I_{s,t}$ distribution obeys the *SIR* model of Eqs. (2.12), where $S_{s,t}$ and $R_{s,t}$ represent the space–time distributions of the population fraction that is susceptible, and recovered, respectively. Particularly, in the special case that one can assume a function $\varphi_{s,t}$ with a smooth shape similar to that of the covariance function of $I_{s,t}$ and $S_{s,t}$,[5] the infecteds' initial condition ($I_{s,0}$) is spatially homogeneous, the population is static while the disease spreads ($q_{s,t} = 1$) and the time-independent kernel that controls population movement across space does not play any role.[6] Due to the above simplified conditions in the present flu study, the generalized *SIR* model of Eqs. (2.12) reduces to

$$\frac{d}{dt}I_{s,t} = (-a'_{s,t} + b'_{s,t}\varphi_{s,t})I_{s,t}$$

$$\frac{d}{dt}S_{s,t} = -b'_{s,t}\varphi_{s,t}I_{s,t} \qquad , \qquad (4.3\text{a–c})$$

$$\frac{d}{dt}R_{s,t} = a'_{s,t}I_{s,t}$$

where, as usual, $a_{s,t} = a'_{s,t}dt$ and $b_{s,t} = b'_{s,t}dt$ are the probabilities that an infected individual recovers and becomes immune, and that the infection transmission occurs during an encounter in the time interval dt involving one infected and one susceptible individual; the initial conditions are $I_{s,t} + S_{s,t} + R_{s,t} = 1$, $S_{s,0} = 1 - I_{s,0}$ and $R_{s,0} = 0$; and the $A_{s,t}$, $B_{s,t}$ and $C_{s,t}$ are functions of $a'_{s,t}$, $b'_{s,t}$ and $\varphi_{s,t}$. The vector $\boldsymbol{\xi}_S$ of the *S-KB* consisted of aggregated observations recorded by physicians at each one of the 21 regions in the French mainland (Fig. 4.4) through the Sentinel project over a time period of 53 weeks (August 1998–August 1999). The corresponding covariance and cross-covariance functions of the space–time flu distributions are shown in Table 4.1 below. In the present study, another way to look at the *BME* equations is that they generate a stochastic solution of the differential *SIR* Eqs. (4.3) that — compared to the standard solutions of general differential equations — has the unique feature of being able to

---

[5]For example, $\varphi_{s,t}$ may be chosen to be a monotonically decreasing function of time t with sufficient flexibility to represent the behavior of the population fraction that is susceptible to infection.

[6]Which means that the integral terms in Eqs. (2.12) can be neglected.

Figure 4.4: Map of France and its 21 mainland regions.

Table 4.1: (Cross)covariances of infecteds, suceptibles and recovereds.

	$I_{s',t'+1}$	$S_{s',t'+1}$	$R_{s',t'+1}$
$I_{s,t+1}$	$A_t A_{t'} c_{I;s-s',0}$	$-A_t(1+B_{t'})c_{I;s-s',0}$	$A_t C_{t'} c_{I;s-s',0}$
$S_{s,t+1}$	$-A_{t'}(1+B_t)c_{I;s-s',0}$	$(1+B_t)(1+B_{t'})c_{I;s-s',0}$	$-(1+B_t)C_{t'} c_{I;s-s',0}$
$R_{s,t+1}$	$A_{t'} C_t c_{I;s-s',0}$	$-(1+B_{t'})C_t c_{I;s-s',0}$	$C_t C_{t'} c_{I;s-s',0}$

account for several other kinds of available knowledge (hard and soft data, empirical relationships, secondary information) and uncertainty sources (in the composite space–time disease variation, the records of infected individuals etc.). For illustration, Fig. 4.5 displays space–time $\rho_{s,t}$ maps of the French population during the flu season using the *BME–SIR* technique. As is the case with most epidemics, the flu spread follows a pattern that depends on the geographical and environmental conditions, the distribution

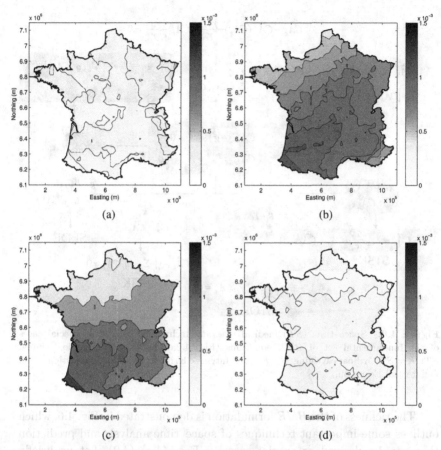

Figure 4.5: Selected maps of French cold *RNI* during week (a) $t = 2$, (b) $t = 6$, (c) $t = 10$, (d) $t = 14$.

and characteristics of the host population, and their cultural behavior. If there is no intervention or change in these conditions, the epidemic tends to repeat itself. Hence, knowledge of various types of epidemics and the conditions under which they occur can be of significant help in managing them, which is also the case with French flu. Then, the role of the medical reasoning metalanguage is to describe the meaning-making resources of images like Fig. 4.5 and image–formula interaction. In the same setting, the space–time predictions of the map are internally valid, i.e. valid for subjects from the underlying population, and, if possible, generalizable to similar populations.

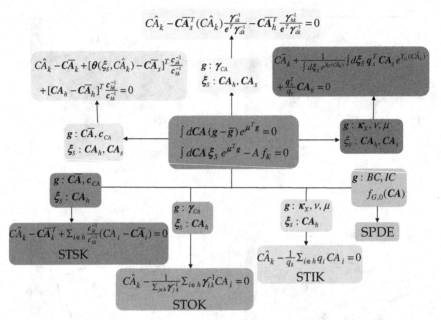

Figure 4.6: Space–time (*ST*) medical mapping techniques derived as special cases of the fundamental set of *BME* equations (Fig. 4.2a); *STSK* = ST Simple Kriging, *STOK* = ST Ordinary Kriging; *STIK* = ST Intrinsic Kriging; *SPDE* = Stochastic Partial Differential Equation.

The richness of the *BME* formulation is demonstrated in Fig. 4.6, which outlines some important techniques of space–time analysis and prediction that can be derived as special cases of Eqs. (4.1)–(4.2). Let us briefly explain a few of the technical terms used in this figure (interested readers are referred to the *BME* literature for a detailed account of the notation and terminology). The *G-KB* is quantitatively represented by the vector g, which may include location–time $CA(p)$ dependence models, natural laws (such as partial differential equations) and empirical relationships (in terms of algebraic equations, charts, graphs etc.). To g also belongs the matrix c_{map} whose elements are covariances between any pairs of space–time points of p_{map} (accordingly, each c_{ij}^{-1} denotes the ij-th element of the inverse matrix c_{map}^{-1}). The *S-KB* includes a dataset CA_d at the space–time data points p_d; CA_d consists of hard data CA_h at points p_h (e.g., exact measurements) and soft data CA_s at points p_s (e.g., expressed in terms of intervals of varying lengths and probabilistic functions of arbitrary shapes). Vector p_{map} includes p_s as well as the map grid points p_k where a space–time prediction $C\hat{A}_k$ is sought, q_i are weight coefficients and θ_i are functions of $\boldsymbol{\xi}_S$ and $C\hat{A}_k$.

Example 4.3: For the purposes of comparison, Fig. 4.2b displays a visualization of $STSK$ that is derived as a special case of the BME visualization (Fig. 4.2a). As before, $C\hat{A}$ is a CA estimate, the KB significance weight vector $\boldsymbol{\mu}$ is reduced to the data significance weight vector $\boldsymbol{\lambda}$ and the site-specific dataset is denoted as S; $\mathbf{c}_S$ is the matrix of covariances between all pairs of S data, and $\mathbf{c}_{CA,S}$ is the matrix of covariances between the estimation point and all data points.

Information gathering for the space–time prediction purposes of Fig. 4.2a is *dynamic* and *non-local*: data gathering at one part of a health system can affect other regions of the system. Another issue that has emerged many times in health studies is that it is impossible to generate accurate predictions of the outcome of knowledge synthesis (say, space–time disease incidence) without an adequate description not only of the individual components of the synthesis (e.g., exposure variables, toxicokinetics, medical diagnostics, population dynamics and confounders), but also of their interconnections and interactions. Visual maps allow health professionals to look for space–time patterns and trends in the hope of discovering something new about the disease (determinants, etiology etc.) (Wang et al., 2008, 2009). So, space–time maps can play a key role in protecting individuals and communities from otherwise undetected toxic exposures and overlooked health risks. Using mapping tools, e.g., health care professionals have managed to link people's respiratory problems with high densities roadway traffic, when these people spend considerable time (home, work) nearby these roadways, thus experiencing high levels of air pollution exposure.

All the above issues are the concern of stochastic medical reasoning in a space–time synthesis context. This is the kind of reasoning that is self-correcting or reflective, including content-dependent (CD) relationships between cause and effect, rather than a highly specialized formal approach. Its cognitive character makes SMR inference possible to argue in a sound manner both about the conditions and limits of a health care professional's knowledge, cognition and empirical support. And so long as these limits are respected, SMR could be the vehicle of disease understanding and spread prediction, in which space–time mapping is viewed as a product of sound and rational reasoning rather than a purely technical affair. Furthermore, when a mapping technique is used, a quantitative test of its performance is needed, so that the health care provider can judge whether the approach is adequate for the particular situation or whether it is better than another mapping technique. Such adequacy tests are related to the "goodness-of-fit" test of a mapping technique that is usually evaluated in the same data, and the "prediction error" test that requires either new data or cross-validation.

4.3. Modeling Space–Time Infectious Disease Spread

Arguably, differences between disease study approaches have to do mainly with knowledge sources they are able to account for, i.e. *S-KB* (the case-specific data available) and *G-KB* (disease mechanisms — biomedical, epidemiological and clinical — and other prior information about the disease). Specifically:

(1) A *statistical* approach may make full use of case-specific information, *S-KB*, but often ignores core knowledge, *G-KB*.
(2) Mainstream *deterministic* and *stochastic* models may take into account the *G*-KB, but usually do not make full use of *S-KB*.
(3) The *state–space* method accounts for *S-KB* (via the observation model) and *G-KB* (via the state model) and generates estimates of disease attributes together with the associated estimation error variance.
(4) The *BME* approach is theoretically more general than the *state–space* method and can incorporate a wider variety of *G* and *S* knowledge sources (via structural and specificatory equations).

Clearly, there are considerable differences between these approaches. *BME*, e.g., does not make any of the restrictive assumptions of the *state–space* method (normality, linearity, second-order statistics) and relies on the rigorous logic of knowledge synthesis under conditions of uncertainty (Yu *et al.*, 2007). The readers may be reminded that in Section 2.5 we reviewed certain representative natural laws and scientific models (*SIR*, human exposure, population dynamics) that occur in a spatiotemporal domain. Indeed, in the real world a disease distribution is a fundamentally spatiotemporal phenomenon, its purely temporal models being merely convenient simplifications. As before, let $I_{s,t}$, $S_{s,t}$ and $R_{s,t}$ represent rates of infected, susceptible and recovered population fractions at time t within a region around location s (i.e., local rates). An *SMR* modeling framework that describes the combined space–time distribution of $I_{s,t}$, $S_{s,t}$ and $R_{s,t}$ is given by the generalized space–time *SIR* model of Eqs. (2.12), which are functions of the population fraction that migrates during a time period, factors that control population movement across space–time and the probabilities of recovery and transmission that can include information about regional topography and local climatic conditions.

The covariances and cross-covariances of $I_{s,t}$, $S_{s,t}$ and $R_{s,t}$, which describe the underlying dependences between space–time points, can be derived from the *SIR* Eqs. (2.12). For illustration, the composite space–time expressions of the covariance of $I_{s,t}$ between points $\boldsymbol{p} = (\boldsymbol{s},t)$ and $\boldsymbol{p}' = (\boldsymbol{s}',t' = t+\tau)$, is:

$$\begin{aligned}
c_{I;s,t,s',t'} = {} & q_{s,t}^2 \{(1-a_{s,t})(1-a_{s',t'})c_{I;s,t,s',t'} \\
& + (1-a_{s,t})b_{s',t'}\mathrm{cov}(\boldsymbol{I}_{s,t}, S_{s',t'}I_{s',t'}) \\
& + (1-a_{s',t'})b_{s,t}\mathrm{cov}(I_{s',t'}, S_{s,t}I_{s,t}) \\
& + b_{s,t}b_{s',t'}\mathrm{cov}(S_{s',t'}I_{s',t'}, S_{s,t}I_{s,t})\} \\
& + q_{s,t}\int_{R^2} du'\kappa_{s'-u',t}\{(1-a_{s,t})(1-a_{u',t'})c_{I;s,t,u',t'} \\
& + (1-a_{s,t})b_{u',t'}\mathrm{cov}(I_{s,t}, S_{u',t'}I_{u',t'}) \\
& + (1-a_{u',t'})b_{s,t}\mathrm{cov}(I_{u',t'}, S_{s,t}I_{s,t}) \\
& + b_{s,t}b_{u',t'}\mathrm{cov}(S_{u',t'}I_{u',t'}, S_{s,t}I_{s,t})\} \\
& + q_{s,t}\int_{R^2} du\kappa_{s-u,t}\{(1-a_{u,t})(1-a_{s',t'})c_{I;u,t,s',t'} \\
& + (1-a_{u,t})b_{s',t'}\mathrm{cov}(I_{u,t}, S_{s',t'}I_{s',t'}) \\
& + (1-a_{s',t'})b_{u,t}\mathrm{cov}(I_{s',t'}, S_{u,t}I_{u,t}) \\
& + b_{u,t}b_{s',t'}\mathrm{cov}(S_{s',t'}I_{s',t'}, S_{u,t}I_{u,t})\} \\
& + \int_{R^2}\int_{R^2} du'du\kappa_{s'-u'}\{\kappa_{s-u,t}(1-a_{u,t})(1-a_{u',t'})c_{I;u,t,u',t'} \\
& + (1-a_{u,t})b_{u',t'}\mathrm{cov}(I_{u,t}, S_{u',t'}I_{u',t'}) \\
& + (1-a_{u',t'})b_{u,t}\mathrm{cov}(I_{u',t'}, S_{u,t}I_{u,t}) \\
& + b_{u,t}b_{u',t'}\mathrm{cov}(S_{u',t'}I_{u',t'}, S_{u,t}I_{u,t})\},
\end{aligned} \qquad (4.4)$$

where $a_{s,t} = a'_{s,t}\tau$, $b_{s,t} = b'_{s,t}\tau$ and $c_{I;s,t,s',t'} = \mathrm{cov}(I_{s,t}, I_{s',t'})$. A practicing health care provider may feel somewhat uncomfortable with Eq. (4.4). Yet, this covariance formula is justifiably long because so is the list of interconnected disease parameters it must take into account, and because of the large amount of information the formula offers concerning disease space–time dependence. Similar expressions can be derived for the covariance of $S_{s,t}$ and the cross-covariance of $I_{s,t}$, $S_{s,t}$ between points $\boldsymbol{p} = (\boldsymbol{s},t)$ and $\boldsymbol{p}' = (\boldsymbol{s}',t+\tau)$. Covariance equations contain a host of valuable information regarding the disease distribution in a space–time domain, taking advantage of the generalized *SIR* model (Section 2.5.1), in which case, arguably, the gain in realism and clarity more than offsets the mathematical complexity. The infecteds covariance model (4.4) could also provide motivation for future experimental work so that some of its parameters are reliably determined *in situ*. No doubt, meaningful model implementation that avoids potential pitfalls requires considerable concrete experience with real situations of population health. Let us consider some special cases of the space–time covariance (4.4).

Example 4.4: In the case of the French flu epidemic of Example 4.2, the corresponding covariance and cross-covariance functions are shown in Table 4.1 with initial condition of the covariance of $I_{s,0}$ and $I_{s',0}$, $c_{I;s-s',0}$. The quantitative expressions in Table 4.1 indicate that the proposed flu covariances and cross-covariances are non-stationary in time. For illustration, the cross-covariance of $I_{s,t+1}$ and $S_{s',t'+1}$ is

$$c_{I,S;s,t+1;s',t'+1} = -A_t(1+B_{t'})c_{I;s-s',0},$$

which depends explicitly on the specific temporal instances considered in the study of the French flu epidemic. As far as health care professionals are concerned, the formulas of Table 4.1 operate on the initial conditions of the flu mapping process such that, if the initial flu conditions convey a scientific fact, so will the covariance and cross-covariance functions obtained by applying these formulas. These formulas properly systematize space–time dependence patterns of flu spread by capturing the consequences in two points, $(\boldsymbol{s},t),(\boldsymbol{s}',t')$, a correlation that arises from local changes in $I_{s,t}$, $S_{s,t}$ and $R_{s,t}$.

Hand–foot–mouth disease (*HFMD*) is the most common infectious disease in China (Li, 2010), which is why there is considerable interest in understanding the spatiotemporal *HFMD* patterns and potential correlations to environmental factors; e.g., Wang *et al.* (2011) have explored *HFMD* and climate associations across Eastern China. The following example refers to a recent *HFMD* study based on the *BME–SIR* approach of Eqs. (2.12) and (4.4).

Example 4.5: Angulo *et al.* (2013) applied the *SIR–BME* method to study the spread of *HFMD* in 145 Chinese counties with relatively high disease incidence. The counties extend between 111°E to 118°E, and 32°N to 37°N (Fig. 2.2a). The data are weekly aggregated *HFMD* rates (infecteds/10,000 people) in the counties over a period of 20 weeks, from the week of September 27–October 3, 2008 ($t = 1$) to that of February 7–13, 2009 ($t = 20$). The observations of the data survey were assumed to be uncertain, and the widths of the associated rate intervals (soft data) were selected on the basis of the recorded national average *HFMD* rates (3.69 in 2008 and 8.68 in 2009, given the corresponding population and reported *HFMD* cases). The initial geographical spread of infecteds, $I_{s,1}$, is obtained from the observed rates at $t = 1$ (no recovered people are assumed at this initial stage). By considering one-week disease duration, the remaining part of the population consisted of susceptibles, $S_{s,t}$. Relocation occurred sparsely during the 20-week study period, which was quantified in terms of a Gaussian kernel (κ_s) of 0.1-bandwidth in Eq. (4.4). This means that on average 97.55% of the population did not relocate during the study period. The recovery and transmission probabilities in Eq. (4.4), $a_{s,t}$, $b_{s,t}$, had initial values $a = 0.1$, $b = 0.4$, and variances $\sigma_a^2 = 0.05$, $\sigma_b^2 = 0.1$. Just as in the flu case (Table 4.1), in the *HFMD* study, the covariances of the $I_{s,t}$ distributions at subsequent times were based on the covariance of the initial $I_{s,1}$ distribution computed from the available data at $t = 1$. The initial covariance had a nugget effect equal to 0.07 (in rate variance units), a spherical shape with a sill equal to 0.07 (rate variance units) and a spatial range equal to 3°. Using this information, space–time maps of the distributions of the $I_{s,t}$, $S_{s,t}$ and $R_{s,t}$ population fractions were generated throughout the 20-week study period. At each consecutive time instance, the *SIR* law (core knowledge) drives the model parameters progressively closer to the values that best interpret the *HFMD* dataset (this process is guided by updating the model with new case-specific information at every time step). Remarkably, the *BME–SIR* approximations of the mean transmission and recovery rates were $b \approx 0.17$, $a \approx 0.21$; i.e., despite the arbitrary initial values ($a = 0.1$, $b = 0.4$), parameter equilibrium was reached relatively fast (within about two weeks).[7] Maps of the predicted $I_{s,t}$ mean distributions were derived for each one of the 20 weeks. Figure 2.2(2) (Chapter 2) displays these means for the counties of interest during

---

[7] Other possibilities exist; e.g., one could use existing data to obtain *SIR*-based regression estimates of the initial parameter values (Wang *et al.*, 2006).

a few selected weeks. The spatiotemporal *BME–SIR* predictions of the infected population rates in Fig. 2.2b account for uncertainty in the *HFMD* observations, and they can provide an informative overview of the disease as it evolves. Yet, successful prediction in conditions of *in situ* uncertainty and space–time variation does not mean perfect accuracy. Accordingly, it was found that the statistical $I_{s,t}$ prediction error ranges, on average, between 0.226 and 0.247 (cases/10,000 people), during the 20-week study period.

In some cases of population health assessment and management, space–time correlation patterns such as the above could provide some light, though not definite, truths about disease causation (or etiology). This is the topic of the section that follows.

4.4. Space–Time Causation Revisited

From the very beginning of civilization, the investigation of the natural world was basically the search for the relevant causes of natural phenomena. For example, in Plato's *Phaedo*, the so-called "inquiry into nature" consisted of a search for the causes of things (why they come into existence, why they go out of existence, why they exist) (Hankinson, 1998). In *Posterior Analytics*, Aristotle states that one has knowledge of a thing only when one has grasped its cause (Mure, 1975). In *Physics*, Aristotle offers his account of the four causes: material cause ("that from which", e.g., the stainless steel of a surgical instrument); formal cause ("what it is to be", e.g., the shape of the instrument is designed to perform specific actions of carrying out desired effects during a surgery or operation); efficient cause ("how it was built", e.g., science and technology was used to build the instrument); and final cause ("the purpose of a thing being built", e.g., to perform the surgery that can improve or save one's life). In the Mohist *Canons* the term "cause" (*ku*) is defined as "what is required for something to happen" (Graham, 1967); also, a noteworthy distinction was made between minor cause ("with this it will not necessarily be so, whereas without this it necessarily will not be so"), and major cause ("with this it will necessarily be so").

A primary concern of medical decision-making is scientific inference, as well as its application to explanation and causation. Undoubtedly, the study of the space–time disease *causation* (or *etiology*) is of great significance in medical sciences, where a health care professional's powers of noticing or perceptual grasp frequently depend upon recognizing what is causally salient and the capacity to respond to the particular clinical or exposure case. New understanding and emerging results are reshaping medical thinking, as is the recognition that establishing realistic cause and effect associations between environmental exposure and population impact is a daunting task. Most professionals acknowledge the necessity of taking causes severely into consideration during the treatment of a disease. Although the spatiotemporal interplay of empiric and epistemic conditions of the population's behavior plays a vital role in disease causation, it has not been satisfactory analyzed in the medical literature. The notion of causation involves the study of a multitude of factors, such as:

(1) *Combined action* of several natural processes possibly contributing to the causation mechanism (some of them may lie outside the boundaries of medicine and epidemiology into environmental, social and cultural domains).
(2) Causation *characterization* in association with specified populations or in connection with the available *KB*s.
(3) Composite *space–time* domain of causation.
(4) Underlying scientific thinking that provides *epistemic support* to causation that is consistent with *in situ* experience.
(5) Causation *hypotheses*[8] that constitute one important category of empirical statements.

Probability plays a key role in disease causation in more than one ways. To demonstrate legally a case of medical malpractice (i.e., a kind of professional negligence), the plaintiff must establish through an expert's medical causation testimony that, to a reasonable probability degree, the plaintiff's injury was caused by the defendant's negligence. The key issues here are "probability" and "causation" and their logical association. Indeed, medical malpractice liability involves a series of probabilistically dependent events. It is usually much easier to disprove than to prove causation. In clinical practice, "treatment" is dependent on "diagnosis", because a patient would not receive treatment for a disease until that disease had been diagnosed. In such cases, an analysis of the joint diagnosis–treatment probability is a vital tool for countering proof of medical causation. In cases of delayed diagnosis the physician (defendant) is allegedly negligent for failing to administer a medical test to a patient (plaintiff) that could have lead to the diagnosis and treatment of a disease at an earlier stage when

---

[8]Hypothesis is an assertion that empirical evidence can either support or reject, but rarely prove.

the odds of successful treatment were considerably higher. The following example discusses a typical malpractice case (Watson, 2010).

Example 4.6: The patient alleges that he would have had a better outcome if the defendants had not negligently failed to diagnose his cancer when it was at stage-one. The patient's expert witness testifies that the defendant's physicians probably would have diagnosed his stage-one cancer if they had ordered additional chest x-rays as required by the standard of care. This testimony supports a jury finding that there was a 51% probability of a stage-one diagnosis, but it does not support a finding of a greater than 51% chance of a better treatment outcome. The reason for this is that the patient's expert testifies that,

"had the patient received treatment when his cancer was at stage-one, he had a roughly 60% chance of living five years or longer".

The above assertions imply that the evidence actually established only that the patient lost a 30.6% probability of survival as a result of the defendant's alleged negligence, i.e.,

"30.6% = a 51% chance of a stage-one diagnosis × a 60% chance of a better outcome with treatment beginning in this stage".

Accordingly, there is no credible proof of medical causation in this case.[9]

Before proceeding further, a comment may be appropriate about epistemic certainty in an etiologic context. If a general assertion a_A is epistemically certain for physician A, then A is justified in being psychologically certain of it; otherwise said, A has a 100% confidence in it, and a_A has an occurrence probability $P[a_A] = 1$. Now, if epistemic certainty is required for a_A to be sufficiently warranted to be a reason A has for suggesting causation, the symptoms and diagnoses the a_A appeals to as reasons for causation may be not reasons after all, because in a professional's daily routine these symptoms and diagnoses lack sufficient warrant most of the time. As a consequence, the view that epistemic certainty is required for a_A to be warranted enough to be A's reason

---

[9]Note that in jurisdictions like California (US), where causation must be proved by a preponderance of the evidence, this case should never get to the jury because the plaintiff cannot prove causation. In jurisdictions like Nevada (US), where plaintiffs may recover damages for the "lost chance" of a better outcome, application of the joint probability of statistically dependent events rule should result in a greatly diminished recovery.

for causation is highly questionable in clinical practice. In real medical reasoning, instead, an assertion can be warranted enough to be a reason a physician has for causation even if that proposition is not epistemically certain for a physician.

While standard logic is descriptive, medical thinking under conditions of uncertainty is creative. This means that, while in the former case in principle one cannot obtain new scientific truths that are not obtainable by means of tautologies, in the latter case causal patterns are derived that are not logical tautologies and can play a central role in causation studies. It was noted earlier that on occasion medical thinking appears to be recursive, taking itself into consideration (e.g., in the process of reasoning about a clinical case the physician would ask him/herself, "What thought am I thinking now?"). At the same time, medical thinking recognizes that causation occurs in a space–time domain, which means that the physical significance of the domain has a key effect on the conception of causation. This is to say, the study of the *disease transmission chain* considers the structure of composite space–time as is determined by the study conditions. The same is valid for the environmental causal chain. Consequently, all case attributes relevant to space–time disease spread are represented as random fields (in the sense of Section 2.2.2).

In sum, the causation (or etiology) of a health effect is generally viewed in a space–time synthesis milieu. In this milieu, both the epistemic and the empiric components of causation are considered stochastic, and, as a result, so is their interplay. Epistemic stochasticity refers to uncertainty due to the physician's incomplete knowledge of the medical case. Empiric has to be conceived as stochastic too, since observed events and measured case attributes linked to disease spread are mediated primarily in cognitive terms (involving justified beliefs, assessments, diagnoses).

4.5. Medical Causation in the *SMR* Inference Setting

According to the preceding analysis, unlike mainstream logic, *SMR* inference is not just an instrument to be used indifferently in clinical or epidemiologic studies, but a theoretical science in its own right that is linked to other scientific fields. The needs of disease causation can put the *SMR* results discussed earlier to good use, keeping in mind that in medical situations one seeks causal facts, not mere accidents of association. The following section briefly describes the lay of the land in medical causation thinking.

4.5.1. Defining the problem

Let $C_{\kappa,p}$ ($\kappa = 1, 2, \ldots, n$) denote possible causes (genetic, environmental, social) of the effect E_p (say, disease spread in a region over time). Disease causation is a highly uncertain affair, which means that mainstream deterministic logic is of little use here. On the other hand, *SMR* inference by definition possesses the flexibility for a realistic description of many causation situations. For a certain school of medical thinking, to make a statement about $C_{\kappa,p}$ is to say that the physician is able to construct an assertion or mind state a_A that adequately accounts for eminent *in situ* characteristics of $C_{\kappa,p}$. Depending on the construction, the assertion may possess different degrees of validity.

The form of physician A's assertions about possible causes ($C_{\kappa,p}$) of disease spread (E_p) may vary according to the context in which the different a_A forms are uttered. What varies are the epistemic standards that A must meet in order for such an assertion form to be valid as part of a triadic formulation of causation. In other words, what varies with context is how well positioned a physician must be with respect to a_A for the latter to count as g_A, b_A, s_A, r_A or k_A. Note that "context" here refers to things like certain features of A's cognitive condition, the objective situation of the medical case ($C_{\kappa,p}$ features, alternatives to the current diagnosis, potential symptoms) and factors of A's psychology, conversational and practical situation. As a result of context-dependence, the truth-value of assertions about causation, made in different contexts, may differ.

Assume that $a_A C_{\kappa,p}$ denotes physician A's assertion that the causes $C_{\kappa,p}$ exist in a region during the time period of interest, and suppose that A needs to study the possibility that $C_{\kappa,p}$ could cause E_p under conditions of *in situ* case uncertainty. The medical decision process may include the rigorous evaluation of relationships like

$$\frac{a_A(E_p \backslash C_{\kappa,p})}{\therefore\ a_A(E_p)}$$
$$P[a_A(E_p \backslash C_{\kappa,p})] > \zeta_a \qquad (4.5\text{a--d})$$
$$P[a_A(E_p \backslash C_{\kappa,p})] > P[a_A(E_p)]$$
$$P[a_A(E_p \backslash C_{\kappa,p})] > P[a_A(E_p \backslash \neg C_{\kappa,p})].$$

The a_A-form of Eq. (4.5a) should be of sufficiently high epistemic strength, and the ζ_a-value in Eq. (4.5b) should be sufficiently large. These relationships constitute causation criteria that can be used both independently of each other or in synergy. In accordance with the analysis in previous sections, the vigor of such criteria surely depends on the a_A-form associated with the case of interest.[10] For instance, the $P[r_A(E_p \backslash C_{\kappa,p})]$ generally presents a causation condition based on the physician's rational state, whereas the $P[b_A(E_p \backslash C_{\kappa,p})]$ offers merely a belief-based assessment. The following is one of a series of examples discussed in the book that aim at familiarizing the readers with the various types of medical causation.

Example 4.7: Physician A studies a case of *prostate cancer* (PC) for Mr. Brenan who belongs in a certain risk group

RG: "63-year-old, African American, with high calcium intake diet, and a sedentary lifestyle".

Based on the relevant database, the physician estimates that $P[a_A(RG \rightarrow PC)] = p = 0.55$. Then, the criterion (4.5a) is interpreted by A based on the *denying the antecedent syllogism* (Table 3.11),

$$\text{Case premises} \quad \frac{a_A(RG \rightarrow PC) \quad (p = 0.55)}{k_A RG}$$

$$\text{Physician's conclusion} \quad \therefore a_A PC \quad (P[a_A PC] = 0.55)$$

Hence, given that Mr. Brenan belongs to RG, in the sense of Eq. (4.5a) the probability that Mr. Brenan has PC is 55%. Note that the same result is obtained if instead of the material conditional one uses the statistical conditional (why?).

As with other reasoning matters, when checking a causation condition a health care professional's calculations must satisfy the mathematical restrictions of probability.

Example 4.8: Physician A studies Ms. Voubrahms' *coronary stenosis* (CS) case. Ms. Voubrahms has been a serious smoker (SS) for many years. Based on previous experience, the physician estimates the following probability values $P_{KB}[CS] = p_1 = 0.40$ $P_{KB}[SS|CS] = p_2 = 0.45$, $P_{KB}[SS|\neg CS] = p_3 = 0.55$. Then the physician uses Bayesian inference to calculate the

---

[10] Readers may recall that $a_A(P[CA_p] = p_A) : p_A = P[a_A CA_p] = P_{KB}[CA_p] \neq P[CA_p] = p$.

posterior probability

$$P_{KB}[CS|SS] = \frac{p_1 p_2}{p_2 p_1 + p_3(1-p_1)} = 0.35.$$

Given that the condition of Eq. (4.5c) is not satisfied, i.e. $P_{KB}[CS|SS] = 0.35 < P_{KB}[CS] = 0.40$, the physician decides that

"Ms Voubrahms' smoking should not be considered a CS causation, in this case".

However, this could be the wrong conclusion, since the probability values selected by the physician are internally inconsistent. Indeed, probability theory shows that these values must be such that $p_1 < 1$ and $p_3 < p_2$ (instead, in the present case $p_3 = 0.55 > p_2 = 0.45$).

Implementation of the causation conditions of Eqs. (4.5a–d) is not always straightforward in practice. This is the focus of the following example.

Example 4.9: Dr. Geroulakos is treating a case of *myocardial infarction* (MI). His patient has high blood pressure (BB). From the published statistics, Dr. Geroulakos knows that

"the probability interval of someone with his patient's characteristics (sex, age, pre-existing health condition, lifestyle) suffering from MI is $P_{KB}[MI] \in [35, 75\%]$".

Moreover, based on the doctor's records,

"the patient's chances of high BB is $P_{KB}[BB] = 0.60\%$".

In order to choose the appropriate course of medication, Dr. Geroulakos needs to calculate the probability of the conditional $MI \backslash BB$ and then compare it with $MI \backslash \neg BB$, according to Eq. (4.5d). The good doctor selects the common health statistics conditional ("$\backslash$" $\equiv$ "$|$") , in which case his inference is as follows,

Case premises
$$P_{KB}[MI] \in [35, 75\%]$$
$$P_{KB}[BB] = 60\%$$

Physicians' conclusion
$$\therefore P_{KB}[MI|BB] = 58.34\%, \quad P_{KB}[MI|\neg BB] \in [0, 100\%]$$

which is uninformative. Hence, Dr. Geroulakos should rather conclude that the chosen inference scheme is not very helpful in the present case.

Each one of the relationships (4.5a–d) relies on a different notion of causation assessment. The first relationship is based on the transition from an assertion state (belief, claim, statement) to an actual state; the second relationship is based on a similar transition but in probabilistic terms; the key notion behind the third and fourth relationships is that the causes raise the probabilities of their effects, all else being equal (e.g., (4.5c) expresses the degree to which $C_{\kappa,p}$ partially imply E_p). Which one of the relationships (4.5a–d) offers a better representation of the clinical situation is a matter of medical decision-making and requires considerable concrete experience with real case studies. Physicians' intuition based on experience plays a central role as regards criteria (4.5a–d), for it allows them to act relatively quickly. In clinical cases that are uncommon for them, the physicians consult each other and analyze each possibility critically in Eqs. (4.5a–d). Actually, the implementation of the relationships in an unstructured domain may suggest interpreting them in a rather hypothetical setting in which possible consequences can be studied. Let us look into some special cases.

Example 4.10: At a first glance, one could rely on epistemic changes introduced by Eqs. (4.5a–d) to study medical etiology. Air pollution and occupational exposures have been linked to a variety of population effects. As noted in various parts of the book, human exposures may include dust, strong fumes, cold air, exercise, inhaled irritants, allergens, pesticides, cigarette smoking and bacterial or viral infections. If the physician suggests that a causal link exists between exposure PE_p and population damage PD_p at location–time p, the conditional in Eq. (4.5c) would be interpreted as a logical counterfactual, " $\cdot \setminus \cdot$ " $\equiv$ " $\cdot \Rightarrow \cdot$ ", so that the following criterion is used,

$$P_{KB}[PD_p \setminus PE_p] \equiv P_{KB}[PE_p \Rightarrow PD_p] > P_{KB}[PD_p].$$

A counterfactual conditional is a conditional statement of what *would be* the case (sickness, death) if its antecedent (exposure) *were* true: if PE_p had not happened, PD_p would not have either. This is to be contrasted with an indicative conditional, which indicates what *is* (in fact) the case if the antecedent *is* (in fact) true.

Classical logic was formulated by Frege and others in an idealized language, which is adequate for mathematical reasoning.[11] There are

---

[11] For mathematicians, if formal logic does not translate our natural language perfectly but plays its intended role in mathematics, so much the worse for natural language.

some oddities, but they do not show usually up in mathematics, and if they do, they can be lived with. However, as we saw in Section 2.6.2, these oddities can be a Procrustean framework, so to speak, when one encounters medical assertions (propositions, beliefs, judgments, claims) about empirical matters. The difference is that in thinking about the empirical world, health care providers frequently assign to these assertions different degrees of confidence that are less than certainty. In the same setting, language continues to play a pivotal role in causation matters too, since, as was stressed in various parts of the book, lexical entities of thought patterns may allow a sentence more than one possible meaning.

Example 4.11: Suppose that PR denotes *man taking the hair-growth drug Propecia*, and SD denotes *sexual dysfunction problems*. Dr. Lalas, an experienced sexologist of the $IASHS$ (Institute for Advanced Study of Human Sexuality), opposes men taking Propecia ($\neg PR$), and, as a consequence, he rejects the option of a

"man taking Propecia thus causing sexual dysfunction problems to himself",[12]

in symbolic terms, for Dr. Lalas the possibility taking $PR \to SD$ is out of the question. According to formal logic, and unknowingly to him, Dr. Lalas apparently has inconsistent views about causation. Because,

"to reject the thought that $PR \to SD$ is to accept $PR \wedge (\neg SD)$,"

since this is the only case in which the possibility that "$PR \to SD$ be applicable in the case of a male" is false. How can Dr. Lalas accept $PR \wedge (\neg SD)$ yet oppose the use of Propecia ($\neg PR$)? Put the other way round, if Dr. Lalas thinks that $\neg PR$, he must think it likely that at least one of the assertions, $\neg PR, SD$ is true. But that is just to think it likely that $PR \to SD$, which is inconsistent. Clearly, it is not just formal logic that seems to fit the patterns of thought of competent, intelligent physicians badly. Due to inadequate use of language, decision-makers would be intellectually disabled: they would not have the power to discriminate between believable and unbelievable conditionals whose premise they think is likely to be false.

The soundness of the stochastic syllogisms used in the above disease etiologic examples generally depends on the internal consistency of the

---

[12]The assertion "*taking* Propecia will cause sexual disfunction to men" (who take it) is not to be confused with "Propecia causes sexual disfunction to men" (in general).

logical process, the adequate understanding of the biological mechanisms underlying the syllogism, and the quality of the case probability estimates involved in the premises of the syllogism. For example, when the representativeness heuristic process (Section 3.3.4) is used in the estimation of these probabilities, possible sources of error include (Sox et al., 2013): (a) the effect of disease prevalence in the parent population (e.g., in a real case even if a Chicago resident's symptoms included a history of intermittent shaking chills, sweats and fever, which fit well with the standard definition of malaria, the disease was unlikely, given its rarity in the American population); (b) the imperfect disease indicators obtained in terms of the cues that make up the standard disease description (e.g., while coughing blood and clinical evidence of leg blood clots are part of the standard description of pulmonary embolism, they are not accurate disease clues, since only about 25% of patients with pulmonary embolism exhibit the first symptom and about 33% exhibit the second); (c) the standard disease predictors used in the case typically all occur together so that more than one predictor does not add information (e.g., that a patient has chest pain that is retrosternal, radiates to the left arm, and is crushing, squeezing and pressure-like essentially does not increase the probability of having coronary artery disease more than one of these symptoms does); (d) misinterpreting random variation in biological measures as response to treatment (e.g., following a single measurement of serum glucose displaying mild hyperglycemia the clinician puts the patient on a diabetic diet that apparently causes the fall of patient's blood sugar in a second test, the clinician wrongly concludes that the patient has diabetes); and (e) mistakenly judging a patient with atypical clinical disease features as highly likely to have a disease with which the physician happens to have a small experience (e.g., although there is no enlargement of the patient's thyroid gland, the physician estimates that the probability of hyperthyroidism is 50% because the patient closely resembles the only two hyperthyroidism cases in the physician's limited practice, ignoring the fact that the chance of a hyperthyroid patient having an enlarged thyroid gland is actually 95%).

4.5.2. *The role of KB and the interpretation of probabilistic causation*

As the readers may recall, a knowledge base KB may include contextual and laboratory information. Contextual clinical information refers to symptoms and signs, clinical setting, past history and medications, and demographic information. Laboratory information contains different kinds of tests and results (e.g., blood sugar tests, enzyme, protein and protein blood tests, flu tests, lipid profile, serum and urine electrolytes, and arterial blood gas results). In principle, science-based etiology integrates biomedical theory and clinical knowledge (both are KB elements). As physicians accrue more experience they increasingly rely on clinical knowledge rather than biomedical science when diagnosing or treating a disease. To put matters in perspective, let us revisit Eq. (4.5) with $\kappa = 1$. As noted earlier, statements such as "C_p causes E_p", or "C_p raises the probability of E_p" are true or false only relative to a particular knowledge base KB (or a population considered on the basis of KB). Generally, one can assert that

"C_p is the cause of E_p with respect to a particular KB".

It turns out that the available or chosen KB can make a big difference and can convert a positive causation factor into a neutral or negative one, or vice versa.

Example 4.12: The value a health care professional A derives for the

"probability of a resident of the city of Granada having a heart attack this year"

is different if the information available to A (A's KB) is the population of 20-year-old residents from what it is if the available KB is the population of 50-year-old residents. Similarly, the estimation of

"Mr. Cornwallis' conditional probability of having a heart attack this year given that he has been a heavy smoker for 20 years",

is different if the available KB is the population of 40-year-olds from what it is if KB is the population of 60-year-olds.

Remarkably, C_p may not be the cause of E_p with respect to a certain KB, but, nevertheless, it can be the cause with respect to another KB'. This situation can be illustrated with the help of a simple arrangement as follows.

Example 4.13: Consider the knowledge base

KB : population of pairs of simultaneous (independent) tosses of two fair coins.

Let C_p denote that the left coin comes up heads (H_-) in a toss of the two coins, and E_p that the right coin comes up H in a similar toss ($_-H$). Clearly,

"H_- is causally neutral for $_-H$ with respect to the above KB",

i.e., $P_{KB}[_-H] = P_{KB}[_-H|H_-] = 0.5$. Next, consider another knowledge base, KB' : population of pairs with half H and matching on left and right coins. In this case,

"H_- is causally positive for $_-H$ with respect to KB'''"

for $P_{KB'}[_-H] = 0.5 < 1.0 = P_{KB'}[_-H|H_-]$.

The above example leads to some interesting conclusions, in particular, in the case of probabilistic causation there should be some conditions imposed on one's choice of a KB. Let us discuss a few of these:

(1) KB may not logically imply the probability value of E_p, or of $E_p|C_p$, or of $E_p|\neg C_p$. Otherwise, some probabilistic causal claims would then be logical truths. This is the case of KB in Example 4.13 above.

(2) KB may not logically imply even the equality or direction of inequality among E_p, $E_p|C_p$ and $E_p|\neg C_p$. Otherwise, some implied probabilistic claims would clearly be wrong.

(3) KB may allow for both possibilities of C_p and $\neg C_p$ being exemplified in KB. Otherwise, if we allow KB to be such that it consists only of C_ps, then it becomes *a priori* impossible for C_p to be causally relevant to E_p(i.e., $P_{KB}[E_p|C_p] = P_{KB}[E_p]$, by definition, as in the KB case in Example 4.13).[13]

Before leaving this section, let us remind our readers that one disease feature can beat out another (according, e.g., to the analysis in Section 3.6), even though both seem reasonable based on the available KB and associated causation. Similarly, two disease treatments $DT^{(1)}$ and $DT^{(2)}$ can be both reasonable choices given the causation assumed, yet $DT^{(1)}$ could be selected for practical (ontic) reasons, such as resource availability and financial costs.

Causation in the sense discussed above looks for a more general mental and analytical comprehension of probability. The question then arises concerning the *interpretation* of the conditional probability $P_{KB}[E_p|C_p]$ that is particularly appropriate for probabilistic causation. Naturally, a probability interpretation should satisfy certain conditions, such as: it must be *coherent*

---

[13] As another example, in a population in which everybody smokes, we still want to be able to say that smoking is a cause of lung cancer.

(i.e., it satisfies the basic mathematics of probability calculus), *ascertainable* (i.e., its values must be determinable) and *contextual* (meaningful in the specified case setting). Within this methodological framework, two major probability causation interpretations are:

(1) *Ontic* interpretation: Probability causation is an objective feature of the actual phenomenon, e.g., the proportion of sampled C_ps are observed to be linked to E_ps. It is independent of human conceptions and descriptions. Probability causation is expressed in object language that relates events, and the validity of arguments about it is empirical. The objective interpretation is closely related to the *biophysical* interpretation of $P_{KB}[E_p|C_p]$, which is the degree to which C_p yields E_p as determined by biological, physical, clinical etc. factors (i.e., in biophysical probability the probabilistic facts are fixed by biophysical factors).

(2) *Epistemic* interpretation: Probability causation is often viewed as the degree to which C_p partially entails E_p, given the investigator's cognitive condition. It is expressed in metalanguage that relates assertions about events, and the validity of arguments about probability causation is logical. A special case is the *subjective* interpretation that is dependent on human conceptions and personal descriptions, i.e., probability is interpreted as the investigator's degree of belief in E_p, given one's own understanding of C_p.

The above interpretations are not without their critics. Critics of subjectivism argue that the subjective probability interpretation (considered, e.g., in a certain kind of Bayesian inference) lacks the resources to distinguish "uncertainty due to lack of knowledge" from "uncertainty that no possible increase in knowledge could eliminate". Bayesians frequently assume that all uncertainty is subjective, although it is not clear how subjectivism entails indeterminism. The above discussion prepares the readers for the analysis to be presented next.

4.5.3. Causation: Epistemic vs. non-epistemic

The focus of this part of the book is on the *warrant* required for, say, an environmental exposure or clinical case assertion to be a reason for causation. On a relevant note, the *SMR* method of translating a physician's mind assertions into space–time probabilities (Section 3.8) may be useful in certain etiologic settings, such as to establish a criterion for whether or not a medical assertion is an *epistemic cause* of another assertion. In many cases, this may be done with the help of the following quantitative criterion:

$$(a_A C_p \triangleright a_A E_p) \leftrightarrow P_{KB}[E_p \backslash C_p] \geq P_{KB}[E_p \backslash \neg C_p], \qquad (4.6)$$

where "$\triangleright$" denotes "is an epistemic reason to believe" or "is an epistemic cause for". That is, $a_A C_p$ is an epistemic cause for $a_A E_p$ iff the conditional probability of E_p given C_p is greater than that of E_p given $\neg C_p$. More precisely, a health care professional needs to epistemically determine under what conditions $a_A C_p$ has the status of "being warranted enough" for a professional to be a reason to assert $a_A E_p$ (e.g., given a belief about exposure to assert its health effect). Naturally, there are several possibilities as to why such a warrant is a sufficient reason for a physician to justify causation, such as that a warrant requires epistemic certainty or it is subject to pragmatic uncertainty. These possibilities are linked to different kinds of epistemic relations (knowledge, true belief, justification etc.) a physician must bear to C_p via the a_A-form in Eq. (4.6) for C_p to be warranted enough to be a cause for E_p.

Example 4.14: A health care professional frequently deals with assertions and their epistemic justification. The assertion

"radiation exposure RE_p at location–time p exceeds a certain threshold θ"

is an epistemic reason to believe that

"population mortality PM_p at p is above a certain critical level ϕ" only if

$$P_{KB}[(PM_p > \phi) \backslash (RE_p > \theta)] > P_{KB}[PM_p > \phi \backslash \neg (RE_p > \theta)]. \qquad (4.7)$$

That is, the statement concerning exposure is an epistemic reason to believe the cause–effect association, in the probabilistic sense above.

The strength of an epistemic cause is directly affected by the particular form of the assertion state a_A (i.e., g_A, b_A, s_A, r_A, k_A), which reflects the quality of evidence and warranted confidence (e.g., how confident a physician ought epistemically to be). Clearly, k_A is linked to the highest level of epistemic strength, and g_A is linked to the lowest. A knowledge state k_A is quite demanding and the most difficult for a physician to achieve. The readers may recall (Section 3.1.3) that physicians are led to grant that they know something even though they clearly do not satisfy anything beyond the truth or true belief conditions of knowledge. Moreover, there are standard contexts in which physicians will claim that they know something, even when it is utterly clear that they only possess a true belief.

On the other hand, an assertion about C_p is considered a *non-epistemic* reason to believe something about E_p if this reason is over and above the extent to which the probability inequality above is valid. Generally, non-epistemic reasons make an assertion less likely to be the case than epistemic reasons that are science-based. The reader must keep in mind that the above case (Example 4.14),

$$C_p > \theta \text{ is an essential reason for } E_p > \phi$$

underlies a natural process; whereas the case

$$a_A(C_p > \theta) \text{ is an epistemic reason for } a_A(E_p > \phi)$$

represents a human reasoning process. The natural process can be valid without the epistemic process being so. On a relevant note, the form $a_A C_p$ is valid iff A asserts that C_p, it is in fact C_p, and A meets certain epistemic standards. The way these standards are selected varies. Standards are determined on the basis of A's cognitive context as well as on C_p's real context (C_p's features, practical facts about C_p's environment etc.). Whether

"physician A asserts that the scan detects a malignant tumor"

is valid or not may be determined by the physician's own epistemic context and/or by practical facts about the scan's own context (technical features etc.).

Example 4.15: While in medical practice the fact that taking the drug hexamethonium is a *natural cause* for an individual's lung being seriously damaged, a Johns Hopkins medical researcher's state of mind did not include a similar *epistemic cause* that related beliefs about the drug and lung damage. That is, the following reasoning mode was absent in the researcher's mind state:

"an individual takes drug *hexamethonium*"

is an epistemic reason to believe that

"individual's lung is seriously damaged".

The result of this absence of an epistemic condition was the tragic death of a healthy 24-year-old woman in an asthma experiment (Bor and Pelton, 2001).

The above should make it obvious that the study of disease etiology demands that competent clinicians possess more than logical and analytical skills. They should also exhibit critical reflection, continuously assess data and methodology, challenge assumptions when necessary and use creative thinking to guide their decision-making.

4.5.4. *Some remarks regarding the form of the causation conditional*

In all the above studies, the practitioner needs to decide the *form* of the conditional $P_{KB}[\cdot\backslash\cdot]$ that best fits the etiologic situation of interest, which is not always a trivial matter (as we saw, e.g., in Sections 2.6 and 3.3.2). An issue worth studying is under what causal case conditions the "$\cdot\backslash\cdot$" could be seen as a material conditional ($\cdot \rightarrow \cdot$), an equivalence conditional ($\cdot \leftrightarrow \cdot$), a counterfactual (or subjunctive) conditional ($\cdot \Rightarrow \cdot$) etc. The choice, of course, will depend on the specific case objective features and the clinician's understanding of it. When making this choice, one must keep in mind a few facts:

(1) The calculus of probability is not that of possibility (e.g., that all Granada citizens will catch a cold in June is perfectly possible, but not realistically probable), which naturally implies that the notion of *probable* causation should not be confused with that of *possible* causation.
(2) It is well established that $P_{KB}[E_p\backslash C_p]$ is not necessarily $P_{KB}[E_p|C_p]$; i.e., *evidential relevance is not causal relevance*, which means that correlation is not necessarily causation.

As a matter of fact, there are two fundamental differences between the notions of correlation and causation, namely, causation is asymmetric and content-dependent.[14] Based on a concept initially proposed by Hempel that replaces the deductive–nomological mode by an inductive–statistical one, causality is sometimes manifested in terms of an almost perfect correlation between human exposure and health effect. This approach is

---

[14]The matter has already been discussed in Sections 2.4.2 and 2.5.4.

not free of some serious complications, nevertheless. One problem with correlation-based techniques is that while correlation is a symmetrical relation, causation is nonsymmetrical; e.g., if cigarette smoking is correlated with cancer, then cancer is correlated with smoking, but the fact that cigarette smoking causes cancer does not imply that cancer causes smoking. A similar technique seeks to determine associations by comparing separate maps of health effects and environmental exposures across geographic areas or over time, using multivariate statistics to establish correlations between health effects and exposures. There are certain problems with this technique, as well. Several cases have been reported in the scientific literature in which the technique led to incorrect conclusions (Krewski et al., 1989).

As mentioned earlier (Section 2.5.4), a difference between correlation and causation is that the former is CI, whereas the latter is CD. According to the correlation approach, C_p may be seen as a cause of the health effect E_p if the correlation between the two, $Corr_{C,E}$, is high. This example describes situations where the approach is useful, as well as some other situations where the approach is utterly useless.

Example 4.16: Assume that $DI_p(=3,6,9,\ldots$ in suitable units) denotes the incidence values of a certain disease at Los Angeles, during a specified time period t; and $PC_p(=1,2,3,\ldots)$ represents pollutant concentration at Los Angeles during the same period. Clearly, the statistical $Corr_{PC,DI}$ is high, which, from the correlation viewpoint, implies that exposure PC_p could be possibly seen as the cause of disease DI_p, assuming that substantive links between PC_p and DI_p can be indeed established. On the other hand, let us assume that $MP_p(=0.5,1,1.5,\ldots$ in suitable units) now represents the stock market prices of some entity at Hong Kong during period t. In numerical terms, $Corr_{MP,DI}$ also turns out to be high. Furthermore, let $HG_p(=0.25,0.50,0.75,\ldots)$ measure the portion of Beijing high schools during period t with average student grades reaching the highest education standard. Again, one finds a high $Corr_{HG,DI}$. Should Hong Kong stock market prices or Beijing high school statistics be viewed as causes for the Los Angeles disease incidence? Certainly not, even if the corresponding correlations happen to be very high.

In light of the above results, the relationship expressed by $Corr_{C,E}$ is CI, since its values do not depend on the substantive meaning of C_p and its physical relation to E_p. Therefore, statistical correlation cannot express any substantive connections that could be characterized as "C_p-E_p" causation.

In a more general setting, one could argue that while classical semantics is concerned with the study of the content of a medical document (that includes text, graphs, images, equations) to understand its meaning medical reasoning often deals with contextual aspects of meaning in particular cases. Medical documents carry rich semantics, but only health care experts can really take advantage of them. Meaningful reasoning renders rich inferencing and reasoning capabilities, as well as a framework to integrate heterogeneous *KB*s. Without the semantics being well understood, the study of disease causation would rely on the presence of words, symbols, equations and plots rather than the concepts standing behind them. As a result, decision-making may retrieve and process too much irrelevant data or fail to identify all the relevant documents in the medical *KB* available.

4.5.5. Stochastic causal inferences

Section 3.4 studied the basics of stochastic inferences in medical decision-making. As central components of scientific inquiry, stochastic inferences are closely linked to causation analysis. In fact, certain of the stochastic syllogisms of Table 3.11 are directly applicable in disease causation investigations.

Example 4.17: Let DS be a symptom considered as the cause of the tropic disease TD. A physician needs to find out the impact of Ms Moneypenny not experiencing DS on her chances of having TD. In this case, the physician may chose to use the stochastic denying the antecedent inference of Table 3.11,

$$
\begin{array}{ll}
\text{Case premises} & \dfrac{P_{KB}[DS \to TD] = p_1}{P_{KB}[\neg DS] = p_2} \\
\text{Physician's} & \\
\text{conclusion} & \therefore P_{KB}[\neg TD] \in [1 - p_1,\ \min\{1 - p_1 + p_2, 1\}]
\end{array} \quad (4.8)
$$

It is instructive to perform a sensitivity analysis of inference (4.8). For numerical illustration purposes, Table 4.2 presents a few combinations of p_1, p_2 values and the corresponding $P_{KB}[\neg TD]$ values. For example, one

Table 4.2: Numerical sensitivity analysis of inference Eq. (4.8).

p_1	p_2	$P_{KB}[\neg TD]$	p_1	p_2	$P_{KB}[\neg TD]$
0.8	0.2	[0.2, 0.4]	0.4	0.2	[0.6, 0.8]
0.8	0.4	[0.2, 0.6]	0.4	0.4	[0.6, 1]

may observe that if $p_1 > p_2$, the width of the interval $P_{KB}[\neg TD] \in [1 - p_1, 1 - p_1 + p_2]$ is always p_2, whereas if $p_1 \leq p_2$, the $P_{KB}[\neg TD] \in [1 - p_1, 1]$ is independent of p_2.

Of some interest in medical thinking is the probabilification of the *transposition* relation of Table 2.5. Specifically, according to the stochastic transposition syllogism,

$$\begin{array}{ll} \text{Case premise} & P_{KB}[C \to E] = p \\ \text{Physician's conclusion} & \therefore P_{KB}[(\neg E) \to (\neg C)] = p \end{array} \tag{4.9}$$

i.e., the material conditional probability of C (symptom, exposure etc.) causing E (health effect, injury etc.) is the same as the probability of not-E causing not-C. This result is not necessarily true for all forms of conditionals "$\cdot \backslash \cdot$". For instance, while a similar syllogism holds for the equivalence conditional,

$$\begin{array}{ll} \text{Case premise} & P_{KB}[C \leftrightarrow E] = p \\ \text{Physician's conclusion} & \therefore P_{KB}[(\neg E) \leftrightarrow (\neg C)] = p \end{array} \tag{4.10}$$

this is not so in the case of the standard statistical conditional,

$$\begin{array}{ll} \text{Case premise} & P_{KB}[E|C] = p \\ \text{Physician's conclusion} & \therefore P_{KB}[(\neg C)|(\neg E)] = \dfrac{P_{KB}[C]}{P_{KB}[\neg E]}(p-1) + 1 \end{array} \tag{4.11}$$

Incidentally, the readers may observe that if $P_{KB}[C] = P_{KB}[\neg E]$ in Eq. (4.11), then $P_{KB}[\neg C|\neg E] = p$ too.

An interesting situation emerges when the physician needs to calculate the impact of a disease symptom DS_1 in the diagnosis DD relevant to another symptom DS_2. A valid inference here is the *cumulative transitivity* syllogism

$$\begin{array}{ll} \text{Case premises} & \begin{array}{l} P_{KB}[DD|DS_1 \wedge DS_2] = p_1 \\ P_{KB}[DS_2|DS_1] = p_2 \end{array} \\ \text{Physician's conclusion} & \therefore P_{KB}[DD|DS_1] \in [p_1 p_2, \, p_1 p_2 + 1 - p_2] \end{array} \tag{4.12}$$

Note that when $p_2 = 1$ (certainty condition), syllogism (4.12) yields $P_{KB}[DD|DS_1] \in [p_1, p_1 + 1 - 1] = p_1$, which makes sense intuitively: if the chance of symptom DS_2 occurring in the presence of symptom DS_1 is 100%, the probability of DD given DS_1 and DS_2 should not differ from that of DD given DS_1. On the other hand, if $p_2 = 0$, syllogism (4.12) gives $P_{KB}[DD|DS_1] \in [0,1]$, i.e., if the presence of DS_1 excludes that of DS_2, ignoring DS_2 makes the calculated probability of conditioning DD on DS_1 uninformative.

Suppose now that in the medical case under consideration a physician knows the conditional probabilities of two individuals' disease symptoms DS_1 and DS_2 given the presence of a diagnosis DD. The probability sought is the conditional probability of the two symptoms together given the presence of the disease. In light of the interval expressions of conditional probabilities, an inference worth considering is as follows:

Case premises
$$P_{KB}[DS_1|DD] \in [p_1, p_2]$$
$$P_{KB}[DS_2|DD] \in [q_1, q_2]$$

Physician's conclusion
$$\therefore P_{KB}[DS_1 \wedge DS_2|DD] \in [\max\{0, p_1 + q_1 - 1\}, \min\{p_2, q_2\}]$$
(4.13)

In the syllogism (4.13) the meaning of the conclusion in the physician's mind clearly depends on the relative values of the bounds p_1, p_2, q_1 and q_2 of the known conditional symptom probabilities. In particular, let $p_1 + q_1 < 1$:

(1) if $p_2 < q_2$, then $P_{KB}[DS_1 \wedge DS_2|DD] \in [0, p_2]$;
(2) if $p_2 > q_2$, then $P_{KB}[DS_1 \wedge DS_2|DD] \in [0, q_2]$;
(3) if $p_1 = p_2 = p$, $q_1 = q_2 = q > p$,[15] then $P_{KB}[DS_1 \wedge DS_2|DD] \in [0, p]$;
(4) if $q < p$, then $P_{KB}[DS_1 \wedge DS_2|DD] \in [0, q]$ (see, also, Table 3.13, inference of "denying the antecedent").

4.5.6. *The role of secondary case attributes*

In several medical cases, it is possible that the physician's investigation includes *secondary* case attributes (symptoms, signs etc.) that are not related directly to the health effect of interest (disease, injury, poisoning), but rather to a possible causation of the effect. Still, under certain conditions these attributes could provide valuable information about the actual causation. The following examples may throw some light on the issue.

---

[15]That is, each symptom probability interval reduces to a single probability value.

Example 4.18: Let DC be a known direct cause of the health effect HE. Suppose that the particular patient exhibits another, secondary case symptom DS that is related to DC in a probabilistic sense. A physician needs to calculate the likelihood that HE actually occurs given the presence of DS. In this setting, the following hypothetical syllogism is a possibility worth considering by the physician

$$\begin{array}{ll} \textit{Case premises} & \begin{array}{l} DC \to HE \\ P_{KB}[DC|DS] = p \end{array} \\ \textit{Physician's conclusion} & \therefore P_{KB}[HE|DS] \in [p, 1] \end{array} \qquad (4.14)$$

More precisely, if in a clinical case it is generally known that the non-exhibited DC causes HE, and the probability of DC given that the patient exhibits a symptom DS is p, then the physician is informed that the probability of the health effect HE given DS belongs to the range $[p, 1]$. Clearly, the more relevant to DC is the DS (larger p), the higher are the chances that HE occurs.

We conclude this section with another example where the secondary case attributes appear as premise strengthening factors.

Example 4.19: A care professional is confronted with a clinical case where it is known that DC causes HE within a certain probability range. In the presence of the symptom DS, the updated probability is calculated using the inference

$$\begin{array}{ll} \textit{Case premises} & P_{KB}[DC \to HE] \in [p_1, p_2] \\ \textit{Physician's conclusion} & \therefore P_{KB}[(DC \land DS) \to HE] \in [p_1, 1] \end{array} \qquad (4.15)$$

According to inference (4.15), the updated causation probability range $P_{KB}[(DC \land DS) \to HE]$ is larger than the original probability range $P_{KB}[DC \to HE]$ when $p_2 < 1$; whereas the two probabilities coincide when $p_2 = 1$. The readers may notice that a different conclusion is reached if the physician choses the statistical conditional syllogism

$$\begin{array}{ll} \textit{Case premises} & P_{KB}[HE|DC] \in [p_1, p_2] \\ \textit{Physician's conclusion} & \therefore P_{KB}[HE|(DC \land DS)] \in [0, 1] \end{array}, \qquad (4.16)$$

which is clearly uninformative. That is, once more, the choice of the conditional type to use in a case can lead to different case prognoses.

4.6. Causation in Terms of Integrative Space–Time Prediction

In the stochastic dynamics milieu of *SMR* inference (Section 3.9.2), disease causation is viewed as a knowledge synthesis affair, in which the medical investigator seeks a causation model that integrates component causes linked with different disciplines in order to reach sound conclusions about the medical case. In the same setting, causes and effects in reality are of definite and finite character, where the cognitive meaning of causation may lie in its potential for improved *integrative space–time prediction* (*ISTP*). This kind of prediction consistently fuses logical and natural laws of spatiotemporal cause and effect. Let D_p and $C_{\kappa,p}$ denote, respectively, disease records and potential causal factors data at a set of locations–times $p = (s, t)$. The *ISTP* approach postulates that the existence of a cause–effect association should lead to disease predictions $D_{p'}$ at locations–times $p' \neq p$ where disease records are not available with a smaller uncertainty when information about both $C_{\kappa,p}$ and D_p are combined than when only D_p records are used (Christakos *et al.*, 2005), i.e.,

$$U_{KB(C_{\kappa,p},D_p)}[D_{p'}] < U_{KB(D_p)}[D_{p'}], \qquad (4.17)$$

where $KB(C_{\kappa,p}, D_p)$ denotes a base that includes information about both $C_{\kappa,p}$ (physical, ecological, climatic and environmental causes) and D_p (disease distribution, epidemic observations etc.) and $KB(D_p)$ is a base that includes D_p data only. In the setting of Eq. (4.17), once initiated, the cause and the effect (disease) can continue to interact in space–time with intensity appropriate to their internal links and possibly affected by external influences. Real case studies based on Eq. (4.17) can be found in Christakos and Serre (2000), Christakos (2005) and references therein.

The *ISTP* approach involves a *CD* causation based on a holistically structured background of meanings that is fundamentally different from a *CI* causation. In the *CD* case, the fact that the specified causes at p imply the disease distribution at p' is due to the specific content of the disease distribution, i.e. the implication will not hold if the meaning of the distribution changes. In the *CI* case, on the other hand, the implication does not depend on the content, i.e., it continues to hold when the meaning of the disease distribution is substituted with a different one. Clearly, *CD* causation establishes a more significant relationship between cause and effect (disease), in that it offers a more intimate connection between the two than *CI* causation does. The content-dependent *ISTP* approach assumes a selective kind of a relationship as it is genuinely linked to the unique meaning of the health effect. Accordingly, physicians do not merely try to establish the presence or absence of causation for its own sake, but whether or not a certain real world objective can be accomplished through this causation. Can the occurrence of $D_{p'}$, e.g., be explained or predicted from one's understanding of the biophysical meaning of $C_{\kappa,p}$ and its connection to the essence of D_p?

In other words, *ISTP* should be implemented *in situ* with the proviso that the "$C_{\kappa,p} - D_{p'}$" link is *CD*, in which the relation $C_{\kappa,p}$ has to $D_{p'}$ exists only by virtue of the particular meaning of $D_{p'}$ and its physical link to the essence of $C_{\kappa,p}$. Our readers may notice that the *ISTP* approach can gain additional support from the fact that in health sciences one finds an interesting association between space–time prediction and causation: occasionally the same epistemic and empiric conditions that generate sound predictions of a phenomenon (before it occurs) are used to obtain meaningful causal explanations of it (after it has occurred), as well. No doubt there are other cases where this symmetry between disease prediction and causation is not valid, i.e. the epistemic and empiric conditions that are used to predict a health effect may not be able to explain it, in which case a different line of thought should be used.

The approach introduced by Eq. (4.17) is distinguished from other causation studies that are also based on predictability. *ISTP* does not have certain of the difficulties of these studies (one such difficulty is linked to the effects of possible explanation–prediction asymmetry, which is the case of deductive–nomological causation studies) (Christakos, 2012). Another interesting feature of the *ISTP* approach is that prediction is understood as a product of teleology of reason, i.e., disease prediction is likely to be more realistic if it is derived via a theory of knowledge. By blending environmental, ecological and biological measurements of potential causes with core and specificatory disease features in a space–time domain, prediction accounts for various sources of uncertain medical knowledge as well as for inter- and intra-subject variations of the specified populations.

In view of the above considerations, disease causation (etiology) is not merely a relation between one or more potential causal factors and a health effect, but a knowledge synthesis process involving potential causal factors, health effects and a specified population (which, in a public health context, usually refers to a group of representative receptors in a composite space–time domain).

4.7. Causation Justification and the Dualistic Opposition

Just as logical justification must conform to logical rules, a physician's causal justification of an *in situ* health effect must conform to scientific standards of evidential support. This thesis can be useful (*inter alia*) in the study of the important issue of justifying the justification of a disease cause. For example, forming a belief that exposure to certain pollutants causes a specific population effect is something that the physician must justify. Naturally, not all belief justifications are equally valid (some doctors do a better job of establishing a belief as justified than others). Which is why, in addition to the justification of a belief (concerning a cause) by means of a given approach, a practitioner often needs to justify the approach itself that was used to develop causal justification.

In the uncertainty of everyday professional practice, how certain a physician needs to be before accepting a disease etiology (say, what form $a_A(E_p \backslash C_p)$ the specified disease causation must have) will depend on how strong the evidence in favor of this assertion is, and how serious a mistake in accepting or rejecting the assertion would be. Then, the question may arise: if $a_A(E_p \backslash C_p)$ is epistemically certain for physician A, could A be justified in being psychologically certain of it, in having a credence or confidence of 100% in it?

The strength of justification of this and many other relevant assertions is a contextually varying matter. Generally, etiologic medical reasoning is a twofold affair:

(1) a belief concerning a cause must be first justified in terms of an approach (logical, analytical, empirical etc.), and
(2) a justification approach itself must be also justified (how good it is compared to other approaches, or if the chosen approach is adequate in justifying a belief about the disease cause).

In the twofold setting above, a causal justification that conforms to scientific standards of evidential support and logical rules is often better justified in public and private medical practice than a causal justification that is supposedly grounded in self-justifying beliefs or beliefs for which there are always other beliefs that make them likely to be valid (in the real world very rarely, if ever, does one finds such beliefs).

No doubt, health care professionals' skills that are necessary for effective reflection and disease etiologic purposes represent a progression from the foundation of self-awareness upon which critical analysis of information supports the ability to make value judgments (cause–effect assessments, diagnoses, prognoses, risk identification, medical treatments), whereas a high level of skill in an unstructured domain seems to require considerable concrete experience with real situations. We will conclude this section by reminding the readers that although large-scale epidemic studies cannot usually offer a definite proof of the actual cause of a disease, on occasion they could provide valuable clues regarding possible disease causes (Bossak and Welford, 2009). In light of the above and similar considerations, creative medical reasoning in the *Age of Synthesis* rather ascribes itself to the motto:

Tomorrow belongs to the surprise.

Of course, there are those who deprive themselves of the surprise experience, since the established system has in place a secure future for them, free of surprises. Yet, life is an effort that deserves a better cause. At this point of our discussion the above motto serves the additional purpose of making the connection with the next chapter of the book.

Chapter 5

Looking Ahead

5.1. An Ibsenian Transformation

The challenges lying ahead require that health care providers develop mind habits such as critical thinking, confidence, contextual perspective, creativity, flexibility, inquisitiveness, intellectual integrity, open-mindedness, perseverance and reflection. In which case, the book's argument is straightforward: how physicians structure their thinking and, in the process, how they interpret the methodological steps they follow, both depend on their *health paradigm*. The latter determines which concepts, models, guidelines and methods can be structured and how their interconnections or differences should be displayed to facilitate disease understanding and medical decision-making in the broad sense. Yet, another book argument is that despite its paramount role in medical thinking, a paradigm is not a sacred and irreplaceable reality. This brings us to a third book argument concerning *intellectual provocation*.

By serving a wider purpose, intellectual provocation enables the introduction of new concepts and arguments into the "marketplace of ideas", opens the minds of health care providers, adjusts their mainstream views, improves their professional practice and allows changes that enrich their culture and social discourse. In certain medical fields a paradigm change is often needed, which may require an Ibsenian transformation on behalf of the care provider. In his theatrical play, *Ghosts*, Ibsen's message is that, "We are traveling with dead weight on board, which we must get rid of". In the case of medical thinking this weight may include a physician's ultimate presumptions, and sometimes even prejudices concerning the meaning of symptoms and the interpretation of a diagnosis methodology. Indeed, medicine frequently needs to get rid of various "dead weights" in the forms of medical dogmas. Such dogmas that have been overturned in light of new evidence and deeper thinking include the following.

Example 5.1: In certain medical circles for years the prevailing dogma was that people must tightly control their blood sugar. But recent studies have shown that the implementation of the dogma of tightly regulating blood sugar can in the end kill more people than it helps. For almost half a century another dogma of the medical profession was that estrogen replacement therapy for post-menopausal women could prevent heart disease and dementia. However, subsequent findings of a set of much better designed, randomized controlled studies have shown that this dogma was, in fact, false. For almost two decades, the prevailing dogma was the supposed benefits of low-fat diets (e.g., carbohydrates were considered a good substitute). Unfortunately, this was yet another false dogma that finally lead to an obesity epidemic and subsequently had to be abandoned.

Also relevant to the above is the argument that many laboratory techniques that are considered highly accurate in one setting may not be so in other settings. A good example is the case of *DNA* techniques. In particular, techniques used to study ancient *DNA* also include considerable amount of uncertainty. Several reports have questioned the reliability claims concerning the recovery of ancient *DNA* (Bryson, 2003). Techniques used to study ancient *DNA* currently contain inherent problems, particularly with regard to the generation of authentic and useful data. Recent studies emphasize that efforts to reduce contamination and artefactual results by adopting authentication criteria that are not foolproof have, in practice, replaced the use of thought and prudence when designing and executing ancient *DNA* studies (Gilbert *et al.*, 2005). Remarkably, this sort of unreliable experimental data lacking sound theoretical support has played an authoritative role in the critical debate concerning the etiology of the Black Death, one of history's deadliest epidemics (Christakos *et al.*, 2005; 2007).

In a similar vein, many thinkers argue that quantitative tools that have for a long time been rather popular among health care professionals can act as "Ibsenian weights" on the professionals' shoulders, if implemented uncritically. An example is health statistics.

Example 5.2: Health statistics have been a very valuable tool in clinical studies. But as with any tool, health statistics have their limitations. One is the ease with which people ignorant of statistical methods can be fooled by the framing of statistics results. There are also inherent limitations, *inter alia*, health statistics fail to capture vital features of space–time change and are not good at predicting fundamental breaks of disease from the past. Specifically, when earlier patterns between dependent and explanatory disease attributes break down, health statistics predictions tend to perform poorly.

As the readers know, it is not unusual that practitioners in various medical and clinical fields form their own conception of the growth of scientific knowledge, neglecting actual science. When the study of the laws of science failed to conform to their models, many practitioners stubbornly held on to their preconceptions. Holding on to the statistics preconception in the face of such anomalies seems to exemplify Kuhn's views on how paradigms operate, this time in medical thinking. Before leaving this section, it is worth reminding our readers that real progress in medical sciences requires the development and preservation of a *space–time for thought* that would make case interpenetration possible, cultivate the mind, feed curiosity and furnish the imagination. Medical scientists understand that they need to be constantly conscious of the implications of using modeling methods in their studies and maintain a high level of responsibility towards the potential victims of their decisions and actions. Moreover, a series of problems have to be overcome that reflect the "growing pains" of success, including increased scrutiny by non-modeler practitioners. The existence of a space–time for thought becomes increasingly necessary since, as many thinkers argue, there is yet no theoretically insightful, radical and intellectually exciting ground being broken in health modeling under uncertain *in situ* conditions and biomedical heterogeneity. Despite the central role of the space–time notion in clinical and medical investigations (disease evolution, etiology, prognosis etc.), its study has been neglected due to strong opposition to basic research with its long-term benefits. This is yet another social symptom of the present generation's indifference for the well-being of future generations.

5.2. *SMR* and Divergence of Rationality in Medical Thinking

Optimally, diagnostic thinking should be explanatory, internally consistent, comprehensive, and should be testable, possessing predictive power. Diagnostic thinking may involve a number of possible schemes:

(1) Much of diagnostic thinking is inductive, i.e., it may derive general conclusions from particular empirical evidence (in the form of observations, physical examinations, lab tests).[1]
(2) In more sophisticated situations, diagnostic schemes are devised by the physician's creative imagination, in which case induction from empirical evidence is used to justify or route the diagnosis.
(3) Some experts argue that the most logically certain diagnostic schemes are those that employ mathematics. From *Susceptible–Infected–Recovered* (*SIR*) models, e.g., one can, via mathematical calculations, make deductive inferences about the space–time distribution of a disease.[2]

An increasing number of studies have shown that rigorous *SMR* is a key component of decision-making, professional accountability and quality patient care. *SMR* uses words and symbols to evoke mental pictures or other sensory experiences and empirical findings. Such imagery may be deployed to conjure up a conceptual synthesis of biophysical, social, statistical, epistemic and linguistic notions; to create a new perspective on something clinically familiar; and to point to additional layers of meaning, as is the case with metaphor and symbolism (quantitative expressions suggesting a range of associations or ideas, and formulas estimating health risk). Each and every one of these topics represent future challenges and opportunities of what lies ahead in medical thinking.

---

[1] This is mostly the case in scientific practice: theories are accepted if they are backed up by sufficient empirical evidence, even if there are no strictly logical grounds for accepting inductive inferences (in the 18th century, David Hume correctly pointed out that no matter how often one observes an eagle taking off, there is no logical reason why it should do so the next time, but, nevertheless, he conceded that this was not a practical issue).

[2] Yet, not all thinking schemes use mathematical concepts and techniques. Darwin's theory of evolution by natural selection, e.g., has immense explanatory power without resorting to mathematics.

Undoubtedly, differences between medical findings occur for a variety of reasons, including genuine population differences, medical term interpretations, experimental fluctuations, laboratory variations, methodological matters, lexical ambiguity and, last but not least, differences in the statistical methods used. *SMR* argumentation seeks to combine these varying elements, reconcile the varying findings of different studies and arrive at a consensus. Certain statistical techniques, (e.g., meta-analysis) seek to combine evidence from different studies, whereas rigorous *SMR* inference is concerned about knowledge synthesis in conditions of location–time uncertainty and its value in medical research.

The abstract mode of space–time synthesis above is a level of reasoning that is removed from the facts of the "here and now", which is the domain of the concrete mode. A feature of the abstract mode is the ability to transfer knowledge from one context to another (say, from physics to biology, or from fluid mechanics to blood flow modeling), whereas the concrete mode specifically needs to consider the data in both contexts. In the context of medical thinking a series of crucial questions arise such as: Why would physicians want to reinterpret mainstream logic in a certain way?

Basically, *SMR*'s answer is that there exists the desire to retain, as far as possible, substantive interpretations of health assessments and medical decision support. Ideally, a physician is trying to develop reasoning in such a way that the justification of an assertion embodies predominantly *CD* arguments (scientific principles, empirical evidence etc.) rather than *CI* ones. Again, a deeper understanding of such matters is part of what lies ahead in medical decision analysis.

Medical reasoning methods need to account for the fact that it is not uncommon that two experts A and B, both of them rational thinkers, reach different conclusions when they are confronted with the same case. There are several possible explanations for this divergence:

(1) experts A and B are provided with different case-specific data about the case;
(2) A and B are provided with the same case-specific data, yet they possess different core knowledge about the case; or
(3) A and B are provided with the same case-specific data and possess the same core knowledge, yet the experts accede to different health paradigms.

SMR dynamics can help health professionals understand the consequences of alternative decisions in various settings. A basic postulate is that in many situations it is more realistic to use terms like "believe", "expect" or "probablify", which offer a range of possibilities (with different likelihoods of occurrence), rather than the term "know" that is usually associated with certainty (the available KB objectively facilitates a high degree of justification for "know"). This allows practitioners to get an improved awareness of how much they do not know about a health situation, in addition to their appreciation of how much they know about it. As such, quantitative medical reasoning helps develop one's abilities to think in several distinct ways and also to explain lucidly these different thought processes. The latter is particularly valuable to health care providers, including the cases in which the mainstream guidelines are process measures of unclear utility, unproven with respect to changing disease outcomes, as well as in their everyday contact with the public. As a result, many decision analysts have dismissed such guidelines as a typically American success-oriented approach, which simply ignores a range of difficult diagnostic questions. Indeed, there are a number of substantive issues to be considered. How do physicians think that they think? How do they justify their justification of a disease diagnosis and treatment? How can an inference method deal with the potential existence of unanticipated knowledge (also, Section 5.3 below), and how can such knowledge be rationally incorporated into the corpus of one's previous beliefs? How can one predict the occurrence of something whose existence one neither knows, nor even suspects? Subjective probability and Bayesian inference would seem at a loss to handle such a problem, given their structure and content.

Lastly, lack of values and logic leads to scientific *nihilism*, which is a reason that SMR in a broad sense refers to a formalism that combines aspects of both stochastic logic and scientific knowledge. Or in a narrow sense, it is seen as a particular kind of logic, namely one that incorporates probabilities in the language or the metalanguage (which, generally, is a system of notation, descriptive terms, explanations etc. for an object language). All the above constitute potentially fruitful domains of research and development in medical decision science.

5.3. Challenges Emerging from the Incompleteness Principle and Unanticipated Knowledge

In all scientific fields, a typical fallacy is that because a certain phenomenon cannot be studied with the available tools, it is sometimes assumed that it does not really exist *in situ*. This fallacy is the result of a lack of appreciation of the gap between the limitations on nature and the limitations upon the particular tools (experimental, mathematical, computational or statistical) one has chosen to describe nature.

Example 5.3: Assume that a mathematical model M_{CA} is used to describe a biomedical phenomenon. It is possible that an attribute CA_p of this phenomenon is actually observed in nature but, nevertheless, the investigator cannot use M_{CA} to describe CA_p adequately. This can usually happen for one of two reasons: either the investigator does not currently possess the tools to solve the M_{CA} equations for CA_p and compare the solution with *in situ* observation (which leaves open the possibility that M_{CA} is an adequate description of CA_p and that the limitation merely lies with the investigator); or M_{CA} can be solved but the solution does not agree with observation (i.e., M_{CA} is a poor description of the specified CA_p, after all).

As a matter of fact, it is a sign of a discipline's maturity that it eventually comes to appreciate its own boundaries and limits. This is often materialized in the form of an *incompleteness principle*. There exist several kinds of incompleteness principles, two of which are mentioned here. A well-known incompleteness principle states that certain issues concerning a theory and its applicability are humanly impossible to resolve in the context of the theory itself (i.e., to use a biomedical theory to solve a clinical problem while satisfying the assumptions and conditions imposed by the theory). This situation has been rigorously exemplified by Gödel's incompleteness theorem and Heisenberg's uncertainty principle, among others. Gödel's incompleteness theorem states, generally speaking, that all sufficiently powerful and consistent logical systems are incomplete. On the positive side, this theorem is valid when one uses the entire system to describe reality. It is possible, however, that the study of the actual medical phenomenon requires only a part of the mathematical system that is complete in itself, in which case Gödel's incompleteness does not pose a problem.

Example 5.4: A spatial epidemiology study of regional disease spread that uses only Euclidean geometry (Section 2.2) is not affected by Gödel's theorem, in which case the Euclidean geometry-based study is logically valid, and also substantively sound, assuming that it is sufficiently realistic (which may not be the case in, say, large-scale health phenomena like epidemics).

Accordingly, physicians should use medical decision analysis to study possible effects of incompleteness in realistic health and clinical environments to avoid the trap of reasoning about the logically unreasonable. Another kind of incompleteness arises during the transition from the *particular* (empirical) to the *general* (theoretical). It is a common approach in disease modeling to try to go beyond a site-specific empirical relationship and derive a more general formulation that applies in other clinical cases or health sites, as well. A mistake often made during the transition process from an empirical relationship to a model of the case or site is to trade away fidelity to the relationship's scientific role in exchange for coherence with a formal modeling setting. In other words, underlying certain formal manipulations is an inadequate reasoning mode, certain matters of which need to be revised (matters of internal consistency, fidelity, rationality or form vs. substance). Accordingly, how to avoid violating principles of *in situ* logic and probability is a field that requires further study in medicine. One must learn to distinguish between theoretical reasoning (which is about assertions) and practical reasoning (which is about actions). What seems theoretically irrational can be perfectly rational for clinical practice purposes, in which case the health study context plays a key role.

Example 5.5: Suspending judgment on whether to give drug D_1 or drug D_2 to a patient due to insufficient information distinguishing the two drugs seems theoretically rational. But if the patient's only hope for survival is that one of the two drugs works, not deciding between D_1 and D_2 may not be a practically rational decision for the doctor to make. In which case, how to rigorously analyze this kind of "forced" decision-making can become an issue with serious medical consequences.

Research in medical reasoning should include both the rigorous formulation of issues such as the above to the extent possible given the current medical understanding, and the development of quantitative methods that assign measures of justification and uncertainty to the formulation.

Another challenge that is relevant to knowledge incompleteness is the so-called *unanticipated knowledge*. The existence of this kind of knowledge has been mostly neglected in current medical decision-making theories. An analogy serves to illustrate the point. According to Pausanias and others, beyond the various monuments dedicated to the 12 gods and numerous smaller deities, the ancient Greeks had erected an altar dedicated to the unknown God (Τώ αγνώστω Θεώ).[3] The ancients continue to make sacrifices upon the altar, although they admittedly knew nothing about the existence or the name of this unanticipated God. *Mutatis mutandis*, one may wonder, how could one erect an appropriate altar for unanticipated knowledge in medical research and clinical practice?

In particular, an important issue rarely discussed by current medical decision models is what the care provider's response should be when novel, unknown or previously unsuspected phenomena occur (new symptoms, unexpected treatment results). In other words, how can such new data be coherently incorporated into the care provider's previous cognitive condition?

Coherence of an old and a new KB does not make sense here, simply because there is no old KB for the new to cohere with, which is why the real issue is how to deal with unanticipated knowledge. Let us recall that the Bayesian approach generally establishes the consistency between pre-test (also called prior in statistics) and post-test (posterior) information. It is not appropriate, however, to use this approach since the new phenomenon was completely unknown *a priori* and, hence, the notion of prior (or old) probability is inconceivable. Undoubtedly, the matter needs further investigation in the context of medical decision support. In doing so, it should be kept in mind that health care providers are not dealing with a phenomenon whose possibility they have considered, but whose probability was judged to be zero. Rather, the care providers observe a new phenomenon whose existence was not even suspected before its occurrence. In other words, it is not that health care providers judge such phenomena impossible — indeed, after the fact, the providers may view them as quite plausible — it is just that beforehand the care providers did not even consider their possibility. On the surface there would appear to be no way of incorporating such new information into the care providers' system of beliefs, other than starting over from scratch and completely reassessing their subjective probabilities. At a deeper level, some thinkers argue for a baroque state of affairs in the theory of unanticipated knowledge consisting of rough and yet imperfectly shaped conceptual pearls, so to speak. In any case, this is an open medical research field of significant decision–theoretic interest.

---

[3] Pausanias, *Periegesis-Attic* 1[4].

5.4. Information Technology-Based Medical Reasoning

Medical investigators are also increasingly confronted with challenges that include both taking advantage of cutting edge technological developments that can be of significant value in clinical practice, and identifying research areas that could potentially improve medical inquiry. Medical reasoning and related technology continue to gradually proceed from a research and standardization context to a concrete and productive *in situ* setting. In various parts of the book, we called the readers attention to the fact that *SMR* is content- and context-dependent reasoning based on the integration of different *KB*s (transcending several disciplines, including medicine, environmental health, geomedicine, biology, psychology, sociology and statistics) under conditions of uncertainty and space–time heterogeneity. Reasoning is quantified in the form of syllogisms and inferences that cover a wide variety of medical studies, clinical cases and population exposure situations. Indeed, the domain of medical knowledge producing the premises of the inferences is vast and often characterized by complex semantic structures and numerous clinical, exposure, environmental, geomedical and other information sources that are not properly structured (a considerable part of these sources is scattered on the Internet).

Figure 5.1 visually describes the computerized setting of medical inferences, which is readily available to the health care professional. It is clear that to assure adequate and effective *in situ* medical decision-making in conditions of uncertainty and time pressure the care professional needs:

(1) a systematic list of sound inferences that is readily available,
(2) an expressive model of *KB*s representation (that would also accommodate for their Web-distributed nature) and
(3) the interactive technology that would allow for the intelligent and productive manipulation and blending of the above inferences and *KB*s with due accuracy, efficiency and speed.[4]

The intention of the visual representation of examples of medical inferences and quantitative arguments (some of which were studied in previous chapters) in Fig. 5.1 is to show that computerized medical reasoning and decision-making could cover in a systematic and comprehensive manner real cases in the field of medicine, clinical practice, environmental health and geomedicine. This includes the meticulous review of old records and medical

---

[4]Various examples of medical inferences have been tabulated in several parts of the book.

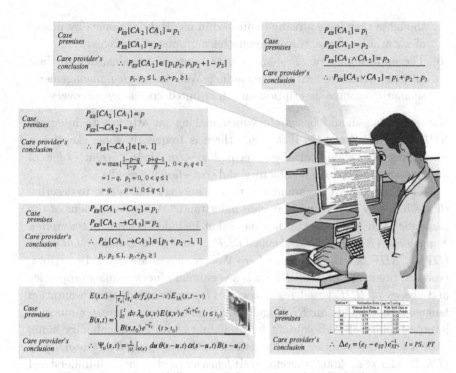

Figure 5.1: A visual illustration of computerized medical inferences.

systems, the appropriate assessment of clinical and exposure evidence, and the adequate performance of physical examinations.

Example 5.6: In order to make an informed decision concerning Ms Zeublouze's health needs, her physician selects from the relevant computerized list an appropriate inference scheme that integrates information about Ms Zeublouze's medical condition, alternative medication courses, other patients' positive outcomes or side effects, recommended operations, what to expect following each procedure, success rates, length of recovery times etc.

Recently, attention has been paid to the possibility that medical decision theory and techniques could benefit from the application of state-of-the-art *information technology* (*IT*) that assists and automates its logical inference, documentation and knowledge interchange needs. This is to be expected for at least two reasons:

(a) It is often highly beneficial that health care experts formally encode in a computer language (some aspects of) the experience, information and

thoughts they have already expressed in medical documents by means of natural language, through equations, graphs and images.
(b) The applicability of *IT* spans a wide range of domains, due to their alleged attractive features, including Web knowledge management, semantic resource description and distributed knowledge discovery.

Naturally, one should be careful when making an attempt to integrate *SMR* with the *Semantic Web*, since there is frequently some skepticism concerning the ability of these technologies in their current stage of development to produce satisfactory and cost-effective results.

The Semantic Web is an information technology that is able to describe things in a way that computers can understand.[5] A key component of the Semantic Web is its markup language that includes *Resource Description Framework (RDF)*, *RDF Schema (RDFS)*, and *Web Ontology Language (OWL)*. *RDF* is a data model where Web resources are defined as triples (subject, predicate, object); *RDFS* is a vocabulary for describing properties and classes of *RDF*-based resources; *OWL* is a knowledge representation language that builds on *RDF* and *RDFS* and facilitates greater machine interpretability of Web content by providing additional vocabulary along with a formal semantic. *OWL*'s semantics derive from *description logics* (*DL*, Baader *et al.*, 2003) therefore *OWL* has attractive and well-understood computational properties. Perfecting the ways the above and other IT conceptual tools can be rigorously integrated into clinical decision support and population exposure assessment constitute promising avenues of future research.

5.5. Social and Cultural Dimensions of Medical Thinking

Many social scientists cite the view that in modern medicine one should not be limited to concepts and observations, but also concerned about processes and evolving activity. They have no use for the idea that medicine is one thing and its context another. They are looking at ideas, their use, and the social process that created and elaborated them, as a single phenomenon. How physicians group themselves, in what ways they publicly disagree and what social significance they claim for medicine are just as revealing as concepts, patterns of thought, forms of physical examination and diagnosis techniques. From this point of view one sees these as part of a single

---

[5] The Semantic Web is an extension of the World Wide Web, in which web information is expressed in a machine-processable form.

manifold, and seeks to understand what makes them one process. A prime example is the close connection between natural process, social process and the vital processes of the human body. In this milieu, the unfairness characterizing many aspects of social life has been seen as a manifestation of the Darwinian notion that random genetic mutation may by chance provide a functional advantage to an organism and so be perpetuated by natural selection.

Consensus and disagreement are two issues of considerable social and cultural importance. For example, the ancient Greek and Chinese cultures differed greatly on the subjects of what one said and how one said it. On the whole, the Chinese valued consensus as much as the Greeks valued dispute. Greek culture encouraged disagreement and disputation in science as in every other field, whereas in China the emphasis remained on consensus. In China scholars understood the risk of favoring alternatives to the current dispensation of power. For Greeks, on the other hand, whatever other purposes it served (political power etc.), disagreement was a tool of competition and progress. It was characteristic for the argumentative nature of Greek medicine that there was widespread disagreement about many health issues. Hippocrates radically disagreed with the Cnidian school on a variety of medical issues. The mind–body debate was to have a long history (e.g., the physical basis of personality and the emotions, how stress is related to heart disease),[6] involving some of the brightest minds of that time, like Alcmaeon, Hippocrates, Plato and Aristotles. In China, on the other hand, people who lived by their knowledge presented their ideas much of the time not to colleagues but rather to their patrons,[7] who, faced with political decisions, seldom showed patience for anything resembling intellectual debate. Similar was the case with teacher–student relations. Plato had deep intellectual disagreements with his own students, the most famous case being that with his best student, Aristotle, whereas in China a teacher and his disciples formed an internally cohesive community that avoided attacking other communities.

Another social dimension is related to the setting of the case. In the previous chapters, most of the medical and exposure assertions were considered in an epistemic setting, which generally makes them more likely

---

[6]This is a debate that touches on the relationship between the mental and the physical, and how the one impinges on the other. This is a relationship that continues to be only partially understood.

[7]Disagreements with other scholars were mostly unimportant by comparison.

to be valid than if they were considered in a non-epistemic setting. The former setting is usually strictly science-based, which is not necessarily the case with the latter setting that may possess a significant social dimension. Non-epistemic reasons for supporting a particular belief may be motivated by financial incentives or political agendas. An example of non-epistemic reasoning and its perils follows.

Example 5.7: Some health investigators who know that an agency only funds research that promotes its chief administrator's views may find it convenient to believe these views, even if they are highly questionable scientifically (say, an administrator argues that tobacco smoking is a healthy habit). Clearly, the reason the investigator uses for believing such views (which in actuality merely supports the administrator's agenda) is non-epistemic, i.e. believing a view because this will assure funding for the investigator does not make it more likely that the view is valid. This is a belief that is based not on scientific rationality but on "practical" rationality, which is the need for funds. In other words, the belief may be pragmatic but is not epistemic (nor is it scientific), nevertheless. Therefore, at the very least any study of the results of such a medical research project should carefully factor in the incentives at work. Under similar circumstances, the health care providers get the feeling that they are working with a somewhat mechanistic process consisting of layers of understanding, emotion, belief and agenda.

It is widely recognized that there are a number of reasoning issues that cannot be addressed on a formal logic basis alone, but which require that the health care providers dig deeper into the ontic and cognitive aspects of the medical case. For example, an interesting question is whether it matters that a wrong diagnosis is the result of a physician's reasoning being irrational given that the wrong diagnosis did not involve irrational reasoning. Furthermore, epidemiologists and medical geographers know that maintaining internal consistency while integrating different knowledge bases and uncertainty sources can exert significant influence in disease mapping.

In an appropriate *SMR* inferences milieu, the connections between vital concepts should be appreciated and incorporated in a health study. For example, if a physician's thinking mode is inconsistent it does not necessarily means that one is being irrational. Lastly, it is not unusual that health modelers use concepts in a different way than the medical experimenters who supply them with data. As noted earlier, the health modeler must be aware of certain pitfalls during the transition process from experimental (empirical) data to general formulation (model).

Yet another social dimension of the medical decision process is due to the possible *legal* consequences of a medical case. In fact, litigation trends indicate that medical causation issues, in particular, are becoming more common. Medical institutions are becoming increasingly complex, with many different specialists and other health care providers commonly involved in patient care. System failures, where different participants in the patient's health care fail to communicate effectively or efficiently, are increasing as health care systems become more complex and diversified (while, at the same time, budgets are tightening and patient loads are increasing). In addition, the cost of medical testing continues to increase, which reduces the likelihood of redundant and prophylactic tests. Together, these circumstances may result in medical causation issues in delayed diagnoses cases becoming more common and more complex. Lastly, causation can become particularly hard to establish when multiple statistically dependent events characterize the causal chain. In the following example, in order to prove medical malpractice liability, the plaintiff attempts to establish through competent expert testimony that, to a reasonable degree of medical probability, the plaintiff's harm was caused by the defendant's negligence (Watson, 2010).

Example 5.8: Ms Charles filed a lawsuit alleging that the negligence of physician A caused a serious delay in the diagnosis of her cancer, thus delaying diagnosis until the cancer was no longer treatable. However, the alleged causal chain of events in this case was a long and complex one. The plaintiff alleged that her physician negligently failed to administer a pap smear, which probably would have been abnormal. Ms Charles further alleged that (a) an abnormal pap smear result would probably have led to a colposcopy, (b) the results of the colposcopy probably would have been unsatisfactory and led to a biopsy and (c) the biopsy probably would have led to a diagnosis of the plaintiff's cancer at a time when she had a 65% chance of being cured. Ms Charles eventually died of cancer. The odds of Ms Charles surviving if she had had a pap smear are determined by multiplying the odds that such a pap smear would have been abnormal by the odds of a follow-up colposcopy being unsuccessful and the odds of her surviving once her disease was diagnosed and treated. The testimony of the plaintiff's expert that a "pap smear would probably have been abnormal" proves only that 51% of pap smears would have been abnormal. And her expert's testimony that the colposcopy required by an abnormal pap smear would "more likely than not" have been unsatisfactory and led to a biopsy is proof that a biopsy would have been required in 51% of the cases where the pap smear was abnormal, i.e., only 26% (= 51% × 51%) of the time. Assuming a biopsy performed in all 26% of such cases would have led to diagnosis and treatment, then 65% of the 26% of biopsy patients would have survived. This means a patient in the plaintiff's condition when the pap smear was allegedly required would survive only 17% (= 26% × 65%) of the time. Viewed from another perspective, the plaintiff's causation evidence established that 49% of women in her condition would have had a normal pap smear and received no further treatment. In addition, 49% of the women with abnormal pap smears would have had a satisfactory colposcopy, so no further diagnostic procedures would have occurred. Finally, 35% of those undergoing biopsy procedures and receiving cancer treatment would have died anyway. In other words, according to the plaintiffs' evidence, 83% (= 49% + 25% + 9%) of the time, a patient's having a pap smear would not change that patient's survival outcome.

Accordingly, defense counsels must be prepared to confront the statistical aspects of expert testimony when such cases arise, which means that these aspects have non-negligible social consequences. Counsel defending against medical causation claims should consider doing all of the following: (i) vigorously examining the plaintiff's experts on the issue of medical causation, and the foundation for their opinions (including specific medical studies); (ii) working with defense experts (including medical experts and, if needed, a statistician) to fully understand whether the plaintiff's causation theory is medically, factually and statistically sound; (iii) if appropriate, move for summary judgment on the issue of medical causation, and/or file a motion *in limine* to bar speculative evidence of causation and (iv) if appropriate, request a pre-trial hearing to test the validity of plaintiff's medical causation evidence.

5.6. Quod Iacet Ante?

It is widely acknowledged that clinical reasoning has lagged behind advances in medical sciences in the last few decades. Improving this situation will be one of the goals and challenges of what lies ahead in medical research. It is not surprising that clinicians (especially the less experienced ones) fall victim to two common fallacies: *round trip fallacy*, i.e., not distinguishing between absence of evidence and evidence of absence (e.g., assuming that no evidence of disease means that there is evidence of no disease); and *narrative fallacy*, i.e., perceiving or imposing causality based on the plausible (e.g., the plausible is viewed as causal). Quality patient care requires that physicians continuously improve both their biomedical knowledge and the way they think about it under conditions of clinical uncertainty and time pressure. A physician must be fully aware of the process behind decision-making, including the complex risk–benefit tradeoffs of tests and treatments (e.g., patient care is difficult in part because of the uncertainty of imperfect diagnostic tests and treatments with unpredictable consequences). Rigorous medical reasoning aids learning and teaching the clinical process. Principles of screening intervention and diagnostic judgment become transparent from an understanding of substantive logic rules, probability conditionals and knowledge synthesis. These scientific reasoning tools can serve as critical templates to share clinical experience, communicate case findings, and resolve differences of opinion concerning controversial medical practices. Advances in the area of synthesis are urgently needed, since techniques for interrogating the complex networks of biomedical knowledge and clinical evidence, and for responding to possible inconsistencies, remain mostly at a craft lore level.

Reasoning schemes need to be improved by means of which physicians are able to correctly convert data into useful knowledge before they can use them, and all this frequently under time pressure. Typically, a datum is converted into knowledge by relating it to the physician's clinical experience, situational environment and other information, so that it leads to case diagnoses, judgments or treatments (which can be potentially used in other contexts or for different purposes). As a simple illustration, the fact that little Maria's body temperature is 39.8° C remains merely a piece of data, but when it is related to the medical knowledge that any temperature above 39° C is considered high fever, then we know that Maria should see a doctor.

Physicians widely acknowledge that in many cases they have only indirect access to a patient's actual health state, i.e., they attempt to infer the health state via indicators or cues (symptoms, physical findings, diagnostic tests) rather than by direct observation of the patient's real state. The key to a safe medical practice is the physician's understanding of how well the standard features of commonly encountered diseases can act as disease predictors in real-world conditions. All the above cues, however, are often imperfect and uncertain, and their interpretation makes clinical sense only when they are considered in the light of physician's prior (or pre-test) knowledge about the patient's state (i.e., before the cues become available). The challenge is to study this problem by assuming that prior knowledge belongs to the G-KB and cues belong to the S-KB, and then derive the posterior (or post-test) probability of the patient's state by means of a clinically meaningful logical synthesis of G and S. Currently, mainstream measures of how well the standard disease features (G) and the clinical cues (S) actually predict the disease are obtained in terms of their likelihood ratios. Obtaining improved measures in an interdisciplinary synthesis environment is another challenge of what lies ahead.

A few decades ago, when computers were thought to be able to solve any problem whatsoever, it was believed that technology could be used to make medical decision systems that will diagnose a disease without a physician (e.g., it was anticipated that all the physicians had to do was to insert the symptoms to the decision system and it will quickly provide them with the correct diagnosis). Beyond technical difficulties, other serious issues eventually emerged that prevented the implementation of the computerized decision systems in the real world. Indeed, even decision systems of limited scope (like MYCIN (Buchanan and Shortliffe, 1984) which was designed to diagnose infectious blood diseases and recommend antibiotics with an anticipated correct diagnosis rate of about 65%) were never used in practice due to legal and ethical issues. Today, the less ambitious goal is to develop diagnostic decision support systems (DDSS) that only offer opinions to physicians who make the final disease diagnosis. In this setting, physicians should not merely *search* for the answer to the patient's health problem in a decision system (e.g., like one performs a Google search), but rather use the system as a tool that can help them *derive* the correct answer. Pattern recognition, for example, can provide valuable information to the physicians, such as interpretations of one-dimensional data (EEG, ECG) and two-dimensional data (detecting cells, tumors or other x-ray abnormalities, MR, tomography); and processing of numerical datasets (e.g., blood test results) and non-numeric information (e.g., patient's history, physician's reports, published literature). Then, it is a matter of professional competence (knowledge, cognitive skills, creativity) for physicians to integrate the pattern recognition outcomes into their reasoning process leading to disease diagnosis and treatment.

At the same time, the high standards of truth when thinking about disease should be held despite physician's professional affiliation, intellectual background and primeval loyalties, and often against the dysfunctional truth-trashing anti-intellectualism that has characterized much of today's Anglo-American world. Research in medical reasoning is concerned with several issues and their synthesis. An important epistemic issue is the relationship between clinical assertion and *truth* (see also Section 2.7.4). The epistemic role typically assigned to truth is to be the norm (goal) of human inquiry, in which case the sense in which truth is a norm of clinical assertion should be carefully spelled out (otherwise it can lead to misunderstandings and the development of deflationist theories of truth). Although the norm of a_ACA is truth (when "physician A asserts that CA," A's yardstick is truth), truth itself is not normative (that "a_ACA is true" is just a fact, a semantic relation between assertion and reality, not a norm). What is normative is the close connection by which case assertion and truth are mutually bound. One may consider a minimalistic meaning: truth is the norm of "A asserts that the disease is contagious" in the sense that "'A asserts that the disease is contagious' is *true* iff the disease *is* contagious." In practice, to make an assertion about CA is to express the physician's (degree of) belief that CA is true, in which case the norm of assertion is knowledge (norm of a_ACA is k_ACA). The supporters of this meaning of truth argue that a physician A who asserts CA for reasons other than the fact that CA is true is irrational in some fashion or A does not have a genuine stance of asserting with regard to CA. The analysis above expresses specific conceptual relations between clinical assertion and truth that are not as trivial as they may seem at first glance, and they surely deserve further investigation in the context of medical decision-making.

Medical decision-makers should be wary of the traps imposed by arguments with a nihilistic ring favored by veriphobes (postmodernists, relativist and others who believe that truth is an empty word; Section 2.7.4). Veriphobes carry their themes to extremes, defending an eliminativistic use of the notion of truth that rests on the claim that an expression like "Cushing's disease occurs predominantly in women" means the same as "'Cushing's disease occurs predominantly in women' is true." This use is problematic for it signals that truth is simply a device physicians can use to speak about case statements and approve them, not a term designating an objective world that transcends the approval of a professional community. For the veriphobe Dr. A it makes no difference whether one says "Chagas disease is caused by *Trypanosoma cruzi (T. cruzi)*" or "'Chagas disease is caused by *T. cruzi*' is true." However, in the former representation of what A says the sentence "Chagas disease is caused by *T.*

cruzi" is merely mentioned, but in the latter representation the sentence appears to be used, which is why the claim that the two representations are equivalent needs careful formulation and may be indefensible: Dr. A might know that "'Chagas disease is caused by $T.$ *cruzi*' is true" without knowing what it means (e.g., A merely finds it in a list of acknowledged medical facts, although A does not know its meaning), and this is different from knowing that Chagas disease is caused by $T.$ *cruzi*. Undoubtedly, whether or not "truth" is an empty word can have crucial consequences in medical reasoning and decision-making. According to the postmodern medicine mode of thought, no biomedical information can ever be strong enough on which to base a "real" clinical opinion and, hence, the possibility of knowledge can be rationalized away using postmodernist conceptions.

Another issue of increasing concern to health care professionals is the effect of climate on population health. For example, several studies have provided evidence of a link between vector-borne disease outbreaks and El Niño-driven climate anomalies, whereas other studies have focused on the effects of the North Atlantic Oscillation on infectious diseases (Morand et al., 2013). Moreover, many scientists consider global climate change as inevitable due to greenhouse gases created by human activities. As noticed in Greer et al. (2008), changes in weather patterns and eco-systems, together with health effects of climate change are expected to be most severe in far northern regions (e.g., the Arctic). North America is also expected to experience marked changes in weather patterns in coming decades (higher temperatures, increased rainfall, droughts and extreme weather events, such as tornadoes and hurricanes). Undoubtedly, changes in climate and associated changes in weather and other environmental exposures can have a serious impact on human health and quality care. Not only do these events directly affect population health (population injuries and displacement due to thermal stress etc.), but they are also likely to cause significant changes in the incidence and distribution of infectious diseases, including vector-borne and zoonotic diseases, water- and food-borne diseases, and diseases with environmental reservoirs. Accordingly, the anticipated nature and direction of these changes pose serious challenges to health care providers and public health agencies.

As an epilogue, the medical profession should make every effort to overcome any nihilism concerning the care providers' ability to become fully aware of their thinking mode and even change it when necessary. Fortunately, there is mounting evidence that health care providers realize that they can drastically improve medical decision-making and clinical judgment if they learn to think about, and understand, their thinking modes, so that they can be at the right mode at the right time, and consciously implement rigorous reasoning concepts and inference techniques in real conditions of uncertainty and time pressure.

Bibliography

Abbey H., 1952. "An examination of the Reed–Frost theory of epidemics". *Human Biology* 24: 201–233.

Adams F., 1939. *The Complete Works of Hippocrates*. Williams & Williams, Baltimore, MD.

Ades A.E., Welton N.J., Caldwell D., Price M., Goubar A. and Lu G., 2008. "Multiparameter evidence synthesis in epidemiology and medical decision-making". *Journal of Health Services Research and Policy* 13: 12–22.

Ahlbom A., Green A., Kheifets L., Savitz D. and Swerdlow A., 2004. "Epidemiology of health effects of radiofrequency exposure". *Environmental Health Perspectives* 112(17): 1741–1754. DOI: 10.1289/ehp.7306.

Alavanja M.C.R., Hoppin J.A. and Kamel F., 2004. "Health effects of chronic pesticide exposure: cancer and neurotoxicity". *Annual Review of Public Health* 25: 155–197. DOI: 10.1146/annurev.publhealth.25.101802.123020.

Alexander J., Divin-Cosgrove C., Faner M.L. and O'Connell M., 2005. "Increasing the knowledge base of asthmatics and their families through asthma clubs along the southwest border". *Journal of American Academy of Nurse Practitioners* 12(7): 260–266.

Al-Jahdali H., Memish Z.A. and Menzies D., 2003. "Tuberculosis in association with travel". *International Journal of Antimicrobial Agents* 21: 125–130.

Allen L.J.S. and Burgin A.M., 2000. "Comparison of deterministic and stochastic SIS and SIR models in discrete time". *Mathematical Biosciences* 163: 1–33.

Alloway B.J., 2005. "Bioavailability of elements in soil". In Selinus O., Alloway B., Centeno J. A., Finkelman R. B., Fuge R., Lindh U. and Smedley P., (eds.). *Essentials of Medical Geology —Impacts of the Natural Environment on Public Health*, Elsevier, London: 347–372.

Andersen, P., 2007. "A review of micronutrient problems in the cultivated soil of Nepal". *Mountain Research and Development* 27: 331–335.

Anderson R.M. and May R.M., 1991. *Infectious Diseases of Humans: Dynamics and Control*. Oxford University Press, Oxford.

Angulo J.M., Yu H.-L., Langousis A., Madrid D. and Christakos G., 2012. "Modeling of space-time infectious disease spread under conditions of uncertainty". *International Journal of Geographical Information Science* 26: 1751–1772. DOI: 10.1080/13658816.2011.648642.

Angulo J.M., Yu H.-L., Langousis A., Kolovos A., Wang J.-F., Madrid D. and Christakos G., 2013. "Integrative 'evidence-law' modeling of spatiotemporal disease spread". *PLOS ONE*, to appear.
Arrivé L., Renard R., Carrat F., Belkacem A., Dahan H., Le Hir P., Monnier-Cholley L. and Tubiana J.-M., 2000. "A scale of methodological quality for clinical studies of radiologic examinations". *Radiology* 217: 69–74.
Baader F., McGuinness D.L., Nardi D. and Patel-Schneider P.F. (eds.), 2003. *The Description Logic Handbook: Theory, Implementation, and Applications*. Cambridge University Press, New York, NY.
Benner P.E., Sutphen M. and Hughes R.G., 2008. "Clinical reasoning, decision-making, and action: Thinking critically and clinically". In Hughes R.G. (ed.), *Patient Safety and Quality: An Evidence-Based Handbook for Nurses*. US Department of Health and Human Services, Rockville, MD: 1–23.
Bennett J.B., 1980. *Rational Thinking: A Study in Basic Logic*. Nelson-Hall Co., Chicago, IL.
Berwick D.M. and Hackbarth A., 2012. "Eliminating waste in US health care". *Journal of the American Medical Association* 307: 15131–516. DOI: 10.1001/jama.2012.362.
Binmore K., Kirman A. and Tani P. (eds.), 1993. *Frontiers of Game Theory*. MIT Press, Cambridge, MA.
Boezen M., Schouten J., Rijcken B., Vonk J., Gerritsen J., van der Zee S., Hoek G., Brunekreef B. and Postma D., 1998. "Peak expiratory flow variability, bronchial responsiveness, and susceptibility to ambient air pollution in adults". *American Journal of Respiratory and Critical Care Medicine* 158(6): 1848–1854.
Bohm D., 1989. *Quantum Theory*. Dover Publications Inc., Mineola, NY.
Bor J. and Pelton T., 2001. "Hopkins faults safety lapses". *The Baltimore Sun*. July 17 2001.
Bossak B.H. and Welford M.R., 2009. "Did medieval trade activity and a viral etiology control the spatial extent and seasonal distribution of Black Death mortality?" *Medical Hypotheses* 72: 749–752.
Bryson B., 2003. *A Short History of Nearly Everything*. Broadway Books, New York, NY.
Buchanan B.G. and Shortliffe E.H., 1984. *Rule-based Expert Systems: The MYCIN Experiments of the Stanford Heuristic Programming Project*. Addison-Wesley, Reading, MA.
Burke W., Fesinmeyer M., Reed K., Hampson L. and Carlsten C., 2003. "Family history as a predictor of asthma risk". *American Journal of Preventive Medicine* 24(2): 160–169.
Campbell T.C., 1967. "Present day knowledge on aflatoxin". *Philippine Journal of Nutrition* 20: 193–201.
Carnap R., 1962. *Logical Foundations of Probability* (2nd ed.). University of Chicago Press, Chicago, Il.
Chen C.-C., Wu C.-F., Yu H.-L., Chan C.-C. and Cheng T.-J., 2012a. "Spatiotemporal modeling with temporal-invariant variogram subgroups to estimate

fine particulate matter PM2.5 concentrations". *Atmospheric Environment* July 2012(54): 1–8.

Chien L.-C. and Bangdiwala S., 2012. "The implementation of Bayesian structural additive regression models in multi-city time series air pollution and human health studies". *Stochastic Environmental Research and Risk Assessment* 26(8): 1041–1051.

Chien L.-C., Yang C.-H. and Yu H.-L., 2012b. "Effects of Asian dust storm on spatiotemporal distribution of children clinic visits of respiratory diseases in Taipei". *Environmental Health Perspectives* 120(8): 1215–1220.

Chmielewski J., 1962. "Notes on early Chinese logic (I)". *Rocznik Orientalistyczny* 26(1): 7–22.

Chmielewski J., 1963a. "Notes on early Chinese logic (II)". *Rocznik Orientalistyczny* 26(2): 91–105.

Chmielewski J., 1963b. "Notes on early Chinese logic (III)". *Rocznik Orientalistyczny* 27(1): 103–121.

Chmielewski J., 1965a. "Notes on early Chinese logic (IV)". *Rocznik Orientalistyczny* 28(2): 87–111.

Chmielewski J., 1965b. "Notes on early Chinese logic (V)". *Rocznik Orientalistyczny* 29(2): 117–138.

Chmielewski J., 1966. "Notes on early Chinese logic (VI)". *Rocznik Orientalistyczny* 30(1): 31–52.

Chmielewski J., 1968. "Notes on early Chinese logic (VII)". *Rocznik Orientalistyczny* 31(1): 117–136.

Chmielewski J., 1969. "Notes on early Chinese logic (VIII)". *Rocznik Orientalistyczny* 32(2): 83–103.

Choi K.-M., Yu H.-L. and Wilson M.L., 2008. "Spatiotemporal statistical analysis of influenza mortality risk in the State of California during the period 1997–2001". *Stochastic Environmental Research and Risk Assessment* 22(1): 15–25.

Christakos G., 2010. *Integrative Problem-Solving in a Time of Decadence*. Springer, New York, NY.

Christakos G., 2012. "Space–Time stochastic modelling of human exposure". In El-Shaarawi A.H. and Piegorsch W. (eds.), *Encyclopedia of Enviromentrics* (2nd ed.). J. Wiley & Sons Ltd, Chichester, UK: 2530–2535.

Christakos G., Bogaert P. and Serre M.L., 2002. *Temporal GIS*. (With CD-ROM). Springer-Verlag, New York, NY.

Christakos G. and Hristopulos D.T., 1998. *Spatiotemporal Environmental Health Modelling: A Tractatus Stochasticus*. Kluwer Academic Publishing, Boston, MA.

Christakos G. and Kolovos A., 1999. "A study of the spatiotemporal health impacts of ozone exposure". *Journal of Exposure Analysis & Environmental Epidemiology* 9(4): 322–335.

Christakos G., Olea R.A., Serre M.L., Yu H.-L. and Wang L.-L., 2005. *Interdisciplinary Public Health Reasoning and Epidemic Modelling: The Case of Black Death*. Springer-Verlag, New York, NY.

Christakos G. and Serre M.L., 2000. "A spatiotemporal study of exposure–health effect associations". *Journal of Exposure Analysis & Environmental Epidemiology* 10(2): 168–187.

Cordingley G. (2005). "Using probability in medical diagnosis — a headache example". December 26 2005. Available online: http://EzineArticles.com/118440.

Crawford-Brown D., 1997. *Theoretical and Mathematical Foundations of Human Health Risk Analysis*. Kluwer, Boston MA.

Croskerry P., 2003. "The importance of cognitive errors in diagnosis and strategies to prevent them". *Academic Medicine* 78: 1–6.

Croskerry P., 2009. "A universal model of diagnostic reasoning". *Academic Medicine* 84: 1022–1028.

Croskerry P., Abbass A. and Wu A., 2010. "Emotional issues in patient safety". *Journal of Patient Safety* 6: 199–205.

Croskerry P. and Nimmo G.R., 2011. "Better clinical decision making and reducing diagnostic error". *Journal of the Royal College of the Physicians of Edinburgh* 41: 155–62.

Crosson F.J., 2012. "Change the microenvironment: delivery system reform essential to controlling costs". *Modern Healthcare and The Commonwealth Fund* April 27 2009. Available online: http://www.commonwealthfund.org/Content/Publications/Commentaries/2009/Apr/Change-the-Microenvironment.aspx.

Daley D.J. and Gani J., 1999. *Epidemic Modelling*. Cambridge University Press, Cambridge, UK.

Davenhall B. 2012. *Geomedicine: Geography and Personal Health*. Esri, Redlands, CA.

Dawson N.V. and Cebul R.D. (1992). "Use of formal methods in medical decision-making: A survey and analysis". *Medical Decision Making* 1992(12): 298–306.

Deckers J. and Steinnes E., 2004. "State of the art on soil- related geo-medical issues in the world". In Sparks D.J. (ed.), *Advances in Agronomy*. Elsevier, Doordrecht: 1–35.

Demyanov V., Soltani S., Kanevski M., Canu S., Maignan M., Savelieva E., Timonin V. and Pisarenko V. (2001). "Wavelet analysis residual kriging vs. neural network residual kriging". *Stochastic Environmental Research and Risk Assessment* 15(1): 18–32.

Dettmer H.W., 2007. *The Logical Thinking Process: A Systems Approach to Complex Problem Solving*. Quality Press, Milwaukee, WI.

Dhondt S., Xuan Q.L., Van H.V. and Hens L., 2011. "Environmental health impacts of mobility and transport in Hai Phong, Vietnam". *Stochastic Environmental Research and Risk Assessment* 25(3): 363–376.

Diamond G.A. and Forrester J.S., 1979. "Analysis of probability as an aid in the clinical diagnosis of coronary-artery disease". *New England Journal of Medicine* 300: 1350–1358.

Dorland W.A.N., 2007. *Dorland's Medical Dictionary for Health Consumers*. Saunders/Elsevier Inc., Philadelphia, PA.

Edholm O.G., Adam J.M. and Fox R.H., 1962. "The effects of work in cool and hot conditions on pulse rate and body temperature". *Ergonomics* 5: 545–556.

Eggleston P.A., 2009. "Complex interactions of pollutant and allergen exposures and their impact on people with asthma". *Pediatrics* 123: S160–S167. DOI: 10.1542/peds.2008-2233F

Ekstrom G. and Akerblom M., 1990. "Pesticide management in food and water safety: international contributions and national approaches". *Reviews of Environmental Contamination and Toxicology* 114: 23–55.

Evans J.S.B.T., Handley S.H. and Over D.E., 2003. "Conditionals and conditional probability". *Journal of Experimental Psychology: Learning, Memory, and Cognition* 29: 321–355.

Fagin R., Halpern J., Moses Y. and Vardi M., 1995. *Reasoning about Knowledge.* MIT Press, Cambridge, MA.

Fenner-Crisp P.E., 2001. "Risk assessment and risk management: the regulatory process". In Kreiger R. (ed.), *Handbook of Pesticide Toxicology.* Academic Press, San Diego, CA: 681–690.

Fenske R.A., 1997. "Pesticide exposure assessment of workers and their families". *Occupational Medicine* 12: 221–237.

de Finetti B. (1974). *Theory of Probability* Vol. 1. Wiley, New York, NY.

de Finetti B. (1975). *Theory of Probability* Vol. 2. Wiley, New York, NY.

Friedman T.L., 2010. "Root canal politics". *The New York Times,* May 9 2010.

Fulda J.S., 1989. "Material implication revisited". *The American Mathematical Monthly* 96(3): 247–250.

Furuva H., 2007. "Risk of transmission of airborne infection during train commute based on mathematical model". *Environmental Health and Preventive Medicine* 12(2): 78–83, DOI: 10.1007/BF02898153.

Gauderman W.J., Vora H., McConnell R., Berhane K., Gilliland F., Thomas D., Lurmann F., Avol E., Kunzli N., Jerrett M. and Peters J., 2007. "Effect of exposure to traffic on lung development from 10 to 18 years of age: a cohort study". *The Lancet* 369(9561): 571–577.

Gilbert M.T.P., Bandelt H.-J., Hofreiter M. and Barnes I., 2005. "Assessing ancient DNA studies". *Trends in Ecology & Evolution* 20(10): 541–544.

Gilhooly K.J., McGeorge P., Hunter J., Rawles J.M., Kirby I.K., Green C. and Wynn V., 1997. "Biomedical knowledge in diagnostic thinking: the case of electrocardiogram (ECG) interpretation". *European Journal of Cognitive Psychology* 9: 199–223.

Gill G.V., Redmond S., Garratt F. and Paisey R., 1994, "Diabetes and alternative medicine: cause for concern". *Diabetic Medicine* 11: 210–213.

Gnedenko B.V. and Khinchin, A.Ya., 2010. *An Elementary Introduction to the Theory of Probability.* Dover Publications Inc., Mineola, NY.

Graber M., 2003. "Metacognitive training to reduce diagnostic errors: Ready for prime time?" *Academic Medicine* 78: 781.

Graham A.C., 1967. "Chinese logic". In Edwards P. (ed.), *The Encyclopedia of Philosophy* Vol. IV. Macmillan, New York, N.Y.

Grawert A., 2009. "The fundamental meaning of 'medical uncertainty': Judicial deference to selective science in Gonzales v. Carhart". *Legislation and Public Policy* 12: 379–408.

Greer A., Ng V. and Fisman D., 2008. "Climate change and infectious diseases in North America: the road ahead". *Canadian Medical Association Journal* 178(6): 715–722.

Griffith D.A. and Christakos G., 2007. "Medical geography as a science of interdisciplinary knowledge synthesis under conditions of uncertainty". *Stochastic Environmental Research and Risk Assessment* 21(5): 459–460.

Groopman J., 2007. *How Doctors Think*. Houghton Mifflin, New York, NY.

Groopman J., 2010. "What is missing in medical thinking". *Bulletin of the American Academy* Winter 2010: 53–58.

Gummer B., 2009. *The Scourging Angel — The Black Death in the British Isles*. The Bodley Head, London.

Gupta U.C. and Gupta S.C., 2005. "Future trends and requirements in micronutrient research". *Communications in Soil Science and Plant Analysis* 36: 33–45.

Gutiérrez R., Roldán C., Gutiérrez-Sánchez R. and Angulo J.M., 2005. "Estimation and prediction of a 2D lognormal diffusion random field". *Stochastic Environmental Research and Risk Assessment* 19(4): 258–265.

Habermas J., 1995. *Moral Consciousness and Communicative Action*. MIT Press, Cambridge, MA.

Hacking I., 1975. *The Emergence of Probability*. Cambridge University Press, Cambridge.

Haining R., Law J., Maheswaran R., Pearson T. and Brindley P., 2007. "Bayesian modelling of environmental risk: example using a small area ecological study of coronary heart disease mortality in relation to modelled outdoor nitrogen oxide levels." *Stochastic Environmental Research and Risk Assessment* 21(5): 501–509.

Hankinson J.R., 1998. *Cause and Explanation in Ancient Greek Thought*. Oxford University Press, Oxford.

Hannay D., 2002. "Oral history and qualitative research". *The British Journal of General Practice* June 2002: 515.

Hansen C., 1983. *Language and Logic in Ancient China*. University of Michigan Press, Ann Arbor, MI.

Hansen C., 1998. "Logic in China". In Craig E. (ed.), *Routledge Encyclopedia of Philosophy*. Routledge, New York, NY.

Harbsmeier C., 1998. "Language and Logic — Part I". In Needham J. (ed.), *Science and Civilisation in China* Vol. VII. Cambridge University Press, Cambridge.

Harris P., Brunsdon C. and Fotheringham A.S., 2011. "Links, comparisons and extensions of the geographically weighted regression model when used as a spatial predictor". *Stochastic Environmental Research and Risk Assessment* 25(2): 123–138.

Hartzband P. and Groopman J.E., 2008. "Off the record — Avoiding the pitfalls of going electronic". *The New England Journal of Medicine* 358: 1656–1658.

Helwade D.R. and Subramanyam A., 2010. "Spatial prediction using bivariate exponential distribution/" *Stochastic Environmental Research and Risk Assessment* 24(2): 271–281.

Hintikka J.K., 1962. *Knowledge and Belief*. Cornell University Press, Ithaca, NY.

Hocutt M., 1974. "Aristotle's Four Becauses". *Philosophy* 49: 385–399.

Holt A., Bichindaritz I., Schmidt R. and Perner P., 2005. "Medical applications in case-based reasoning". *The Knowledge Engineering Review* 20: 289–292.

Hovell M.F., Meltzer S.B., Wahlgren D.R., Matt G.E., Hofstetter C.R., Jones J.A., Meltzer E.O., Bernert J.T. and Pirkle J.L., 2002. "Asthma management and environmental tobacco smoke exposure reduction in Latino children: a controlled trial". *Pediatrics* 110(5): 946–956.

Hsu C.C. and Ho C.S., 2004. "A new hybrid case-based architecture for medical diagnosis". *Information Sciences, Informatics and Computer Science: An International Journal* 166(1–4): 231–247.

Hunter J.B., Guernsey de Zapien J., Papenfuss M., Fernandez M.L., Meister J. and Giuliano A.R., 2004. "The impact of a promotora on increasing routine chronic disease: Prevention among women aged 40 and older at the US-Mexico Border". *Health Education and Behavior* 31(4): 18S–21S.

Hurley P.J., 2012. *A Concise Introduction to Logic*. (11th ed.). Wadsworth, Boston, MA.

Innocent P.R. and John R.I., 2004. "Computer aided fuzzy medical diagnosis". *Information Sciences: an International Journal* 162(2): 81–104.

Ioannidis J.P.A., 2005. "Contradicted and initially stronger effects in highly cited clinical research". *Journal of the American Medical Association* 294: 218–228.

Jacquez J.A. and O'Neill P., 1991. "Reproduction numbers and thresholds in stochastic epidemic models I. Homogeneous populations". *Mathematical Bioscience* 107: 161–186.

Jeffreys H., 1961. *Theory of Probability*. Clarendon Press, Oxford.

Jenicek M., 2011. *Medical Error and Harm: Understanding, Prevention and Control*. Productivity Press, New York, NY.

John R.I. and Innocent P.R., 2005. "Modeling uncertainty in clinical diagnosis using fuzzy logic". *IEEE Transactions on systems, Man, and Cybernetics* 35(6): 1340–1350.

Jones W.H.S, 1953. *Hippocrates*. Loeb Classical Library, London and Cambridge.

Kahn C.E. Jr., Roberts L.M., Wang K., Jenks D. and Haddawy P., 1995. "Preliminary investigation of a Bayesian network for mammographic diagnosis of breast cancer". *Proceedings of the Annual Symposium on Computing Applied to Medical Care* 1995: 208–212.

Kane S.A., 2009. *Introduction to Physics in Modern Medicine*. CRC Press, Boca Raton, FL.

Kant I., (1922). *Critique of Pure Reason*. Norman Kemp Smith (trans.). Macmillan, London.

Karnon J., Brennan A. and Akehurst R., 2010. "Decision modeling to inform decision-making: Seeing the wood for the trees". *Medical Decision Making* 30: E20–E22, DOI:10.1177/0272989×10364245.

Kassirer J.P., 2010. "Teaching clinical reasoning: Case-based and coached". *Academic Medicine* 85: 1118–1124.

Kassirer J.P. and Kopelman R.I., 1989. "Cognitive diagnostic errors: Instantiation, classification, and consequences". *American Journal of Medicine* 86: 433–441.

Kassirer J.P., Wong J.B. and Kopelman R.I., 2009. *Learning Clinical Reasoning*. Williams and Wilkins, Philadelphia, PA.

Klein G.A., 1998. *Sources of Power: How People Make Decisions*. MIT Press, Cambridge, MA.

Klein J.T., 1996. *Crossing boundaries: Knowledge, Disciplinarities, and Interdisciplinarities*. University Press of Virginia, Charlottesville, VA.

Koch T. and Denike K., 2007. "Certainty, uncertainty, and the spatiality of disease: a West Nile Virus example". *Stochastic Environmental Research and Risk Assessment* 21(5): 523–531.

Kohn L.T., Corrigan J.M. and Donaldson M.S. (eds.), 1999. *To Err is Human: Building a Safer Health System*. National Academy Press, Washington DC.

Kolmogorov A.N., 1933. *Grundbegriffe der Wahrscheinlichkeitsrechnung*. Springer, Berlin.

Kolovos A., Angulo J.M., Modis K., Papantonopoulos G., Wang J.-F. and Christakos G., 2013. "New model-based covariances for spatiotemporal environmental health analysis: mapping the spatiotemporal spread of French flu". *Environmental Monitoring & Assessment* 185: 815–831. DOI: 10.1007/s10661-012-2593-1.

Kompridis N., 2006. *Critique and Disclosure*. MIT Press, Cambridge, MA.

Kottegoda N.T. and Rosso R., 1997. *Statistics, Probability for Civil and Environmental Engineers*. McGraw-Hill, New York, NY.

Krewski D., Wigle D., Clayson D.B. and Howe G.R., 1989. "Role of epidemiology in health risk assessment". *Recent Results in Cancer Research* 120: 1–24.

Kulikowski C., 2000 "Artificial intelligence in medical decision making: History, evolution, and prospects". In Bronzino J.D. (ed.), *The Biomedical Engineering Handbook: 2nd Edition*. CRC Press LLC, Boca Raton, FL.

Läg J., 1978. "Oversikt over geomedininske problemstillinger med endel eksempler fra norske undersokesler" (with English summary). *Norsk Veterinartidsskr* 90: 621–627.

Läg J. (ed.), 1990. *Geomedicine*. CRC Press, Boca Raton, FL.

Layman C.S., 2005. *The Power of Logic*. (3rd ed.). McGraw-Hill, New York, NY.

Lear J., 1980. *Aristotle and Logical Theory*. Cambridge University Press, Cambridge.

Lee G. and Rogerson P., 2007. "Monitoring global spatial statistics". *Stochastic Environmental Research and Risk Assessment* 21(5): 545–553.

Lele S. and Norgaard R.B., 2005. "Practicing interdisciplinarity". *Bioscience* 55: 967–975.

Lennox J., 2011. "Aristotle's Biology". In Edward N. Zalta (ed.), *The Stanford Encyclopedia of Philosophy* (Fall 2011 ed.). Available online: http://plato.stanford.edu/archives/fall2011/entries/aristotle-biology.

Lesgold A., Rubinson H., Feltovich P., Glaser R., Klopfer D. and Wang Y., 1988. "Expertise in a complex skill: diagnosing X-ray pictures". In Chi M.T.H., Glaser R. and Farr M. (eds.), *The Nature of Expertise*. Lawrence Erlbaum Association, Hillsdale, NJ: 311–342.

Li L., 2010. "Review of hand, foot and mouth disease". *Frontiers of Medicine in China* 4: 139–146.

Liao Y.-L., Wang J.-F., Guo Y.-Q. and Zheng X.-Y., 2011a. "Risk assessment of human neural tube defects using a Bayesian belief network". *Stochastic Environmental Research and Risk Assessment* 24(1): 93–100.

Liao Y.-L., Wang J.-F., Wu J.-L., Wang J.-J. and Zheng X.-Y., 2011b. "A comparison of methods for spatial relative risk mapping of human neural tube defects". *Stochastic Environmental Research and Risk Assessment* 25(1): 99–106.

Lin Y.-P., Cheng B.-Y., Chu H.-J., Chang T.-K. and Yu H.-L., 2011. "Assessing how heavy metal pollution and human activity are related by using logistic regression and kriging methods". *Geoderma* 163: 275–282.

Lindley D.V., 1982. "Scoring rules and the inevitability of probability". *International Statistics Review* 50: 1–26.

Liu L.-M., Lo K.-C. and Wu J.-T., 1996. "A probabilistic interpretation of 'If-Then'". *The Quarterly Journal of Experimental Psychology* 49(A): 828–844.

Lucas P.J.F., 2001. "Expert knowledge and its role in learning Bayesian networks in medicine: an appraisal". In Quaglini, S., Barahona P. and Andreassen S. (eds.), *Artificial Intelligence in Medicine*. Springer-Verlag, Berlin: 156–166.

Luo M. and Opaluch J.J., 2011. "Analyze the risks of biological invasion". *Stochastic Environmental Research and Risk Assessment* 25(3): 377–388.

Malmesbury M.B. (ed.), 1839. *The English Works of Thomas Hobbes of Malmesbury*. J. Bohn, London.

Malomo A.O., Idowu O.E. and Osuagwu F.C., 2006. "Lessons from history: Human anatomy, from the origin to the renaissance". *International Journal of Morphology* 24(1): 99–104.

Marshall B.J. and Warren J.R., 1983. "Unidentified curved bacilli on gastric epithelium in active chronic gastritis". *The Lancet* 321(8336): 1273–1275. DOI:10.1016/S0140-6736(83)92719-8.

Matalene C., 1985. "Contrastive rhetoric: an American writing teacher in China". *College English* 47(8): 789–808.

McConnell R., Berhane K., Gilliand F., London S., Vora H., Avol E., Gauderman W.J., Margolis H.G., Lurmann F., Thomas D.C. and Peters J.M., 1999. "Air pollution and bronchitic symptoms in southern California children with asthma". *Environmental Health Perspectives* 107(9): 757–760.

McGlynn E.A., Asch S.M., Adams J., Keesey J., Hicks J., De-Cristofaro A. and Kerr E.A., 2003. "The quality of health care delivered to adults in the United States". *The New England Journal of Medicine* 348: 2635–2645.

McLaughlin K., Novak K., Rikers R.M. and Schmidt H.G., 2010. "Does applying biomedical knowledge improve diagnostic performance when solving electrolyte problems?" *Canadian Medical Education Journal* 1(1): e4–e9.

Mendelsohn R.L., 1987. *Basic Logic*. Prentice Hall, Englewood Cliffs, NJ.

Mikler A.R., Venkatachalam S. and Ramisetty-Mikler S., 2006. "Decisions under uncertainty: a computational framework for quantification of policies addressing infectious disease epidemics". *Stochastic Environmental Research and Risk Assessment* 21(5): 533–543.

Miller R.A., 1994. "Medical diagnostic decision support systems – Past, Present, and Future". *Journal of the American Medical Informatics Association* 1(1): 8–27.

Miller R.A. and A. Geissbuhler, 2009. "Diagnostic Decision Support Systems". In Berner E.S. (ed.), *Clinical Decision Support Systems: Theory and Practice*. Springer, New York, NY.

Modell S.M., 2010. "Aristotelian influence in the formation of medical theory". *The European Legacy: Toward New Paradigms* 15(4): 409–424.

Moeller D.W., 2011. *Environmental Health*. (4th ed.). Harvard University Press, Cambridge, MA.

Montgomery K., 2006. *How Doctors Think: Clinical Judgment and the Practice of Medicine*. Oxford University Press, New York, NY.

Morand S., Owers K.A., Waret-Szkuta A., McIntyre K.M. and Baylis M., 2013. "Climate variability and outbreaks of infectious diseases in Europe". *Sci. Rep.* 3: 1774; DOI:10.1038/srep01774.

Mure G.R.G., 1975. "Cause and because in Aristotle". *Philosophy* 50: 356–357.

Nguyen H.T., Mukaidono M. and Kreinovich V., 2002. "Probability of implication, logical version of Bayes' theorem, and fuzzy logic operations". *Proceedings of International IEEE Conference on Fuzzy Systems* FUZZ-IEEE 1: 530–535.

Niknam K. and Niknam S., 2008. "A mathematical description of physician decision-making". *Proceedings of Second International Conference on Bioinformatics and Biomedical Engineering* ICBBE 2008: 581–584. DOI: 10.1109/ICBBE.2008.141.

Nikovski D., 2000. "Constructing Bayesian networks for medical diagnosis from incomplete and partially correct statistics". *IEEE Transactions on Knowledge and Data Engineering* 12(4): 509–516.

O'Mathúna D.P. and McCallum D., 1996. "Postmodern medicine: miracle or menace?" *Today's Christian Doctor* 27: 28–32.

Oliver M.A., 1997. "Soil and human health: A review". *European Journal of Soil Science* 48: 573–592.

Oliver M.A., 2004. "Soil and human health: geomedical aspects in relation to agriculture". In Steinnes E. (ed.), *Geomedical Aspects of Organic Farming*. The Norwegian Academy of Science and Letters, Oslo: 16–32.

Orfanos C.E., 2007. "From Hippocrates to modern medicine". *JEADV* 21: 852–858. DOI: 10.1111/j.1468-3083.2007.02273.x.

Over D.E. and Evans J.S.B.T., 2003. "The probability of conditionals: The psychological evidence". *Mind & Language* 18: 340–358.

Palmer C.L., 2001. *Work at the Boundaries of Science*. Kluwer, Dordrecht.

Pawlowsky-Glahn V. and Buccianti A., 2011. *Compositional Data Analysis: Theory and Applications*. Wiley, Hoboken, NJ.

Peng F. and Hall W.J., 1996. "Bayesian Analysis of ROC Curves Using Markov-chain Monte Carlo Methods". *Medical Decision Making* 16: 404–411. DOI: 10.1177/0272989X9601600411.

Porter R., 1997. *The Greatest Benefit of Mankind. A Medical History of Humanity from Antiquity to the Present*. Harper Collins, London.

Redelmeier D.A., 2005. "The cognitive psychology of missed diagnoses". *Annals of International Medicine* 142: 115–120.

Redelmeier D.A., Koehler D.J., Liberman V. and Tversky A., 1995. "Probability judgment in medicine: Discounting unspecified possibilities". *Medical Decision Making* 15: 227–230.

Reiczigel J., Brugger K., Rubel F., Solymosi N. and Lang Z., 2010. "Bayesian analysis of a dynamical model for the spread of the Usutu virus". *Stochastic Environmental Research and Risk Assessment* 24(3): 455–462.

Reyna V.F., 2008. "Theories of medical decision-making and health: an evidence-based approach". *Medical Decision Making* 28: 829–833.

Reynolds J.H., Hobart R.L., Ayala P. and Eischen M.H., 2005. "Clean Indoor Air in El Paso, Texas: A Case Study". *Preventing Chronic Disease: Public Health Research, Practice, and Policy* 2(1). Available online: http://www.cdc.gov/pcd/issues/2005jan/04_0065.htm.

Riley E.C., Murphy G. and Riley R.L., 1978. "Airborne spread of measles in a suburban elementary school". *American Journal of Epidemiology* 107: 421–432.

Roberts M.G., 2004. "Modelling strategies for minimizing the impact of an imported exotic infection". *Proceedings of the Royal Society B* 271: 2411–2415. DOI: 10.1098/rspb.2004.2865.

Sampson W., 2000. "Postmodern medicine". *The Scientific Review of Alternative Medicine* 4(1): 18–20.

Schiff G., Hasan O., Kim S., Abrams R., Cosby K., Lambert B.L., Elstein A.S., Hasler S., Kabongo M.L., Krosnjar N., Odwazny R., Wisniewski M.F. and McNutt R.A., 2009. "Diagnostic error in medicine: analysis of 583 physician-reported errors". *Archives of International Medicine* 169: 1881–1887.

Selinus O., Alloway B., Centeno J.A., Finkelman R.B., Fuge R., Lindh U. and Smedley P. (eds.), 2005. *Essentials of Medical Geology — Impacts of the Natural Environment on Public Health*. Elsevier Academic Press, London, UK.

Schwartz S. and Griffin T., 1986. *Medical Thinking: The Psychology of Medical Judgment and Decision-Making*. Springer-Verlag, New York, NY.

Sharpe R.M. and Irvine D.S., 2004. "How strong is the evidence of a link between environmental chemicals and adverse effects on human reproductive health?" *British Medical Journal* 328: 447–451.

Sivin N., 1995. *Medicine, Philosophy and Religion in Ancient China: Researches and Reflections*. Variorum, Aldershot.

Smith R., 2012. "Aristotle's Logic". In Zalta, E.N. (ed.), *The Stanford Encyclopedia of Philosophy* (Spring 2012 ed.). Available online: http://plato.stanford.edu/archives/spr2012/entries/aristotle-logic.

Sox H.C., Higgins M.C. and Owens D.K., 2013. *Medical Decision Making*. John Wiley & Sons, Ltd., New York, NY.

Steinnes E. (ed.), 2004. *Geomedical Aspects of Organic Farming*. The Norwegian Academy of Science and Letters, Oslo.

Steinnes E., 2009. "Soils and geomedicine". *Environmental Geochemistry and Health* 31: 523–535.

Tamerius J.D., Wise E.K., Uejio C.K., McCoy A.L. and Comrie A.C., 2006. "Climate and human health: synthesizing environmental complexity and uncertainty". *Stochastic Environmental Research and Risk Assessment* 21(5): 601–613.

Tampellini D., Rahman N., Lin M.T., Capetillo-Zarate E. and Gouras G.K., 2011. "Impaired β-amyloid secretion in Alzheimer's disease pathogenesis". *Journal of Neuroscience* 31(43): 15384–15390. DOI: 10.1523/JNEUROSCI.2986-11.2011.

Tan J., Zhu W., Wang W., Li R., Hou S., Wang D. and Yang L., 2002. "Selenium in soil and endemic diseases in China". *The Science of the Total Environment* 284: 227–235.

Tan W.-Y., 2000. *Stochastic Modeling of AIDS Epidemiology and HIV Pathogenesis*. World Scientific Publishing Co., Singapore.

Tanaka M., Volle M.A., Brisson G.B. and Dion M., 1979. "Body temperature and heart rate relationships during submaximal bicycle ergometer exercises". *European Journal of Applied Physiology* 42(4): 263–270.

Thayer W.C., Griffith D.A. and Diamond G.L., 2006. "Medical geography as a science of interdisciplinary knowledge synthesis under conditions of uncertainty". *Stochastic Environmental Research and Risk Assessment* 21(5): 459–460.

Tidman P. and Kahane H., 1999. *Logic & Philosophy: A Modern Introduction*. (8th ed.). Wadsworth, Boston, MA.

Tiefelsdorf M., 2007. "Controlling for migration effects in ecological disease mapping of prostate cancer". *Stochastic Environmental Research and Risk Assessment* 21(5): 615–624.

Tilikidis A., 1999. *The Basic Theory of Traditional Chinese Medicine*. Protoporia, Athens.

Toth F.L. (ed.), 2011. *Geological Disposal of Carbon Dioxide and Radioactive Waste: A Comparative Assessment*. Springer, New York, NY.

Tsipouras M.G., Voglis C. and Fotiadis D.I., 2007. "A framework for fuzzy expert system creation-application to cardiovascular diseases". *IEEE Transactions on Biomedical Engineering* 54(11): 2089–2105.

Tversky A. and Koehler D.J., 1994. "Support theory: a non-extensional representation of subjective probability". *Psychological Review* 101(4): 547–567.

Ugarte M.D., Goicoa T., Etxeberria J. and Militino A.F., 2012. "A P-spline ANOVA type model in space-time disease mapping". *Stochastic Environmental Research and Risk Assessment* 21(5): 615–624.

Ullman D.G., 2006. *Making Robust Decisions*. Trafford Publishing, Victoria, BC.

Volz E. and Meyers L.A., 2009. "Epidemic thresholds in dynamic contact networks". *Journal of the Royal Society Interface* 6(32): 233–241. DOI: 10.1098/rsif.2008.0218.

Walker B., Stokes, L.D. and Warren R., 2003. "Environmental factors associated with asthma". *Journal of the National Medical Association* 95(2): 152–166.

Wang J.-F., Christakos G., Han W.G. and Meng B., 2008. "Data-driven exploration of 'spatial pattern-time process-driving forces' associations of SARS epidemic in Beijing, China". *Journal of Public Health* 30(3): 234–244. DOI: 10.1093/pubmed/fdn023.

Wang J.-F., Guo Y.S., Christakos G., Yang W.Z., Liao Y.L., Li Z.J., Li X.Z., Lai S.J. and Chen H.Y., 2011. "Hand, foot and mouth disease: spatiotemporal transmission and climate". *International Journal of Health Geographics* 10(25). DOI: 10.1186/1476-072X-10-25.

Wang J.-F., Hu M.-G., Xu C.-D., Christakos G. and Zhao Y., 2012. "Estimation of citywide air pollution in Beijing". *PLOS ONE* 8(1): e53400. DOI: 10.1371/journal.pone.0053400.

Wang J.-F., Liu X., Christakos G., Liao Y.-L., Gu X. and Zheng X.-Y., 2009. "Assessing local determinants of neural tube defects in the Heshun region, Shanxi province, China". *BMC Public Health* 10(52). DOI: 10.1186/1471-2458-10-52.

Wang J.-F., McMichael A.J., Meng B., Becker N.G., Han W.G., Glass K., Wu J., Liu X., Liu J., Li X. and Zheng X., 2006. "Spatial dynamics of an epidemic of severe acute respiratory syndrome in an urban area". *Bulletin of the World Health Organization* 84(12): 965–968.

Wang T., 2012. "Chinese medicine needs the baptism of science". *Zhongguo Zhong Xi Yi Jie He Za Zhi* (in Chinese). 32(8): 1014–1022.

Watson H.T., 2010. "Determining whether medical causation is established using statistical analysis". *Law Journal Letters — Medical Malpractice Law & Strategy* February 2010. Available online: http://www.lawjournalnewsletters.com/issues/ljn_medlaw/27_5/news/153291-1.html.

Weiner D.A., Ryan T.J., McCabe C.H., Kennedy J.W., Schloss M., Tristani F., Chaitman B.R. and Fisher L.D., 1979. "Exercise stress testing: correlations among history of angina, ST-segment response, and prevalence of coronary-artery disease in the Coronary Artery Surgery Study (CASS)". *New England Journal of Medicine* 301: 230–235.

Weiss K.B., Gergen P.J. and Hodgson T.A., 1992. "An economic evaluation of asthma in the United States". *New England Journal of Medicine* 326: 862–866.

Weisser U., 1989. "Das corpus hippocraticum in der arabischen medizin". *Sudhoffs Arch Z Wissenschaftsgesch Beih* 27: 377–408.

West R.W. and Thompson J.R., 1997. "Models for the simple epidemic". *Mathematical Biosciences* 141(1): 29–39.

Wiwanitkit V., 2008. "PM10 in the atmosphere and incidence of respiratory illness in Chiangmai during the smoggy pollution". *Stochastic Environmental Research and Risk Assessment* 22(3): 437–440.

Wright W.E., Peters J.M. and Mack T.M., 1982. "Leukaemia in workers exposed to electrical and magnetic fields". *The Lancet* 2(8308): 1160–1161.

Van Dalen D., 2001. "Intuitionistic logic". In Gobble L. (ed.), *The Blackwell Guide to Philosophical Logic*. Blackwell, Oxford: 224–257.

Vasilescu N., Badica M. and Munteanu M., 1997. "A fuzzy model for simulation and medical diagnosis". *Studies in Health Technology and Informatics* 43 (Pt. B): 638–641.

de Vet H.C.W., Terwee C.B., Mokkink L.B. and Knol D.L., 2011. *Measurement in Medicine*. Cambridge University Press, Cambridge.

Vickers A.J. and Elkin E.B., 2006. "Decision curve analysis: a novel method for evaluating prediction models". *Medical Decision Making* 26: 565–574. DOI: 10.1177/0272989X06295361.

Yao J.F. and Yao J., 2001. "Fuzzy decision-making for medical diagnosis based on fuzzy number and compositional rule of inference". *Fuzzy Sets and Systems* 120(2): 351–366.

Yearwood J. and Pham B., 2000. "Case-based support in a cooperative medical diagnosis environment". *Telemedicine Journal* 6(2): 243–250.

Yoon P.W., Scheuner M.T., Muin J. and Khoury M.J., 2003. "Research priorities for evaluating family history in the prevention of common chronic disease". *American Journal of Preventive Medicine* 24(2): 128–135.

Yu H.-L., Chen J.-C. and Christakos G., 2009. "BME estimation of residential exposure to ambient PM10 and ozone at multiple time-scales". *Environmental Health Perspectives* 117(4): 537–544.

Yu H.-L., Chien L.-C. and Yang C.-H., 2012. "Asian dust storm elevates children's respiratory health risks: a spatiotemporal analysis of children's clinic visits across Taipei (Taiwan)". *PLOS ONE* 7(7): e41317. DOI: 10.1371/journal.pone.0041317.

Yu H.-L., Chiang C.-T., Lin S.-T. and Chang C.-K., 2010. "Spatiotemporal analysis and mapping of oral cancer risk in changhua county (Taiwan): an application of generalized bayesian maximum entropy method". *Annals of Epidemiology* 20(2): 99–107.

Yu H.-L. and Christakos G., 2006. "Spatiotemporal modelling and mapping of the bubonic plague epidemic in India". *International Journal of Health Geographics* 5(12). Available online: http://www.ij-healthgeographics.com.

Yu H.-L., Kolovos A., Christakos G., Chen J.-C., Warmerdam S. and Dev B., 2007. "Interactive spatiotemporal modelling of health systems: The SEKS-GUI framework". *Stochastic Environmental Research & Risk Assessment* 21(5): 555–572. Special issue on *Medical Geography as a Science of Interdisciplinary Knowledge Synthesis under Conditions of Uncertainty*, Griffith D. and Christakos G. (eds.).

Yu H.-L., Lin Y.-C., Sivakumar B. and Kuo Y.-M., 2013. "A study of the temporal dynamics of ambient particulate matter using stochastic and chaotic techniques". *Atmospheric Environment* 69: 37–45. DOI: 10.1016/j.bbr.2011.03.031

Yu H.-L. and Wang C.-H., 2010. "Retrospective prediction of spatiotemporal distribution of PM2.5 in the metropolis area: a case study in Taipei (Taiwan)". *Atmospheric Environment* 44(25): 3053–3065.

Yu H.-L. and Wang C.-H., 2013. "Quantile-based bayesian maximum entropy approach for spatiotemporal modeling of ambient air quality levels". *Environmental Science & Technology* 47(3): 1416–24. DOI: 10.1021/ es302539f.

Yu H.-L., Wang C.-H., Liu M.-C. and Kuo Y.-M., 2011b. "Estimation of fine particulate matter in Taipei using landuse regression and bayesian maximum entropy methods". *International Journal of Environmental Research and Public Health* 8(6): 2153–2169.

Yu H.-L., Yang S.-J., Yeh H.-J. and Christakos G., 2011a. "A Spatiotemporal climate-based model of early dengue fever warning in southern Taiwan". *Stochastic Environmental Research and Risk Assessment* 25(4): 485–494.

Zeiss H., 1931. "Geomedizin (geographische Medizin) oder medizinische Geographie". *Münchner Medizinischen Wochenschrift* 78: 189–201.

Index

abstract mode, 24
absurd statement, 55
acquired immune deficiency syndrome, 86
acute myocardial infarction, 152
affective dispositions to respond (ADR), 40
affirmed consequent to conditional, 107
aflatoxin, 3
Age of Synthesis, 1, 292
agricultural crops, 8
AIDS, 86
Alcmaeon, 305
algorithmic decision-making (ADM), 33
alopecia, 26
Alzheimer's disease, 57
ambiguity, 59
analogical argumentation, 114
anatomic pathology laboratory, 12
anchoring, 121
ancient DNA, 294
anti-depressant Prozac, 151
antinomy ($\alpha\nu\tau\iota\nu o\mu\acute{\iota}\alpha$), 15
apoplexia, 26
argumentation process, 166
Aristotelian logic, 14
Aristotelian principles of logic, 15
Aristotle's (logical) dictum, 1, 42, 60, 268, 305
assertion, 66
    relation to mind state, 67
assertion form, 76

asthma, 8
Avicenna, 2
axiomatic–deductive scheme, 65
axiomatic method, 12

bacteria clostridium botulinum, 123
base rate fallacy, 219
Bayes rule, 167
Bayesian inference, 174
Bayesian Maximum Entropy (BME), 253
Bayesian networks, 33
behavioral analysis, 69
Beijing, 69
belief, 68, 144
benign or malignant histology, 89
biological markers, 8
biomarkers, 8
biomedical knowledge, 29, 46
biopsy, 307
biostatistics, 249
Black Death, 35, 65, 183, 294
blood pressure, 19
blood sugar, 294
BME–SIR approach, 255, 258, 266
BME–SIR equations, 256
body temperature, 19, 69
body temperature (BT) rhythms, 44
bony thorax, 54
Boole, George, 60
botulism disease, 123
bubonic plague, 87
burden-response curve, 6

carbohydrates, 294
carcinogenic substance, 3
cardiac arrhythmia, 103
cardiac diagnoses, 54
case, 253
case assertion
    types, 19
        belief, 19
        guess, 19
        knowledge, 19
        rationalization, 19
        sustained belief, 19
case attribute, 20, 51, 76
case-based reasoning approach, 34
case communication, 58
case individuality, 56
case likelihood ratios, 240
case prognosis (CP), 29, 39, 211
case-specific or specificatory
    knowledge base, 45
CAT, 43
causal medical conditionals, 100
causally neutral, 279
causally positive, 279
causation (or etiology), 268
causation assessment, 275
causation characterization, 269
causation conditional, 283
causation criteria, 272
causation hypotheses, 269
causation justification, 291
cause and effect, 263
cause–effect association, 281
central nervous system, 43
Chagas disease, 311
chain rule of formal logic, 15
chikungunya disease, 137
childhood liver cancer, 3
China, 3, 4, 9, 10, 52
Chinese medicine, 9, 11, 12
Chinese semantics, 15
cholecystitis, 182
cholesterol, 19
Chrysippus, 14
circadian rhythmicity, 19
classical truth table (CTT), 80, 176

climate, 312
climate change, 312
clinical assertion and truth, 311
clinical checklists, 13
clinical data, 9
clinical medicine, 1
clinical practice, 13
clinical symptomatology, 26
Cnidian school, 305
cognitive biases, 40
cognitive condition, 153
cognitive disposition to respond
    (CDR), 40
cognitive errors, 39
cognitive favorability, 244
cognitive situation, 154
colposcopy, 307
common cold, 20
common sense, 31
communication uncertainty, 60
comparative anatomy, 17
composite space–time domain of
    causation, 269
computed axial tomography, 43
computerized medical inferences, 303
conceptual source, 59
concrete mode, 24
conditional disjunctivity, 109
conditional excluded middle, 108
conditional probability, 166, 279
conditional, 117
confirmation strength (CS), 213
conjunction confirmation principle
    (CCP), 216
conjunction disconfirmation principle
    (CDP), 217
content-dependency (CD), 61
content-dependent, 83, 96, 166
content-dependent conditional, 100, 102
content dependent syllogisms, 100
content-independent (CI), 61, 166
content-independent assertions, 96
content independent conditionals, 100
content-independent connectives, 83
continuum, 19

contraposition, 91
conversational (dialogical) connective, 80
conversational or commonsensical interpretation, 81
conversational or dialogical conditional, 99
core or general knowledge base, 38, 45
coronary stenosis, 273
Corpus Hippocraticum (Ιπποκρατική Συλλογή), 9, 10
counterfactual (subjunctive) conditionals, 108, 230, 283
covariance, 265, 266
covariance function, 259
critical reflective thinking, 32
cross-covariance, 266
Cushing's disease, 311
cytological examination, 150

Danaides, 31
Darwin, Charles, 17
Darwinian notion, 305
data, 45
de Finetti's scoring rules, 183
De Materia Medica (Περί 'Υλης Ιατρικής), 10
decision-making, 12
decision-making methods, 30
decision support, 2, 12
deductive–nomological causation, 290
degree of acceptance of disease diagnosis, 213
degree of cellular dysplasia/anaplasia, 19
degree of covering, 214
dementia, 294
dengue fever, 7
denying the antecedent syllogism, 273
derivative assertions, 131, 132
description logics (DL), 304
deterministic, 264
diabetes, 277
diagnoses set, 205
diagnosis confirmation, 31

diagnosis multiplicity principle (DMP), 216
diagnosis procedure standard (DP std), 28
   of common diseases, 28
diagnostic decision support systems (DDSS), 310
diagnostic error, 25
diagnostic procedure (DP), 28
diagnostic process, 40
diagnostic testing, 12
dialogical assertions, 84
diet, 3
differential diagnosis, 31
Dioscurides, 10
disability, 26
disconfirmation degree, 214
disease, 26
   cure, 55
   treatment, 55
disease causation (etiology), 22, 272, 290
disease covariance, 49
disease diagnosis (DD), 9, 26, 39, 211
   types, 27
      admitting, 27
      clinical, 27
      computer-aided, 27
      differential, 27
      discharge, 27
      dual, 27
      exclusion, 27
      laboratory, 27
      nursing, 27
      prenatal, 27
      principal, 27
      radiology, 27
      remote, 27
      self, 27
disease diagnostic performance, 46
disease features, 21
disease incidence, 19, 20
disease indicators, 277
disease predictors, 277
disease prevalence, 277
disease risk assessment, 160

disease spread, 48
disease symptoms (DS), 20, 26, 39
disease transmission, 21
disease transmission chain, 271
diseases prognostic scoring, 30
disjunction connective, 82
Dissections (Ανατομών), 16
divergence of rationality, 296
Divine Farmer's Materia Medica (Shennong Chang Bai Cao), 11
DNA, 294
dual-process approach, 31
dualistic opposition, 291
dysfunction, 26

ear infections, 20
Ebola disease, 169, 174
Ebola test, 169
eco-systems, 312
El Niño, 312
electrocardiographic (EKG), 152
electromagnetic field, 88
elliptic arguments, 95
embryology, 17
empirical aspects, 254
empirical models, 63
empirico-logical investigation, 17
empirico-logical process, 30
environmental causal chain, 271
environmental exposure, 8
environmental health, 1, 5, 117
environmental tobacco smoke, 88, 91
epidemic spread, 2
epidemiologists, 55
episodic memory, 46
epistemic, 280
epistemic adequacy, 164
epistemic cause, 282
epistemic certainty, 270
epistemic interpretation, 280
epistemic justification, 281
epistemic logic, 53, 122
epistemic stochasticity, 271
epistemic strength, 63, 67, 209, 243
epistemic support, 67, 269
epistemic uncertainty, 55

epistemologies, 17
equivalence conditional (or biconditional), 117, 167, 230, 283
Equivalence Conditional Maximum Entropy (ECME), 253
equivalence conditional probability, 172
erythema, 26
esophageal diagnoses, 54
esophageal disease, 70
estrogen replacement therapy, 294
etiologic medical reasoning, 291
etiology, 61
Euboulides, 105
Euclidean geometry, 299
evidence, 22
evidence-based medicine (EBM), 144
evidential relevance, 283
exanthema, 26
excluded middle principle, 15
exclusion strength (ES), 214
exotic infection, 21
expansion, 232
experimental method, 12
experimentalists, 23
expert judgment, 34
expert knowledge, 34
explanation–prediction asymmetry, 290
exposure causal chain, 92
exposure indicators, 8
exposure risks, 9
exposures, 7
externally available health data, 5
extrapolating, 119, 121

fallibilism, 81
false positive, 174
falsehood implies any assertion, 107
flu (influenza), 20
flu epidemic, 121
formal, 153, 183
formal logic oddities, 179
formal or classical logic, 60
formalists, 24
Frege, Gottlob, 60

French flu, 258, 261, 266
frequency exclusion principle (FEP), 217
frequency inclusion principle (FIP), 216
frequency of co-occurrence, 212
fundamental mapping equations, 253
fuzzy reasoning, 34

Gödel, Kurt, 60, 299
Galen, 2, 17
gastric ulcer, 71
gastroesophageal reflux disease, 133
Gaussian kernel, 87, 267
generalized SIR (Susceptible–Infected–Recovered) model, 86, 259, 264
generalizers, 23
geographical distributions, 52
geographical medicine, 5
geomedicine, 1, 5
giardiasis, 252
glomerular filtration rate, 29
glucose levels, 19
Gongsun Long, 15
great vessels disease, 54
Greece, 10
Greek, 9
Greek logic, 15
Greek medicine, 11, 12
greenhouse gases, 312
Groopman, Jerome E., 13, 23
growth law, 88
guesses, 68

H9N7 avian influenza virus, 135
hand–foot–mouth disease (HFMD), 51, 52, 266
Hangzhou city, 3
hard data, 63
Harvey, William, 17
health care, 26
health care professionals, 12
health care provider, 3, 68, 76
health care standards, 9
health damage, 6

health dataset
    converted into information, 64
    related to medical knowledge, 64
health dynamics, 235
health information infrastructure, 141
health judgment, 12
health paradigm, 293
health risk assessment, 151, 163
health risks, 4
health standards, 19
health statistics, 62, 233, 249, 295
health status, 18
    continuum, 18
    indicators, 18
    individual, 18
    population, 18
    spatiotemporal structure, 18
heart disease, 294
heart rate, 19, 41
Helicobacter pylori, 71
hemoglobin levels, 19
hemorrhagic disease (HD), 218
Hepatitis E, 16
herbal medicine, 11
Hermagoras ($E\rho\mu\alpha\gamma\acute{o}\rho\alpha\varsigma$), 74
hexamethonium, 282
high-quality care, 39
higher cumulants, 86
Hippocrates, 1, 9, 305
holistic approach, 9
Hua Tuo, 10
Hui Shi, 15
human circulatory systems, 32
human effect, 6
human exposure, 20
human immunodeficiency virus (HIV), 87
Hume, David, 162, 296
Huntington's disease, 245
hyperglycemia, 277
hyperkalemia, 29
hypertension prevalence, 57
hyperthyroidism, 277
hypothesis generation, 31
hypothesis testing, 31
hypothetical syllogism, 288

Ibsenian transformation, 293
in situ logic, 53, 61
in situ uncertainty, 13
incompleteness principle, 298, 299
indeterminism, 155
India, 52
indicative conditional, 104
indicators, 19
individuality, 57
infected, 51, 256
infectious disease, 20, 49, 312
inference rules, 164
influenza, 99
information, 45, 205
information entropy, 185
information technology (IT), 303
information technology-based medical reasoning, 302
informational perspective, 66
informativeness, 208
insights, 12
intake rate, 88
integrative space–time prediction (ISTP), 289
intellectual provocation, 293
intensive care unit (ICU), 12, 165
internally consistent, 24, 199, 200
internally generated health care records, 5
interval probabilities, 200
intuitionistic logic, 72
intuitive inferences, 31
ischemic pain, 150

justification, 133
justification degree, 244
justifying the justification, 94

Kashin–Beck disease (KBD), 9
kernel bandwidth, 256
Keshan disease (KSD), 9
kidney transplant, 111
knowledge, 45
knowledge base (KB), 44, 67, 68
knowledge continuity, 146

knowledge synthesis (KS), 36, 38, 62, 264
knowledge theory, 145
knows, 68
Kuhn, Thomas S., 295

lab results, 9
language, 42, 43
language–metalanguage association, 44
Lao-Tsi, 131
large-scale health studies, 2
legal consequences, 307
legal disputes, 56, 57
lepra, 26
leuke/leukoderma, 26
leukemia, 8
lexical ambiguity, 169
lichen, 26
lifestyle factors, 3
linguistic source, 59
litigation, 307
livestock, 8
local climatic conditions, 265
location–time coordinates, 20, 47, 48
location–time metric, 48
logical connectives, 61, 79
    conjunction, 61
    disjunction, 61
    entailment, 61
    negation, 61
logical equivalence, 156
logics
    content-dependent, 18
    context-dependent, 18
low-fat diets, 294
lung cancer, 168, 182
lung diagnoses, 54
lung mass, 168

malaria, 132, 252
Manchester score, 30
map interpretation, 252
marketplace of ideas, 293
Marshall, Barry J., 57, 173

Master Mo, 15
material conditional, 104, 117, 167, 283
material conditional maximum entropy (MCME), 253
material conditional probability, 172
materialistic perspective, 66
maximum infected fraction, 258
medical case assertion, 19
medical causation, 61, 269, 271
medical conditionals, 164
medical connectives, 79
medical decision-making, 12, 25, 30, 33
medical dialectics, 36
medical ethics, 112
medical expertise, 147
medical guidelines, 13
medical inferences, 79, 131
medical interventions, 161, 164
medical journaling, 58
medical reasoning, 39, 53
medical reasoning errors, 40
medical records, 9
medical sciences, 1
medical syllogism, 89
medical testing, 174
melancholia, 26
melatonin, 19
metalanguage, 42–44, 66, 104
metalanguage interpretation of medical probability, 159
methodological quality, 38
metric, 47
microbes, 20
Ming-Chia, 15
minimum mean squared error (MMSE), 249
modal logic, 66
mode of thinking, 23
modelers, 23
modes of thinking, 23

modus ponens, 187, 193
modus tollens, 187, 193
Mohists, 15
  Canons, 15, 105, 268
  school, 15
mononucleosis, 20
monotonicity, 108
morbidity, 12
morbidity measures, 19
mortality, 12, 19
mortality measures, 19
Mozi, 15
multiple possible worlds, 65
MYCIN, 310
myocardial infarction, 274

narrative fallacy, 309
natural cause, 282
natural laws, 49, 63, 85, 117
negated antecedent to conditional, 107
negligence of physician, 307
New York City, 6
nihilism, 298, 312
non-egocentric individualism, 46
non-epistemic, 280, 282, 306
non-Hodgkin lymphoma, 8, 30
non-local, 263
non-monotonic medical reasoning, 238
non-self-referential (NSR) assertions, 244
non-separable location–time metric, 48
non-stationary in time, 266
normal breathing rate for air, 88
normative aspects, 254
North Atlantic Oscillation, 312

object language, 44, 66
object language interpretation of probability, 159
oedema, 26
On Animals (Περί Ζώων), 16
ontic interpretation, 280
ontic uncertainty, 54, 60

oral cancer, 70
over-extending, 119
ozone (O_3), 6

pachyderma, 26
pap smear, 307
paradoxes, 105
Parkinson's disease, 219
particularizers, 23
pathophysiological (causal) reasoning, 173
Pausanias, 300
people's environment, 3
peptic ulcer, 57, 172
Περί την Ιατρικήν, 11
pericardial or pleural diagnoses, 54
personal experience, 187
pesticides, 8
Phaedo, 268
Philadelphia, 7
physical examination, 9, 55
physician's cognitive situation, 160
Physics, 268
Φ-function, 89, 92, 187
P_{KB}-approach, 243
place, 253
plague mortality, 52
Plato, 145, 268, 305
Platonic conditions, 145
platonism, 145
pleonastic premises, 95
$PM_{2.5}$, 70
PM_{10}, 70
pneumonia, 20
poliomyelitis, 47
poliosis, 26
pollutant particles in the air, 88
population cohorts, 6
population exposure, 2, 48
population fraction, 256
population health, 54
population mortality, 50, 57
population ratio of new infecteds, 258
population risks, 8
possible worlds, 51
post-menopausal women, 294

Posterior Analytics, 268
posterior or post-test probability, 162
postmodern decision analysis, 138
postmodernists, 311
pragmatic uncertainty, 134
pre-existing medical condition, 6
premise strengthening or strengthening of the antecedent, 108, 199, 204, 233, 288
presentational source, 59
prevalence, 19
principles of medical practice (PMP), 215
prior or pre-test probability, 162
probabilistic causation, 277, 279
probability, 30, 148, 205
probability calculation, 181
probability calculus, 169
probability conditionals, 170
probability density function (PDF), 253
probability dynamics, 235
probability functions, 187
probability interpretations, 148, 157
probability intervals, 201
probability of infection transmission, 256
probability of recovery, 256
probability rules, 155
probability theory, 44
probameter, 181
professional assertions, 89
prognosis, 22
Propecia, 276
prostate cancer, 273
psora/psoriasis, 26
public health, 25, 39, 232
public health finances, 31
public health management, 160
published experience, 187
pulmonary embolism, 277
purely data-driven analysis (PDD), 62
Pyrrhonism, 81

Index

quality assessment, 18
quality of evidence, 208
quantitative modeling, 85

radioactive cloud, 88
radioactive repositories, 22
ranking degree (RD), 212
ranking of assertion forms, 71
rationality, 144
rationalizes, 68
RDF Schema (RDFS), 304
reasoning, 39
recognition primed decision (RPD), 32
reconstruction of argumentation styles, 94
recovered population, 51, 256
recovered population fraction, 258
rectal temperature, 41
rectal temperature–heart rate, 41
Reed–Frost model, 65
referred pain diagnoses, 54
regional disease spread, 299
regional topography, 265
regression mapping, 254
relative risk, 115
reliability, 41
representativeness heuristic, 181
Resource Description Framework (RDF), 304
respiratory diseases, 7
retinal detachment, 82
revision, 232
Riemann–Minkowski metric, 48
risk identification, 292
robust decision (RD) methods, 33
round trip fallacy, 309
Russell, Bertrand, 60

S/TRF, 38, 53, 63, 231
scientific models, 85
scientific theory, 63
score function, 183
S_e, 9
secondary case attributes, 287
secondary evidence, 63

self-awareness, 292
self-contradiction, 107
self-reference, 105, 239
self-referential (SR) assertions, 244
semantic ambiguity, 59
semantic interoperability, 30
semantic memory, 46
semantic uncertainty, 60
Semantic Web, 304
semantically distinct, 118
semantics, 15
sexual dysfunction, 276
Shannon information theory, 211
simplex triangle technique, 256
site-specific knowledge base, 38
skin cancer, 116
SMR dynamics, 297
social and cultural dimensions, 304
social dimension, 305
socioeconomic adversities, 5
soft or uncertain information, 63
soil pollution, 8
sophistry, 113
space–time, 20
space–time causation, 268
space–time dependence, 19–21, 48
space–time disease mapping, 250
space–time disease spread, 22
space–time distance, 48
space–time domain, 47
space–time infectious disease spread, 264
space–time intrinsic kriging (STIK), 262
space–time metric, 49
space–time milieu, 47
space–time ordinary kriging (STOK), 262
space–time probabilities, 148
space–time simple kriging (STSK), 262
spatial epidemiology, 299
spatiotemporal phenomenon, 21
spatiotemporal random field S/TRF, 38, 51
spatiotemporal structure, 20, 21

species population, 88
specificity, 170
standard logical relations, 164
state of mind, 63, 66
state of nature, 63
state–space method, 264
statistical approach, 264
statistical conditional, 167, 172, 230
stereotypes, 121
stochastic causal inferences, 285
stochastic medical reasoning (SMR), 16, 18–20, 35, 36, 41, 47, 53, 61, 62, 67, 80, 85, 250, 272
stochastic models, 264
stochastic partial differential equation (SPDE), 262
stochastic syllogisms, 187
stochastic truth table (STT), 176
stochastic truth-value (STV), 154
stochastics, 18
Stoic logic, 14
structural space–time, 20
styles of professional practice, 2
subjective belief, 78
subjective uncertainty, 77
substansivists, 24
substantive, 183
substantive aspect, 153
substantive conditional, 99, 100, 229
substantive information, 206
support theory, 182
survey data analysis, 151
susceptible, 21, 51, 256
susceptible fraction, 258
Susceptible–Infected–Recovered (SIR), 296
sustainably believes, 68
symptom detection, 31
symptom necessity principle (SNP), 217
symptom's confirmation or disconfirmation strength, 211
symptoms multiplicity principle (SMP), 217
symptoms set, 205
syntactic ambiguity, 59

syntactically equivalent, 118
syntactics, 15
syntax, 43
synthesis, 2, 12
systems-based thinking, 41

Taipei, 70
Taiwan, 70
Talmud, 67
tautology, 109
technical information, 206
Temporal Geographical Information Systems (TGIS), 21, 250
test sensitivity, 170
The Inner Canon of Huangdi (Huangdi Neijiang), 10
theory of knowledge, 17, 164, 184
theory–practice interface, 25
therapeutic interventions, 25
therapy decisions, 12
thinking styles, 14
three Qs, 74
thyroid, 277
thyroid cancer, 3
time, 253
time pressure, 13
tonsillitis, 20
toxicokinetic laws, 6
transposition relationship, 101
transtubular potassium gradient, 29
Treatise on Cold Pathogenic and Miscellaneous Diseases (Shangan Za Bing Lun), 10
triadic case formula, 75, 153
truth-value (TV), 80
Trypanosoma cruzi (*T. cruzi*), 311
tumor malignancy, 19
tumor malignity, 66

U_{KB}-approach, 243
ultraviolet radiation, 115
unaccounted uncertainty, 40
unanticipated knowledge, 298, 300, 301
uncertain mind states, 53
uncertainty, 12, 13, 19, 21, 54, 77, 205

uncertainty factors, 76
uncertainty inequality, 78, 237
uncertainty reasoning, 236
understanding, 45
uninformative, 202
uninformative inference, 199, 204
uninformative priors, 202
uniqueness principle (UP), 216
unknowable, 160
updating, 232

vagueness, 59
vector-borne disease, 312
venous thrombosis, 150
veriphobes, 311

Warren, J. Robin, 57, 173
water- and food-borne diseases, 312
weather patterns, 312
Web Ontology Language (OWL), 304
West Nile virus, 7

Yellow Emperor's Inner Canon (Huangdi Neijiang), 10

ζ_a-approach, 243
Zhang Zhongjing, 10
zoonotic diseases, 312

Printed in the United States
By Bookmasters